The Post-Screen Through Virtual Reality,
Holograms and Light Projections

MediaMatters

MediaMatters is an international book series published by Amsterdam University Press on current debates about media technology and its extended practices (cultural, social, political, spatial, aesthetic, artistic). The series focuses on critical analysis and theory, exploring the entanglements of materiality and performativity in 'old' and 'new' media and seeks contributions that engage with today's (digital) media culture.

For more information about the series, see: www.aup.nl

The Post-Screen Through Virtual Reality, Holograms and Light Projections

Where Screen Boundaries Lie

Jenna Ng

Amsterdam University Press

Cover photo: Nick Holliman

Cover design: Coördesign, Leiden
Lay-out: Crius Group, Hulshout

ISBN   978 94 6372 354 1
e-ISBN   978 90 4855 256 6
DOI   10.5117/9789463723541
NUR   670

Printed and bound by CPI Group (UK) Ltd, Croydon, CR0 4YY

"The true is what he can; the false is what he wants."
Madame de Duras

For Ackbar – teacher extraordinaire, and the true godfather

Table of Contents

Acknowledgements   11

Introduction   15
    Post-Screen Media: Meshing the Chain Mail   15
    Eroding Boundaries in the Contemporary Mediascape   24
    Why Boundaries Matter   29
    Chapter Outlines   36
    The Post-what?   39

1   Screen Boundaries as Movement   51
    Re-placing the Screen: Play and Display, Appearance and
    Dis-Appearance   51
    Screen Boundaries: Physical and Virtual, and of the Movement
    Betwixt   56
    Metaphors for the Screen   61
    Crossing Screen Boundaries: Love, Pleasure, Information, Trans-
    formation   67
    Interactivity and the Moveable Window   74
    Screen Boundaries Across Dimensions   75

2   Leaking at the Edges   79
    Protections and Partitions   79
    Rupturing Screen Boundaries   84
    Interplay between Fictional and Factual Threat   88
    Leaking at the Edges: The Merging of the Amalgamated Real   92
    Virtual Co-location in Real-time... and in the Era of Covid-19   99
    The Screen Boundary Against the Algorithm   102
    Screen Boundaries in Flux   105

3   Virtual Reality: Confinement and Engulfment; Replacement
    and Re-placement   107
    "Multitudes of Amys"   107
    On Immersion (Briefly)   110
    The Affective Surround: The Two Vectors of Immersion   114
    The Post-Screen Through VR (1): Confinement and Engulfment   123
    The Post-Screen Through VR (2): Replacement and Re-placement   127
    The Danger Paradox   133

VR as Immersion: Travel, Escape, Fulfilment 137
VR as Inversion: Witness, Empathy, Subjectivity 145
Defeated by the Ghosts 151

4 Holograms/Holographic Projections: Ghosts Amongst the
Living; Ghosts of the Living 155
How We See Ghosts, or, In Love with the Post-Screen 155
Ghosts in the Media: Re-inventing the Afterlife 158
The Post-Screen Through Holograms/Holographic Projections 168
Holographic Projections (1): Ghosts Amongst the Living – Limbo
Between Deadness and Aliveness 170
Holographic Projections (2): Ghosts of the Living – Vivification
of the Virtual Real 174
A Funny Thing Happened on the Way to Substitution 186

4A (Remix) True Holograms: A Different Kind of Screen; A
Different Kind of Ghost 189
Screens and Ghosts, or, the Window and the Guy in the Basement 189
True Holograms 191
A Different Kind of Screen: Brains, Nerves, Thought 195
A Different Kind of Ghost: "A Memory, A Daydream, A Secret," or,
Digital Apparitions 199

5 Light Projections: On the Matter of Light and the Lightness of
Matter 207
The City Rises 207
The Light Rises, or, Light as the Matter of Light 209
Cities of Screens 215
Light Projections (1): Light that Dissolves and Constructs... and
of Latency 223
Light Projections (2): Walls that Fall Apart... and Re-Form 228
Light Projections (3): Particles that Gain a Body... and Transform 235
Projection Mapping (1): The Image that Devours Structure; the
Voracity that is a Media History 238
Projection Mapping (2): The Exterior that Reveals; the Permanence
that Fades 247
The Ground Beneath Our Feet 251

Conclusion/Coda   253
   Postscripts to the Post-Screen: The Holiday and the Global Pan-
   demic   253
   Twin Obsessions (1): Difference   256
   Twin Obsessions (2): The Gluttony   263
   The Post-Screen in the Time of Covid-19   270

Index   275

Acknowledgements

This book took 7 years to complete, and so its rivers of gratitude run deep; this acknowledgements page can thus only go so far. But there are some whose shadows are cast particularly long across this book and so this page starts with them. The Leverhulme Trust, which funded me as an early career fellow and thereby is not only a literal kickstarter to this work, but also a decisive gateway to many other beginnings; and my students at Cambridge in our "Coding the Frame: Space and Time in Digital Media" classes – Alison Fornell; Conor McKeown; Andrew Pel; Milosz Paul Rosinski – whose many lively discussions and ideas germinated the frame into the post-screen.

Writing requires a room of one's own in every sense, and I had the good fortune of sitting in some wonderful ones. I benefited hugely from various discussions as a visiting scholar at the Centre for Liberal Arts and Social Sciences (CLASS) in Nanyang Technological University, Singapore, including audience comments to an early presentation of this book at the kind invitation of the Asia Research Institute (ARI), National University of Singapore. I was also hosted as a visiting researcher by the Bill Douglas Cinema Museum, University of Exeter, which provided invaluable resources on early media whose consultations not only make appearances in this book, but also shaped much of my thinking for the post-screen. I also benefited enormously from the many questions and comments from and discussions with audiences at talks I gave presenting readings of this book's drafts at the University of Bergen; Lingnan University, Hong Kong; University of Southampton; Goldsmiths University of London and the *Cinema Experiences: Immersive Pasts and Futures* event in Singapore. I am particularly grateful to Liew Kai Khiun; Jihoon Kim; Chua Beng Huat; Phil Wickham; Øyvind Vågnes; Asbjørn Grønstad; Teju Niranjana; Seth Giddings; Jussi Parikka; Rachel Moore; and Lucy Bolton for their generous invitations and conversations which opened my mind and spurred on so many more inspirations.

If there needs to be any proof that writing requires both space and time, this book is it. I thank my Department of Theatre, Film, Television and Interactive Media at the University of York for granting me two terms of research leave in 2020, both of which were indispensable to the completion of this work and during which I finished this book. I am also grateful for the Department's support of two undergraduate research assistants, Joe Lamyman and Junge Shi, as well as to Joe and Junge themselves for their meticulous and assiduous work in hunting out numerous obscure case studies (many of which made it into the book) and looking up theoretical references; in how you were exemplary students, so were you as my research assistants. There are also many in the University who have given me invaluable support in numerous ways: my Interactive Media colleagues are truly the best team in the world and they light up my life at work in so many ways. I am particularly grateful to Nick Jones, Richard Carter and Jandy Luik for reading drafts of the book and providing invaluable comments, all of which made the book better. I also thank colleagues in the Department and the University for their support and mentoring, particularly Marian Ursu, David Barnett, Kristyn Gorton, Helen Hills, Judith Buchanan and Duncan Petrie. You have all inspired me.

Roads in academia are especially long, and I am also grateful to so many colleagues and mentors in the academy over the years who have, one way or another, given and shown me moral support, advice, encouragement, friendship, laughter and kindness to get through the tough times – this one's particularly for Chris Newfield; Keith Wagner; Andrew Webber; John Rink; Christos Lynteris; Simon Mills and Simon Popple. Acknowledgement should also be made of data from the Newcastle Urban Observatory as shown on this book's cover image, photographed by Nick Holliman at The Hive in Curtin University, Perth, Western Australia. I also thank David Theo Goldberg; Nishant Shah; Fred Turner; and Andrew Prescott for not only being especially inspiring colleagues, but also their comments, reviews and ideas from discussions over the years (and the world!) which have all fed directly and indirectly into this book. My biggest thanks, though, goes to Ackbar Abbas for his readings (and re-readings...and more re-readings) of this book's many drafts, for his constant illuminations to me on its possibilities, and for his inimitable schooling over all these years on insight, patience, argument, writing, thinking, lucidity and paradoxes; for these reasons, this book is dedicated to him.

Academia is but one part of my life, and I am always thankful for that and those that I have after school: I also thank my other writing family, namely, the Bi'an network – and in particular Mary Cooper, Yan Ge, and Jeremy Tiang – for writing with me and, in its essence, teaching me how to write. Yours are lessons which genuinely turned the key to my writing – and completing – this book, and which I still apply every time I fire up the laptop. The stalwarts in my life – Lee Guan Jyh; Grant Jenkins; James G. Barrett – who get (and deserve) a mention in every one of my books; this is your mention here: you know how important you are to me. Jim also walked the long path of this book with me, reading multiple drafts and providing enormously helpful comments every time. In no small way, he, too, saw me through this book, as he has done with so much else.

I am also very grateful to Amsterdam University Press for their support of this book project, and particularly Maryse Elliott, Julie Benschop-Plokker, Louise Visser, Chantal Nicolaes and Mike Sanders for all their professionalism and work in getting this book... well, onto the books, and also to the series editors and the two manuscript reviewers for their careful readings, encouraging support and incisive feedback, all of which unquestionably made the work better.

My final – lucky last – thanks are unreservedly for Nick H, my partner in life, love, light...and crime. Every day with you in it renews my strength, and every writing day, no matter its pain and problems, resets in shared joy and laughter. The ghosts in every chapter are for you.

A few parts of this text have been previously published, but they have been re-contextualized and/or substantively extended here. Parts of Chapters 4 and 5 appear in "Surface, Display, Life: Re-thinking the Screen from Projection to Video Mapping," *Archives of Design Research*, 27:1 (2014): 72-91; and parts of Chapter 2 appear in "She Crawls Out of the TV, or, On the Gendered Screen," *Media Fields Journal: Critical Explorations in Media and Space*, No. 14 (July 2019), http://mediafieldsjournal.org/she-crawls/.

Introduction

Abstract

This Introduction presents the context for the book's argument of the post-screen, namely, an argument for a state of critical attention to the delimitations of screen media and the ensuing problematizations of relations between image and object; an intensifying evolution of the virtual and its role in defining media consumers and their realities; and an era of screen media marked by the disappearances of boundaries of differentiation between subject and object; and a point in media history. The central query of the post-screen lies in the growing imperceptibility and instability of screen boundaries. Where these thresholds begin to disappear is also where the need arises to re-question the definitional states of the actual and the virtual, and the renewed contestations for dominance between them.

Keywords: post-screen; boundaries; *La Condition Humaine*; Bazin; Baudrillard; hunger

Post-Screen Media: Meshing the Chain Mail

Screens offer a seemingly endless supply of information, but the true value of the page is not what it allows us to know, but how it allows us to be known.
~ Jonathan Safran Foer[1]

The frame descended at the end, capping a mysterious drama. Minutes earlier, a flash mob had appeared inside a shopping centre in Breda as an ensemble of characters in seventeenth century dress. They re-enacted various scenes: a thief clutching his spoils and fleeing with guards in hot pursuit; two military figures marching into the square at the head of a

---

1   Jonathan Foer, "Technology is diminishing us," *The Guardian* online, December 3, 2016, https://www.theguardian.com/books/2016/dec/03/jonathan-safran-foer-technology-diminishing-us.

Ng, J., *The Post-Screen Through Virtual Reality, Holograms and Light Projections. Where Screen Boundaries Lie*. Amsterdam: Amsterdam University Press 2021
DOI: 10.5117/9789463723541_INTRO

cavalry; a dwarf scurrying along while shooing the crowd; a girl picking up her skirts and running after a squawking chicken.

The performance turned out to be an ingenious publicity stunt for the 2013 re-opening of the Rijksmuseum in Amsterdam, which had closed for a decade-long renovation.[2] For the skit's denouement, the actors assembled in the central space of the shopping centre. They settled into approximate correspondences with the postures exhibited by the characters in Rembrandt's *De Nachtwatcht*,[3] thus re-creating the museum's arguably most famous painting. Once everyone was in place, the concluding touch arrived: a rectangular construction, bearing the museum's opening and sponsorship notices, descended from the ceiling and came to rest around the actors.

There lie boundaries.

The dropping of the frame is not just a clever ending to an eye-catching publicity skit. It also demonstrates a fluidity, an almost casual instability to the visibility of boundaries as structures which control and organize the scene's meaning and content. As the frame falls, the actors are no longer a motley crew of performers. Instead, they visually echo a famous symbol of the museum. The frame further demarcates the mall's space, relatively homogeneous until that moment: it differentiates the *here* of the performers, and *there* of the shoppers; the *here* of the painting, and *there* of the mall. "Where boundaries lie" thus embroils dual meanings of the word "lie": the first as establishment in laying down positions of separations and differentiations; the second as slippage and trickery exposed in the whimsy of partitioning – one moment a perplexing public drama, the next a meaningful sign.

Snagged in these cross-hairs of demarcation and deception, the fluid fluctuations of boundaries agitate and muddy the site of the image against its surroundings, and renew contestations between reality and representation. As contemporary screen media today present increasingly immersive and ubiquitous image worlds amidst changing visibilities and perceptibility of

2   An online video of the stunt can be viewed at ING Nederland, "Onze helden zijn terug," April 1, 2013, YouTube video, 1:25, https://www.youtube.com/watch?v=a6W2ZMpsxhg. The publicity exercise was a big success: as of writing, the video has registered more than 7.9 million hits and created abundant media coverage. On the latter, see, as one instance, "Flashmob recreates Rembrandt painting in Dutch shopping centre – video," *The Guardian* online, April 5, 2013, http://www.theguardian.com/world/video/2013/apr/05/flashmob-rembrandt-amsterdam-shopping-video.
3   Rembrandt van Rijn, *De Nachtwacht* (*The Night Watch*), 1642, oil on canvas, 363 cm x 437 cm, Rijksmuseum.

screen boundaries, these contestations intensify in complexity and with heightened stakes. As what is image becomes indistinguishable against the viewer's actual surroundings, its unsettling re-visits how we might think about truth and authenticity; actuality and virtuality; art and life. As screen boundaries shift and lie, we confront a new regime of relations between images and reality, images and viewers, viewers and screens. A new imagination for images arises, and a new space of definitions and understandings emerges – *the post-screen.*

***

You don't talk, you watch talk shows. You don't play games, you watch game shows.
Travel, relationships, risk; every meaningful experience must be packaged and
delivered to you to watch at a distance so that you can remain ever-sheltered,
ever-passive, ever-ravenous consumers who can't free themselves to rise from their
couches, break a sweat, and participate in life. ... Grab your snacks, watch your
screens, and see what happens. You are no longer in control.
~ Dialogue line from *Incredibles 2*[4]

The use of screens as the villain's weapon of choice in a film as mainstream and family-oriented as the 2018 Disney-produced computer-animated superhero film, *Incredibles 2*, is remarkable. In the film, the villain uses screens to hypnotize people into carrying out her nefarious bidding, which works well for her as screens are ubiquitous (appearing in shop windows, studio broadcasts and so on) and portable (where they can be placed over a person's eyes like goggles). What is remarkable is how the film, itself ironically a mega-blockbuster exhibited on multiple screens across the globe,[5] so effectively leverages the ominousness of screen displays against the all-encompassing reliance and wholly accepting relationship viewers have with screens today. Given the film's success, this ambivalence appears as an experience everyone from young children to adults *worldwide* may sense and understand.

The villainous ubiquity and mesmerism of screens in *Incredibles 2* are signs of current times. Screens are omnipresent today. They appear in

---

4   *Incredibles 2*, directed by Brad Bird (2018; Los Angeles, CA: Disney, 2018), DVD.
5   Worldwide, the film eventually generated more than US$1.2 billion in ticket sales, with a little more than half of that coming from international markets: see the box office numbers for *Incredibles 2* at https://www.boxofficemojo.com/release/rl2071758337/.

manifold contexts. They are the main vehicles for defining contemporary relations between viewers and representations, commandeering their primary engagements with the "age of the image."[6] The average home contains multiple screens by way of television sets, personal computers, laptops and, increasingly, smart displays. Screens pervade a myriad of public spaces such as banks, hospitals, schools, offices, train stations, bus stops, shopping malls and museums, as just a few examples.[7] Large screens rigged up outdoors magnify live events such as concerts and sporting matches; giant screens, up to ten metres wide, convert into outdoor cinema.[8] Small screens appear in public transport, including planes, trains and subways. With a genealogy reaching back to 1960s war-time equipment,[9] screens acquire increased, if controversial, relevance today in the context of drone warfare.[10] Personal devices, such as mobile phones, smartwatches, tablets and GPS finders, present mobile screens. In turn, these screens make the multiple connections which characterize twenty-first century living – to mobile webpages through the Internet; to users through activity tracking apps; to other users on social media; to objects via the Internet of Things; and, increasingly, to augmented and virtual realities through games and

6   This is becoming an oft-used phrase today, though I trace its first mainstream use to Amy E. Schwartz's article, "The Age of the Image," in *The Washington Post*, February 8, 1997, A21, where she discusses the phenomenon in relation to imaging women. More recently, the phrase has emerged to refer to the power of twentieth (and twenty-first) century images in commanding stories, advertising, news and understanding of the world: see, for instance, Stephen Apkon's book, *The Age of the Image: Redefining Literacy in a World of Screens* (New York: Farrar, Straus and Giroux, 2013); and BBC4's television programme, *Age of the Image*, episodes 1-4, featuring James Fox, aired March 6, 10, 17 and 24, 2020.

7   On deconstructing how screens divide public and private spaces, see Anna McCarthy, *Ambient Television: Visual Culture and Public Space* (Durham and London: Duke University Press, 2001).

8   Usually set up in unique locations, outdoor cinema in recent years has taken off as a phenomenon: see Rob Walker, "Jaws at a swimming pool, Gladiator at a castle: how outdoor cinema seduced Britain," *The Guardian* online, April 22, 2018, https://www.the-guardian.com/film/2018/apr/22/outdoor-cinema-britain-boom, which declared Britain "in the middle of a boom in outdoor cinema." (np) More recently, due to social distancing rules in the wake of the Covid-19 pandemic, reports are that open-air cinema on large screens outdoors has become even more popular: see Sam Jones, Kate Connolly and Robert Tait, "'Demand is huge': EU citizens flock to open-air cinemas as lockdown eases," *The Guardian* online, May 29, 2020, https://www.theguardian.com/world/2020/may/29/demand-is-huge-eu-citizens-flock-to-open-air-cinemas-as-lockdown-eases.

9   See Charlie Gere, "Genealogy of Screens," *Visual Communication* 5(2) (2006): 141-152.

10   See Grégoire Chamayou, *A Theory of the Ione* (New York: The New Press, 2015).

other applications.[11] Indeed, "[t]he world we encounter is increasingly a screened world."[12]

As "material technologies,"[13] screens constitute critical hardware components of any media apparatus. But they are also *facilitating technologies* which host, mould and define relations between viewers and the worlds of text, images and representation. Giuliana Bruno writes of screens not only as surfaces, but surfaces which carry out significant work of such facilitation, where they "positively shape our culture, generating contact, connectivity, and communication."[14] Dudley Andrew, too, in a different discussion, describes the screen as a containment of the imprints of reality not as a heedless storage, but as "the ultimate interface between human viewer and world viewed"[15] which hosts substantive terms of engagement between image and reality. Here Branden Hookway's discussion of *the interface* "as a form of relation" also comes to mind, in particular how he describes the interface's essence as "not in the qualities of an entity or in lineages of devices or technologies, but rather in the *qualities of relation between entities*." (emphasis added)[16] In these terms, then, of relations and facilitation, the screen becomes *an exemplar of boundaries*, whose surface and edges establish, police and maintain critical differentiations between virtual and actual realities, art and life, image and viewing subject. It *cuts* between each of them, to use Anne Friedberg's imagery of the "ontological cut," a term she takes from Victor Stoichita who had used it to refer to the

11   There is increasing imbrication and interplay between physical and virtual worlds, where people navigate their physical world through screens: see, for instance, the phenomenal success of *Pokémon Go*, an augmented reality game played through one's mobile phone or tablet, which at the height of its popularity in May 2018 had a reported 147 million monthly active users worldwide: Mansoor Iqbal, "Pokémon GO Revenue and Usage Statistics (2020)," *Business of Apps* online, March 24, 2020, https://www.businessofapps.com/data/pokemon-go-statistics/. Alternatively, they shut out the physical world in favour of the screen world, such as by wearing virtual reality (VR) headsets on public transport: see Brian Krassenstein, "Virtual Reality is Finally Here – Already Annoying People On Public Transportation," *IR.net* online, April 6, 2016, http://ir.net/news/virtual-reality-headsets/124116/virtual-reality-public-train/.

12   L.D. Introna and F.M. Ilharco, "On the Meaning of Screens: Towards a Phenomenological Account of Screenness," *Human Studies* 29 (2006): 57-76, 58.

13   Sean Cubitt, "Current Screens," in *Imagery in the 21st Century*, eds. Oliver Grau and Thomas Veigl (Cambridge, MA: MIT Press, 2011): 21-36, 21. Note, though, that his argument is to leverage the material construction of screen technologies into a discussion of screens as "normative technologies," in particular "to express the nature of public life… rearticulate it." (21)

14   Giuliana Bruno, "Surface Encounters," *e-flux journal* 65, May-August (2015), np.

15   Dudley Andrew, *What Cinema Is! Bazin's Quest and its Charge* (Hoboken, NJ: Wiley-Blackwell, 2010), 69.

16   Branden Hookway, *Interface* (Cambridge, MA: MIT Press, 2014), 4.

demarcation between the portable panel painting and the wall. Friedberg writes:

> Like the frame of the architectural window and the frame of the painting, the frame of the moving-image screen marks a separation – an 'ontological cut' – between the material surface of the wall and the view contained within [the frame's] aperture.[17]

However, emerging media technologies today, such as virtual reality (VR), diminish the force of that cut by seeking to eliminate the presence of the screen and the visibility of its boundaries. Such contemporary screen media thus signals another era: the arrival of *post-screen media*. Like the frame dropped around the actors in Breda who re-created *De Nachtwatcht*, the boundaries of post-screen media are similarly arbitrary and volatile in their appearance and disappearance. This fluidity reinvigorates questions about the screen, prompting re-examination about not only what the screen is, but also how it demarcates and what it stands for in relation to how we understand the actualities of our world in, outside and against images. In formulating the post-screen, the following questions form the central concerns in this book: *in the wake of imperceptible or unstable screen boundaries, how do their imperceptibility or instability change the relations between image and viewer? As those separations diminish, how do we, as viewers, understand our realities and our relations with those realities?*

These conditions of the screen as facilitation and interface thus inform this book's mission, namely, to think of the screen not so much as an entity in itself, but in terms of its, as Hookway puts it, "qualities of relations between entities."[18] Via a series of contemporary media technologies, the book examines this state of collapsing screen boundaries and their ramifications on the relations between image and reality as might be beckoned by the post-screen. As Janet Murray observes, "[p]art of the early work in any medium is the exploration of the border between the representational world and the actual world."[19] My own exploration of the post-screen border through this book will take the form of four arguments. They are neither discrete nor chronological, but more akin to meshed interlinks like chain mail.

---

17   Anne Friedberg, *The Virtual Window: From Alberti to Microsoft* (Cambridge, MA: MIT Press, 2009), 5 and repeated at 157.

18   Hookway, *Interface*, 4.

19   Janet H. Murray, *Hamlet on the Holodeck: The Future of Narrative in Cyberspace* (Cambridge, MA: MIT Press, 1997), 103.

The first and most straightforward argument is that the manifestations of screens in contemporary media seek to diminish, if not erase, the viewer's perceptual differentiations between the actual reality they live in and the virtual reality of the image that they experience. This argument focuses on how contemporary media technology is changing the visibility of screens and thereby the nature and perceptibility of their boundaries. This shift obscures the "ontological cut" which marks out difference, so that screens move *from being spaces of difference to spaces of indifference.*

I argue that this re-positioning gives rise to what I call post-screen media, whose fluid appearances and disappearances of screens are not only about their technological or aesthetic thinking, but also contain deeper implications for our understanding of the relations between images and reality. Elizabeth Grosz had noted similar issues of the diminishing boundary with respect to the computer screen:

> Can the computer screen act as the clear-cut barrier separating cyberspace from real space, the space of mental inhabitation from the physical space of corporeality? What if the boundary is more permeable than the smooth glassy finality of the screen? What if it is no longer clear where matter converts into information and information is reconfigured as matter or representation?[20]

Eroding screen boundaries is thus not just about the blurring of differences, but, as Grosz's questions show, opens up substantive issues of the real and queries the changing natures of virtuality, actuality, corporeality, matter, information and representation.

In turn, the issue of these changing natures forms the basis of the second argument. As with other reality-shifting tenets of the contemporary mediascape, such as the viral circulation of social media, "post-truth" cycles of "fake news" and mis/disinformation, and hyperrealistic immersive simulations, within the changing nature of screens also lie changing equations between truth, lies, representation and illusion.[21] Such shifts have resonated throughout the history of visual media from perspective painting to photography to cinema. To that extent, the increasingly complex

---

20   Elizabeth Grosz, *Architecture from the Outside: Essays on Virtual and Real Space* (Cambridge, MA: MIT Press, 2001), 87.

21   These are also issues bound up with understanding complex processes of mediation between current and earlier media forms. See Maria Engberg and Jay Bolter, "The aesthetics of reality media," *Journal of Visual Culture* 19(1) (2020): 81-95.

relations and imbrications between the virtual and the actual are issues all media negotiate to varying degrees.

However, the ubiquity of screens today imposes a qualitatively different structure of realities, namely, *a* mise-en-abîme *virtualization of the virtual* – a piecemeal building of the virtual upon the virtual. Where nearly 20 years ago as images and spectacle disappeared into digital immateriality and computer code, and as Arthur Kroker wrote of how "ours is a culture signified by the triumph of virtuality,"[22] post-screen media today add to that victory in its slippage and trickery of screen boundaries. *Where screen boundaries lie* is thus not only about where the image's borders and demarcations are established, but also about the screen boundary as the instrumentation of an intense virtualization that does not tell the truth. At the heart of the double entendre is thus a system of trickery entwined with omnipresent displays of images, made possible only out of the sheer ubiquity of screens. The second argument thus re-shapes these virtual realities of the post-screen, drawing from them a new imagination of relations with and definitions of the real.

To Grosz's what ifs, I add a few more key concerns: what if an audience no longer cares about screen boundaries? What if they become inured to the erosion of boundaries between reality and simulacra, and indifferent to distinguishing between them? What if they desire representation to the extent of *wishing* for that erasure?[23] These questions prompt the third argument, which addresses how the changing nature of virtuality out of disappearing screens also points to *the changing nature of affect and subjectivity.* As media objects are consumed, so are their consuming subjects reconfigured and affected. This concern is thus also a critical attention to understanding ourselves as beings in increasingly intertwined actual and representational realities. With minds and bodies bombarded with and in constant absorption of burgeoning quantities of media through expanding bandwidths of information, screens change as do, and with, their viewers.

The fourth argument is effectively the hanging of the mail, which is to thread the first three arguments into a provocation of *imagining the post-screen.* Some problematics of this imagination will be elaborated in the final pages of this introduction, but, for now, imagining the post-screen

22   Arthur Kroker, "The Image Matrix," *ctheory.net*, published 20 March, 2020, http://ctheory. net/ctheory_wp/the-image-matrix/.

23   The character of Cipher from *The Matrix* (directed by The Wachowskis (1999; Burbank, CA: Warner Home Video, 1999), DVD), comes to mind here: in the film, Cipher chooses to live in his computer-generated matrix of reality, despite his awareness of its falseness.

may be articulated as the following concerns: *a state of critical attention to the delimitations of screen media and the ensuing problematization of relations between image and object; an intensifying evolution of the virtual and its role in defining media consumers and their realities; an era of screen media marked by the disappearances of boundaries of differentiation between subject and object; and a point in media history.* As with media in general historically, screen media today has become an inevitable interlocutor of life: what comes through on our laptops, PCs and mobile phones enables, facilitates, solicits, causes, results in, directs and shapes virtually every human action from wrangling wages to waging war, and virtually every emotion from anger to grief to compassion. With the constant interpolation of the screen in everyday life, the liminality of the screen boundary signifies an expanding and increasingly fluid space not just for watching, but for living itself. Imagining the post-screen, then, is wrapped up with this ubiquity of screens to the point of their invisibility or imperceptibility, yet with continued substantive impact not only on our relationships with images, but also on our lives, ways of living and understandings of ourselves.

In that respect, the post-screen marks a point in media history, which, *cf* media's history, is not about the history of media, but about history *and* screen media, or the correlation between media invention and significant cultural, social and political changes. Recall, for instance, the impact of the camera obscura in the eighteenth century with respect to perspective;[24] photography in the nineteenth century on the role of automatism; or cinema in the twentieth century on the meaning of documentation. These are just a few examples of media as "a discursive object – an object to think with,"[25] as is the screen today. The post-screen thus also points to a discourse in how the erosion of screen boundaries exposes the in-between-ness in the gap of the border – that area of the middle – as a different epistemological space. As John Durham Peters writes, "things in the middle, like spines and bowels, often get demeaned, but they too deserve their place in our

---

24   See Lee W. Bailey, "Skull's Darkroom: The Camera Obscura and Subjectivity," in *Philosophy of Technology: Practical, Historical and Other Dimensions*, ed. Paul T. Durbin (Dordrecht: Kluwer Academic Publishers, 1989): 63-79.

25   Martin Lister, Jon Dovey, Seth Giddings, Iain Grant and Kieran Kelly, *New Media: A Critical Introduction*, 2nd edition (London and New York: Routledge, 2009), 110. In their use of that phrase, the authors immediately reference Jonathan Crary's discussion of the camera obscura (in *Techniques of the Observer* (Cambridge, MA: MIT Press, 1992), 25-66) as presumably an exemplar of thinking about media as such a discursive object.

analysis."[26] They deserve their place because, per Grosz again, the middle are spaces for transformation, *because* they are in the middle:

> The space of the in-between is the locus for social, cultural, and natural transformations: it is not simply a convenient space for movements and realignments but in fact is the only place—the place around identities, between identities—where becoming, openness to futurity, outstrips the conservational impetus to retain cohesion and unity.[27]

This fourth argument is thus about how that middle is becoming less noticeable, yet in its diminishment remains more potent than ever as a transformative space for shifts in attention regarding how we know, perceive and become aware of our lived realities. Hence, the question to ask need not always be "what is the truth"; as relevant a question is: "what truth do we care to know or perceive, and what does that say about how we live?" In a sense, that query is also a holistic one asked of all humanities work, which is yet another mission of in-between-ness: as David Theo Goldberg puts it, the humanities is really that "of translating ourselves...to ourselves."[28] The key to imagining the post-screen is to articulate a critical attention that points squarely to re-visiting that query. Or to take Foer's wording in the opening quotation of this introduction, trusting that value lies not in what the information allows us to know, but how it allows us to be known.

Eroding Boundaries in the Contemporary Mediascape

This book will situate its discussion of the post-screen around three groupings of screen media, identified as key exemplars for their various intriguing subversions of screen boundaries particularly in contemporary instantiations: Virtual Reality (VR; chapter 3); holographic projection (chapter 4) and true holograms (chapter 4A); and light projections (chapter 5). Their examples, chosen for their substantive illustrations of the meanings of the post-screen, will traverse across a relatively wide range, drawing primarily out of the moving and still image (paintings; photography; films; television; video games; mobile apps), but also

---

26   John Durham Peters, *The Marvellous Clouds: Toward a Philosophy of Elemental Media* (Chicago, IL: The University of Chicago Press, 2015), 50.

27   Grosz, *Architecture from the Outside*, 90.

28   David Theo Goldberg, "Deprovincializing Digital Humanities," in *Between Humanities and the Digital*, eds. Patrik Svensson and David Theo Goldberg (Cambridge, MA: MIT Press, 2015): 163-171, 165.

from screen media out of concerts; museum installations; advertising; fashion shows; architecture and political spheres (rallies, protests and expressions of activism). By no means, though, is this range exhaustive or meant to be so; notably, technical fields, such as military, scientific and medical applications of screens, have been omitted not because of their inapplicability to the post-screen, but their contextual referencing to the disappearance of screen boundaries is not as clear. The post-screen is not only a phenomenon across multiple screen applications, but also a substantive statement on media and its connections to contemporary changing conditions of truth and reality, expressions that are evidenced with greater clarity through screen works in some spheres as compared to others. Similarly, these groupings of exemplars do not imply the post-screen as a new phenomenon limited to "new" media. Numerous historical instantiations show such practices to be as old as on-screen display itself. Early cinema exhibitors, for example, projected images of historical figures on screens as part of multimedia entertainment experiences even as they concealed their boundaries through various engineering feats and optical trickery. These "older" media will likewise be threaded through the book alongside their "newer" counterparts.

At the same time, the impetus of the post-screen is undoubtedly the ceaseless screen innovations of image display and boundaries, each crop-ping up at trade shows to trumpet their status as the latest gadgets on the market. For instance, "3D hologram fans" advertised at trade shows in 2019 and 2020 create "screens" out of rapidly rotating LED fans. These images do not appear imprinted or projected on any sort of surface resembling a conventional screen. Rather, strips of LED pixels attached to (usually four) fan blades are lit by a control unit as the blades spin, tricking the observer's brain into seeing the image as not only a whole, but also volumetric. These effects are due to the near-invisibility of the fast-spinning fan blades creating a see-through space for the image to take the illusion of a three-dimensional form.[29] At the 2020 Consumer Electronics Show (CES) in Las Vegas and one of the largest, if not the largest, trade shows in the industry, Samsung presented, among other products, "a new 8K bezel-less TV,"[30] or in more

29   Andrea James, "Watch These 3D Jellyfish Holograms Created With Fans," *BoingBoing*, September 21, 2020, https://boingboing.net/2020/09/21/watch-these-3d-jellyfish-holograms-created-with-fans.html.
30   Ivan Mehta, "Samsung unveils a bezel-less 8K TV and a rotating TV at CES," *TNW* online, January 6, 2020, https://thenextweb.com/plugged/2020/01/06/samsung-unveils-a-bezel-less-8k-tv-and-a-rotating-tv-at-ces/. Despite the headline, the report then states that the TV actually has "a barely-visible 2.3 mm thick bezel" (np), which contradicts its headline proclamation of the screen being "bezel-less."

hyperbolic reportage, "the world's first ever frame-less TV."[31] Optical illusion is no longer the name of the game here – the categorical absence of visible boundaries around the screen in the clear light of day announces the industry's unambiguous ambition to blend the virtuality of the image ever more seamlessly with the actuality of its surroundings. In this respect, a deliberate media archaeology[32] of screens' long history of paradoxical revelation and concealment will also be one of this book's key frameworks in discussing the post-screen's subversion of screen boundaries.

Across the wider mediascape, the erosion or elimination of boundaries between art and its surroundings further resonates with the post-screen's thesis of disappearing boundaries and encroaching virtualization. The location of art is not only everywhere but seamlessly so, augmenting and adding layers to multiple processes of constant virtualization. Take, for example, the general containment of paintings within their frames. Much of landscape painting, as one genre amongst many, is about the boundaries of the frame around the painting that, as Bernard Comment puts it, "give them shape."[33] Comment quotes famous painters, such as Leon Battista Alberti and Pierre-Henri Valenciennes, to emphasize the role of the frame in how it "designate[s] the representation"[34] in their paintings as a specific view through a window. Alberti, in particular, famously asked for the painting to be seen as if out of "an open window through which the story can be viewed."[35] Valenciennes described the canvas as "the aspect of nature that is circumscribed by the frame, always creating the effect of a window."[36] As Comment concludes: "It is therefore the frame that denotes that a work of art is what it is."[37]

Yet, eventually – perhaps even inevitably, if we take the viewpoint of a kind of post-screen determinism – even the frame is abolished. Instead, virtual reality floods the viewer's eye. Arthur Danto, for instance, in his argument on "contemporary art" in the 1990s as marking an end to an era

31   James Pero, "Samsung is set to unveil the wold's first ever bezel-less TV next week at the Consumer Electronics Show in Las Vegas," *Mail Online*, December 31, 2019, https://www.dailymail. co.uk/sciencetech/article-7841467/Samsung-set-unveil-worlds-bezel-free-TV-week-CES-Las-Vegas.html. Again, though, strictly speaking, the TV is not bezel-less (see footnote 30).

32   See Jussi Parikka's instructive book on media archaeology as method, *What is Media Archaeology* (Malden, MA; Cambridge: Polity Press, 2012), explaining, among others, its "excavating the past in order to understand the present and the future." (2)

33   Bernard Comment, *The Painted Panorama* (New York: Harry N. Abrams, Inc., 2000), 99.

34   Comment, *The Painted Panorama*, 99.

35   As quoted in Comment, *The Painted Panorama*, 99. The Albertian window in relation to the screen will also be discussed in greater detail in chapter one.

36   As quoted in Comment, *The Painted Panorama*, 99.

37   Comment, *The Painted Panorama*, 99.

of modern art, describes a "generation" of such art ("of which the Museum of Modern Art is the great exemplar") as "defined in formalist terms": "Nothing was to distract from the formal visual interest of the works themselves. *Even picture frames were eliminated as distractions...: paintings were no longer windows onto imagined scenes, but objects in their own right......*" (emphasis added)[38] Other genres subvert the formal containment of art in more elaborate ways, such as land art from the 1960s and 1970s which sited art in remote locations by sculpting the land itself with its natural materials, bypassing the traditional confinement of art in a frame that sets it apart against its surroundings.[39] Arguably, Marcel Duchamp's readymades in the 1910s, by presenting as art ordinary manufactured objects which he sometimes modified and sometimes not, already rubbed out the boundaries between art and the real world, if only by upending the definitions and parameters by which each became one or the other.

Shifting boundaries between artifice and life may also be seen in other, if more oblique, instantiations. There are many examples out of diverse contexts; a couple to illustrate our purposes here will suffice. For instance, in the 1960s, Richard Schechner, with the Performance Group, founded and performed what Schechner later termed "environmental theatre"[40] – a "non-frontal, spectator-incorporative theatre" that aimed to eliminate the distinction between conventional audience and stage territories.[41] On sets designed to deliberately encroach on the audience's space, the actors have greater space and flexibility of interaction with the audience. They are subsequently able to "incorporate the spectator in some way within the performance and to diminish the sense of aesthetic distance."[42] These experimentations with space, started by Schechner but since taken up and further developed by other performance groups, thus erase, even abandon,

---

38   Arthur Danto, *After the End of Art: Contemporary Art and the Pale of History* (Princeton, NJ: Princeton University Press, 1998), 16. Danto centres his arguments of this "post-historical museum" (16) around the subversion of the logic of the painting's frame – "the architecture of the altarpiece, the installation in which a painting is set like a jewel." (xii) Paintings are no longer situated within them, but take on multiple different frameworks, such as other media forms (e.g. sculpture, installations, film) or other prescriptions of space (e.g. fictive space).

39   See *Land, Art: A Cultural Ecology Handbook*, ed. Max Andrews (London: RSA, 2006); or Suzaan Boettger, *Earthworks: Art and the Landscape of the Sixties* (Berkeley, CA: University of California Press, 2002).

40   See Richard Schechner, "6 Axioms for Environmental Theatre," *The Drama Review: TDR* 12, no. 3 (Spring, 1968): 41-64.

41   "Environmental theatre," *The Oxford Companion to Theatre and Performance*, ed. Dennis Kennedy (Oxford: Oxford University Press, 2010), 190.

42   Ibid.

the theatre's conventional boundaries which separate art from life, or audience from performance (and performers), as usually signified via theatre architecture such as stage, proscenium and stage curtains.

A second example is the genres of twenty-first century interactive mobile narratives which integrate physical and narrative spaces, such as what has been called Locative Narratives or Locative Literature, whose stories are told through media assets attached to physical spaces.[43] An example is [murmur], a "documentary oral project" whose creators collected recordings of stories and memories about specific neighbourhoods in Toronto and made them accessible to the public through posted signs bearing a telephone number for people to call.[44] The result, as Jeremy Hight puts it, is that "stories are written in and read in motion in the physical world itself."[45] As with the examples described above, these mobile genres bypass their traditional frameworks – in this case the book, which normally defines the ontological borders for narrative, at least for the Western canon.[46] The boundaries within which a literary text exists, is authored and read thus become less certain, more fluid and more contingent on movement and the location of the body in public space. As Hight suggests, this shifting of boundaries disrupts "form and completion and the fetishistic notion of a work as a singular set artefact and architecture."[47] What signifies as textual literature is now spread across the landscape, a layer of fictional reality fused with its environment, its boundaries indistinguishable and no longer defined via any specific textual frame.

A newly virtualized virtual reality propagated by visual and narrative media today thus emerges out of this volatile interchanging between the

43   For a wide range of examples in diverse contexts demonstrating the practice of storytelling on mobile media, see *The Mobile Story: Narrative Practices with Locative Technologies*, ed. Jason Farman (New York; Abingdon: Routledge, 2014).

44   The [murmur] project website, at http://murmurtoronto.ca/about.php/ (as of June 2020) is unfortunately defunct, but a detailed description of the project can be found at the Canadian Film Centre website, accessed June 11, 2020, http://cfccreates.com/productions/76-murmur.

45   Jeremy Hight, "Locative Narrative, Literature and Form," in *Beyond the Screen: Transformations of Literary Structures, Interfaces and Genres*, eds. Jörgen Schäfer and Peter Gendolla (Bielefeld: Transcript Verlag, 2010): 317-330, 319.

46   Narrative traditions are notably more entwined with place in non-Western cultures, such as Australian Aboriginal narrative systems: see generally, for instance, *Emplaced Myth: Space Narrative and Knowledge in Aboriginal Australia and Papua New Guinea*, eds. Alan Rumsey and James F. Weiner (Honolulu: University of Hawaii Press, 2001). For an extension of Aboriginal narrative systems to new media narratives, see James Barrett, "Virtual Worlds and Indigenous Narratives," in *The Immersive Internet: Reflections on the Entangling of the Virtual with Society, Politics and the Economy*, eds. Robin Teigland and Dominic Power (Basingstoke: Palgrave Macmillan, 2013): 77-91.

47   Hight, "Locative Narrative," 322.

establishment and scuppering of frames and borders, a fluidity that is also the central dual-edged challenge of *where boundaries lie*. In turn, this unsettled state changes the nature of the contestation between the virtual and the actual, where virtuality increasingly encroaches on the actual, revising not only the value of representation but also who we are in relation to representation. This leads us to the next point – why this matters.

Why Boundaries Matter

In the first place, boundaries are difficult spaces – paradoxical, interstitial, liminal. As mentioned, they are a facilitating interface bound by the qualities of relations, defined by what is outside it as much as what is inside it. Jacques Derrida's definition of the *parergon*, appearing in the first section of *The Truth in Painting* and itself an explication of framing and the *passe-partout*, applies well to the boundary's competing contradictions: "neither work (*ergon*) nor outside the work [*hors d'oeuvre*] neither inside nor outside, neither above nor below, it disconcerts any opposition but does not remain indeterminate and it *gives rise* to the work." (emphasis in original)[48] The *parergon* is caught in simultaneous disavowal and affirmation – it exists by not being a part of the object (or *ergon*) as much as by uniting with the *ergon so as* not to be a part of it. In this, it echoes the koan of the doughnut's hole, which exists as an *absence* of edible doughnut ring as much as it does in relation to *being part of* the edible doughnut ring.[49] It is what it is as also

48 Jacques Derrida, *The Truth in Painting*, trans. Geoff Bennington and Ian McLeod (Chicago, IL: University of Chicago Press, 1987), 9. See also Gregory Minissale, *Framing Consciousness in Art: Transcultural Perspectives* (Amsterdam: Rodopi, 2009) for elaboration on Derrida's use of framing – "alongside texts and traditions in order to create a view of their lacunae" – as a discursive figure for revelation, 91.

49 This also brings to mind an oft-quoted verse from Lao Tsu's *Tao Te Ching*, which similarly emphasizes the paradox of what is there against what is not there, how they interrelate to each other, and, most importantly in relation to the *Tao*, understanding their worth against each other:
 Thirty spokes share the wheel's hub;
 It is the center hole that makes it useful.
 Shape clay into a vessel;
 It is the space within that makes it useful.
 Cut doors and windows for a room;
 It is the holes which make it useful.
 Therefore profit comes from what is there;
 Usefulness from what is not there.
From Lao Tsu, *Tao Te Ching*, trans. Gia Fu Feng and Jane English (New York: Vintage Books, 1998), chapter 11, 7.

based on what it is not. Where such a complex space starts to shift in its nature and manifestations, the implications are also bound to be interesting.

On a functional level, boundaries are important because they are defini-tional. Not least due to their inherent ambiguity as being both included and excluded spaces, boundaries define – and are themselves – both beginnings and endings. A frame that surrounds an image marks where the image begins and ends, or the differentiation between what is reality and what is representation. To return to paintings, Rosalind Krauss describes the painting's frame as this "very boundary of the image"; the frame "crops or cuts the represented element out of reality-at-large." What is cropped or cut thus becomes "an example of nature-as-representation, nature-as-sign." Hence, "[t]he frame announces that between the part of reality that was cut away and this part *there is a difference*." (emphasis added)[50]

By being definitional, boundaries also lay down other dictates. They become instructive, even imperative, as they direct what a viewer should look at and what they should ignore. As Dudley Andrew writes: "The frame is the physical embodiment of the bar between image/reality and it marks as well the case that this experience is presented to me by another. I must attend 'there' to the frame and not elsewhere."[51] By marking out objects for a viewer's attention, boundaries facilitate their being seen, and enable them to be seen: "To frame something is to re-present it... Re-presentation invites us to look again; it renders visible."[52] Through such prescription of attention and visibility, boundaries thus also command power in asserting what is and is not important, what deserves and does not deserve the viewer's gaze, what possesses and lacks meaning. Boundaries, as Malcolm McCullough writes, "privilege the contained."[53]

For these reasons, boundaries do intense work. They direct attention, provide meaning, include and exclude, allow and withhold access. To that end, media and media theory have also long been attentive to the ambiguity and the ensuing relational richness of boundaries which contain them. Paintings, photography, literature, television and cinema have all explored, interrogated and played with meanings portended within, without and across their respective frames; many of these discussions will feature

---

50   Rosalind Krauss, *The Originality of the Avant-garde and Other Modernist Myths* (Cambridge, MA: MIT Press, 1985), 115.

51   Dudley Andrew, *Concepts in Film Theory* (Oxford: Oxford University Press, 1984), 43.

52   Karsten Harries, *The Broken Frame: Three Lectures* (Washington, DC: The Catholic University of America Press, 1989), 85.

53   Malcolm McCullough, *Ambient Commons: Attention in the Age of Embodied Information* (Cambridge, MA: MIT Press, 2013), 156.

in the next two chapters. As Anne Friedberg writes, "how the world is framed may be as important as what is contained within that frame."[54] As used in theory, boundaries explicate the nature of media, mapping how media evolves and changes. Jay Bolter and Richard Grusin's argument of remediation, for instance, employs the visibilities and differentiations marked by media's boundaries to underpin their historical account of media change.[55] Across a wide range of media forms, Bolter and Grusin argue how the twin logics of "hypermediacy" and "immediacy" power aesthetic and/or ontological connections between "older" and "newer" media. Respectively, these logics highlight or erase the visibility of those connections. Not unlike the flash mob of Breda, at the heart of Grusin and Bolter's argument is a fluid and competing interplay between the presence and erasure of boundaries. In this sense, the logic of immediacy diminishes the perceptibility of media boundaries so that "the medium itself should disappear and leave us in the presence of the thing represented."[56] Conversely, the logic of hypermediacy emphasizes media boundaries so as to highlight connections to or replacements of other media forms, and to remind viewers of the opacity of media.[57] The shifting of screen boundaries thus delineates the trajectory of media's development, and draws the lines connecting past and present, old and new.

Media theory also rationalizes how screen boundaries form and operate as critical thresholds between image and object to host tensions and transgressions. For example, it is across the screen's boundaries that the *onscreen* (a signified reality visible to the audience) functions as a yin-yang dialectic to the *offscreen* (not visible to the audience).[58] Moreover, like twisted cabling, their realities further entangle across their boundaries to influence and

---

54 Friedberg, *The Virtual Window*, 1.

55 Jay David Bolter and Richard Grusin, *Remediation: Understanding New Media* (Cambridge, MA: MIT Press, 2000).

56 Bolter and Grusin, *Remediation*, 6. This binary division of media in terms of visibility and disappearance echoes Peter Lunenfeld's dialectics of new media, where Lunenfeld identifies two key paradigms of the new computer media, namely, *immersion*, as associated with virtual reality, and *extraction*, as associated with hypertext: see Peter Lunenfeld, "Digital Dialectics: A Hybrid Theory of Computer Media," *Afterimage* (November 1993): 5.

57 Bolter and Grusin, *Remediation*, 31-44.

58 See Noël Burch, *Theory of Film Practice*, trans. Helen R. Lane (Princeton, NJ: Princeton University Press, 1981), 17-31, and in particular how he identifies six "segments" of offscreen space around the film image: offscreen right; offscreen left; offscreen top; offscreen bottom; behind the set; and behind the camera. What is at stake here is how, across boundaries signifying on- and off-screen, the image world is constructed, contained and separated.

affect each other.[59] Tom Gunning's "cinema of attractions" argument,[60] for instance, describes early cinema as technological and visual excitement – or what he ascribes to "an aesthetic of astonishment"[61] – by drawing precisely on the intersections across screen boundaries. Where those boundaries differentiate between virtual and actual realities, they also mark where and how the early cinema spectator's astonishment arose in relation to the incredible (or incredibly mimetic) nature of the illusion they were seeing against their reality. Likewise, the cinematic mode of direct address, referred to here as "characters in movie fictions who appear to acknowledge our presence as spectators; they seem to look at us,"[62] achieves its status of anomalous use precisely due to the pressure of crossing the supposedly inviolable divider between the audience's and the character's world.[63] These are just a few examples of how screen boundaries underpin significant theorizations of evolving relations between mediated and physical realities, reliant on what is within and without the screen's boundaries, and trading off tensions and ambivalences that arise across them.

Moreover, thinking about screen boundaries also leads to a deeper understanding of the object itself – the screen. There may not seem to be much to understand about a screen beyond its technology and engineering[64] – is it not simply a surface filled with light that displays text and images? As Charles Acland puts it: "we just seem to know [a screen] reflexively: a thing that glows and attracts attention with changing images, sounds, and

---

59   For a more extensive analysis of the intertwining between onscreen and offscreen spaces, see Eyal Peretz, *The Off-Screen: An Investigation of the Cinematic Frame* (Stanford: Stanford University Press, 2017), especially Part 1, 15-60.

60   This refers to Tom Gunning's well-known argument of early cinema not as a storytelling medium, but as an exhibitionist cinema "that displays its visibility, willing to rupture a self-enclosed fictional world for a chance to solicit the attention of the spectator": see "The Cinema of Attractions: Early Film, Its Spectator and the Avant-Garde," in *Early Cinema: Space, Frame, Narrative*, eds. Thomas Elsaesser and Adam Barker (London: BFI Publishing, 1990): 56-62, 57.

61   See Tom Gunning, "An Aesthetic of Astonishment: Early Film and the (In)Credulous Spectator," in *Film Theory: Critical Concepts in Media and Cultural Studies*, eds. Philip Simpson, Andrew Utterson and Karen J. Shepherdson (New York; Oxford: Routledge, 2003): 114-133.

62   Tom Brown, *Breaking the Fourth Wall: Direct Address in the Cinema* (Edinburgh: Edinburgh University Press, 2012), x.

63   Referring to Pascal Bonitzer's characterization of this counter-look as the "rupture of the cinematic fiction," Brown similarly notes that, as such "rupture," such address "can only ever be tentative": Brown, *Breaking the Fourth Wall*, 23.

64   There are similar sentiments in other work on this tension between the technical knowledge of a media form and the formation of knowledge out of it, such as that of the book, on which see, for instance, N. Katherine Hayles, *Writing Machines* (Cambridge, MA: MIT Press, 2002).

information."[65] Yet, as Acland also points out, and to that extent echoing calls for the same as set out from the mid-2000s,[66] scholarship, particularly from history, media and critical theory, is needed precisely to forge new ways of understanding the screen beyond being an instrumental technology. As Acland writes, "the mechanical level [of technical specifications, such as screen size, aspect ratio, resolution etc] only gets us so far in our job of actually understanding the related senses, sensibilities, and practices that form as a consequence of media use."[67] Much of such scholarship in recent years have concentrated on re-rationalizing the boundaries and lines of the surfaces to which we commonly designate as screens;[68] in turn, they re-visit our wider visual and media culture in relation to the nature and status of representation in our world. An example of such key thinking for me is Fred Turner's lecture in 2014 on renewed conceptualizations of the screen in terms of its "ubiquity and integration" which mark similarly diminishing screen boundaries.[69] As screens envelop their audiences in their omnipresence, Turner proposes the framework of thinking about screens to shift across various binaries, from "screen" to "surround"; "representation" to "attention"; "production" to "integration"; "reception" to "interaction." What emerges then, in wider terms, is a different screen history, or "the screen history we need."[70] Vivian Sobchack, too, argues for

65   Charles R. Acland, "The Crack in the Electric Window," *Cinema Journal*, 15:2 (2012): 167-171, 168.

66   See in particular Erkki Huhtamo, "Elements of Screenology: Toward an Archaeology of the Screen," *Iconics: International Studies of the Modern Image*, 7 (2004): 31-82, in which he specifically calls for "the creation of a new field of research which would be called 'screenology,'" which would focus not only on "screens as designed artefacts, but also on their uses, their intermedial relations with other cultural forms and on the discourses that have enveloped them in different times and places": 32. He repeats this call in "Screen Tests: Why Do We Need an Archaeology of the Screen," *Cinema Journal*, 51(2) (Winter 2012): 144-148.

67   Ackland, "Crack in the Electric Window," 168.

68   For instance, Acland discusses how "production screen" innovations, such as the "Simulcam" as used by James Cameron for the filming of *Avatar* (2009), has moved the screen "from the endpoint of spectatorship to the position previously occupied by the industry-standard motion picture camera," so that "conventional definitions of monitor, computer, and camera are disrupted. The camera is a screen and the screen is a computer, and all are windows onto a live, virtual performance": "Crack in the Electric Window," 169-170. In other words, the definitional and ontological boundaries of the screen collapse; our understanding of the screen itself changes.

69   Fred Turner, "From Screens to Surrounds" (presentation, Genres of Scholarly Knowledge Production HUMlab conference, Umeå, December 10-12, 2014).

70   See Fred Turner, *The Democratic Surround: Multimedia and American Liberalism from World War II to the Psychedelic Sixties* (Chicago, IL: University of Chicago Press, 2013), which fleshes out the visual landscape that feeds into this screen history in terms of what he terms as the "surround." I pick up again on this sense of the "surround" in chapter 3.

the reconstitution of "what was once a 'screen-scape' into the surround of a systemically-unified, if componentially diversified, 'screen-sphere.'"[71] In this sense, screens, being part of "our lifeworld," become a systemic structure, both enfolding life and affirming "being" within each other. Both arguments disentangle the shifting of screen boundaries to clarify the morphosis of the screen itself, in turn pushing for a larger understanding of it, revising trajectories and taxonomies of its changing forms, structures, functions and purposes. The impact of such work has been both a source of inspiration and an important starter premise for the main threads of enquiry running through this book.

However, by far the most significant importance for boundaries in relation to the thoughts in this book is how they signify relations which bound back *to us* as viewers, so that understanding boundaries becomes, as well, understanding ourselves. Demarcating between art and life, boundaries point to fundamental truths about both, and in the process to qualities of being human in navigating between the two. Of the many cultural expressions on frames and borders, one painting stands out for precisely its sheer pathos in this connection drawn between boundaries and being human: Réne Magritte's (and in this case aptly titled) *La Condition Humaine*.[72] *La Condition Humaine* (and others featuring the same theme, for Magritte was fond of repeating his visual tropes across several paintings) depicts a segment of a landscape portrayed as a near-continuous view, with consistent positioning, as seen both through a window and on a painted canvas set in front of the window. Magritte describes the painting thus: "In front of a window seen from inside a room, I placed a painting representing exactly that portion of the landscape covered by the painting."[73]

71 Vivian Sobchack, "From Screen-Scape to Screen-Sphere: A Meditation *in Medias Res*," in *Screens: From Materiality to Spectatorship – A Historical and Theoretical Reassessment*, eds. Dominique Chateau and José Moure (Amsterdam: Amsterdam University Press, 2016): 157-175, 158.

72 René Magritte, *La Condition Humaine (The Human Condition)*, 1933, oil on canvas, 100 cm x 81 cm, National Gallery of Art, Washington DC. Magritte actually made two similar paintings of this title, *La Condition Humaine I*, 1933, and *La Condition Humaine II*, 1935, the latter of the same dimensions as the first version and currently located at the Simon Spierer Collection in Geneva, Switzerland. Moreover, Magritte repeated *La Condition Humaine*'s theme of artifice against landscape in various other paintings, such as *La Llama de la Cimas (The Call of the Peaks)*, 1943, oil on canvas, 65 cm x 54 cm, The Magritte Museum, Brussels.

73 As quoted in Malcolm Andrews, *Landscape and Western Art* (Oxford: Oxford University Press, 1999), 124.

The key lies in the nuances of the word "exactly."[74] The canvas's depictions of the landscape outside the window very nearly – but do not quite – "exactly" match the view from it. What clearly and deliberately disrupt the painting's otherwise flawless alignment are subtle yet unmistakable indications of the canvas's borders as it rests on its easel – the faint strokes of the canvas outline, its edges topped and tailed by clip and stand, and a clear white strip along its right edge studded with pinheads that fix the canvas in place. In his letters, Magritte explains the painting as an interrogation of how a person sees the world, with its multiple representations indicating a viewer's internal and external representations.

However, art critics such as Renée Riese Hubert go further, reading the painting as a veritable expression of creative failure:

> When Magritte makes his spectator see simultaneously the landscape as a segment of nature and a work of art, he does not primarily deal with the question of aesthetic transformation. He stresses the creating, makes painting unnecessary, turns it into failure.[75]

Read this way, the painting becomes a statement on how "man in relation to both nature and art imprisons himself," where the artist overlooks perspective and "forget[s] that the 'outer' landscape is situated at a certain remoteness, if compared to the proximity of the scene imprinted on the window or the easel."[76] It is a stark announcement of defeat in bridging representation and object, marked by the highlighting of the canvas's edges in what would otherwise have been indeed an "exactly" seamless placing. In short, the boundaries are always there; *the gap always shows*. But the painting is not only about failure, it is also about the desire to seal that gap, control our environments, master artifice and the virtual to the apogee of the real. It is about the broader yearning at play in our mediation of our surroundings

---

74   I take much trouble in qualifying the consistency of the view across canvas and window in *La Condition Humaine* because, to me, how the boundaries of the canvas patently and deliberately break up that consistency are paramount to its meaning and, above all, are crystal clear. It puzzles me why scholars tend to treat the view painted on the canvas without such qualification, such as Andrews, *Landscape*, who declares that "the artificial looks just as real as the scene it represents," and that "the landscape inside the room is indistinguishable from the landscape outside," 124. It is not. Otherwise, elsewhere Andrews also declares these distinguishing features to be part of a Surrealist dream, an "intrusion of something alien," 126, which is an interpretation from another direction altogether.
75   Renée Riese Hubert, "The Other Worldly Landscapes of E.A. Poe and René Magritte," *SubStance*, 6/7(21), (Winter, 1978-1979): 68-78, 72.
76   Ibid.

and our realities. Boundaries thus take on this weight of reference in relation to media and the human condition: in revealing the unbridgeable, the chasms, in turn, expose what we really want and the truths of being human in the failures to attain them. The significance of the boundary in *La Condition Humaine* is thus its revelation of the human condition in the way only boundaries can – in paradox, in riddle, in ascribing to what is there as much as what is not there. The painting is a powerful statement about why boundaries matter and, for that reason, will be a frequent reference in the ensuing chapters of this book. It is not only an inspiration, but also a thoughtful reminder of how, just as the chink in the armour does with weakness, it is the gap in the boundary which exposes truth.

Boundaries are thus important because they are prescriptive in fundamental ways, defining *ergon* against *parergon*; giving rise to the object of attention against what is to be ignored; creating meaning through what they privilege and what they exclude. They underpin significant theory for our understanding of images and realities. They are prime articulations of how, as humans, we seek and fail to master our world, and hence are in themselves fundamental expressions of who we are and what we desire. They are lines drawn in the dust of elemental contestations – human versus nature; art versus life; artifice versus organic; representation versus reality; copy versus original; virtual versus actual. Disrupting boundaries means revising the nature of these battlefields, and waging its wars anew.

Chapter Outlines

The book will proceed as follows. Chapters 1 and 2 first elaborate on specific articulations of screen boundaries via cinema, television, video games and mobile apps, chosen as the main exemplars of screen media in the last hundred years. Each chapter presents a different argument on thinking through the screen in terms of its boundaries, and in particular showcases the paradoxes in their operative frameworks of image against its surroundings: chapter 1 on how screen boundaries display yet conceal the virtual against the actual; chapter 2 on how they separate and partition the image, yet are undermined by various practices and in particular the emergence of interactive media which destabilize their delimitations of screen reality. With readings through theory and examples, particularly from cinema, the two chapters demonstrate the contradictory nature of screen boundaries and the brittleness of their space. These contradictions in turn set up the book's main arguments for the more complex thresholds across virtuality and actuality of the post-screen's disappearing boundaries in contemporary media.

Chapters 3, 4 and 5 formally address the notion of the post-screen through three media technologies: virtual reality (VR); holographic projection; and projection mapping. In particular, they highlight how the erosion of screen boundaries in each media form gives rise to the post-screen, and how each instantiation of the post-screen shifts our understanding of specific concepts in relation to images and representation: the placements of virtual and actual reality (chapter 3); the understandings of bodies, death, life and the afterlife (chapter 4); and the convertibility between materiality, matter, light, energy and mass (chapter 5), each chosen as the most apposite and arresting ideas to emerge from the respective technologies. While these concepts are discussed discretely within the chapters, they combine to colour the shifting real of the post-screen, where the fusion of the virtual and the actual builds a new imagination of both representations and objects, and, in turn, a new kind of media history.

Chapter 3 first situates VR in the context of totalizing media environments, before detailing its erosion of screen boundaries in terms of what I call "the affective surround." In turn, the totalization of reality in VR can be realized via two approaches: immersion and inversion. In this process, the post-screen emerges as renewed imbrications between the actual and the virtual not in terms of the more conventional paradigm of replacement (one for the other), but a more nuanced *re-placement* (one shifted to another) across VR's screen boundaries. This re-placement of the real thus provides a new paradigm in which to consider how actual and virtual realities intertwine in inherent paradoxes across screen boundaries. In turn, this paradigm sheds light on our processes of virtual perception, on remembering and forgetting, and on the dimensional shifts from the physical to the virtual.

Chapter 4 next considers the subverted boundaries of holographic projections as presentations of ghosts and apparitions. It first considers media's long history with death, ghosts and reanimation, tracing four different moments in that entwined trajectory: resurrection; necrophilia; necromancy; and interactivity. The last paves the way for the post-screen of holographic projections to radically relocate our ideas of the afterlife in two ways: the first as ghosts *amongst* the living in a newly nuanced limbo between deadness and aliveness; and the second as ghosts *of* the living, located in a tetravalence of their being here/elsewhere against their actuality/virtuality. Both senses of ghosts thus re-emerge in the post-screen with paradoxical spectralities: one as more alive *when* dead; the other as what I call being *vivified*, or bodies gaining the realness of being alive in their being elsewhere on tetravalent axes of space/time and actual/virtual reality.

Chapter 4 then segues into chapter 4A as what I call a "remix chapter" which contemplates the post-screen through the true hologram, commonly misconceived as or mixed up with the holographic projections of chapter 4. The true hologram is not a projection, but relies on unique technical recording processes and technologies. Nonetheless, pivoting from the argument on ghosts from chapter 4, chapter 4A argues that the hologram may yet be considered an instantiation of the post-screen in terms of the spectral; the argument, though, takes on a very different shade. Rather, the post-screen through the true hologram expresses a different kind of ghost from a different kind of screen: in relation to the latter, an aggregate of brains, nerves and thought; and in relation to the former, the ghosts which emerge are ultimately those from the viewer's own psychology, drawn from as much a different ontology of the world as points or point elements as the viewer's own dreams and hidden secrets. The ghosts of the post-screen through the true holograms thus also *re-place* the living: not ghosts as from the dead or from the living's being of elsewhere-ness, but from the living as re-placed to different levels of introspection and terms of being. These are ghosts which ultimately bound back to ourselves.

Finally, chapter 5 discusses the third instantiation of the post-screen through light projections, specifically advancing its argument on light as giving rise to dynamic interrelations between materiality and immateriality; matter and energy; rigidity and fluidity. As such, light projections translate the boundaries of the image across a variety of surfaces – the urban (e.g. building façades); the amorphous (e.g. water droplets and ash); and the biological (e.g. bodies and faces) – into the post-screen by way of their convertibility between matter, solidity and energy. In this frenzy of disembodiment, the post-screen here thus also sets itself out as part of a culture of gluttony for media, and in particular for images which dissociate themselves completely from the physical realities of their object. The contestation of the actual and the virtual thus takes on a different note here, where it is not just about the totality of the consumption of the image, but a clarion marker of a different chapter of media history: one whose ease of convertibility in the post-screen has also become a signal fire for the politics of the twenty-first century of misinformation, post-truth and shit storms. These ideas, drawn also in a late parallel against the viral contagion of the Covid-19 pandemic which has indelibly marked the world for at least the years of 2020-21, will be summarized in the book's conclusion.

The Post-what?

> *What was separated in the past is now everywhere merged; distance is abolished*
> *in all things: between the sexes, between opposite poles, between stage and*
> *auditorium, between the protagonists of action, between subject and object,*
> *between the real and its double. And this confusion of terms, this collision of poles*
> *means that nowhere – in art, morality or politics – is there now any possibility of a*
> *moral judgement.*
>
> ~ Jean Baudrillard[77]

"The post-what?" enquiry of this section addresses the elephant in the room, which is the exponentially worn groove of the "post" prefix in critical theory. Even criticality is not spared, as seen in Michael Polanyi's coinage of *the post-critical* that designates the shift of critique itself to being "beyond" "critical" sensibility.[78] Across the "post-" lexicon, the *posthuman* – in terms of the enquiry which decentres the human – is perhaps its most deep-rooted term, and in prolific use today across multiple disciplines. Yet, despite (or perhaps because of) its proliferation, even the posthuman itself splinters into various facets of "post-" concepts, as Rosi Braidotti and Maria Hlavajova's 2018 *Posthuman Glossary* demonstrates with their extensive list of "critical terms of posthumanity."[79] This list includes posthuman sexuality and posthumanist performativity, as well as more tangential "post-" tenets in the posthuman scope, such as postdisciplinarity, postanimalism, postglacial, postimage and postmedieval. In recent years, yet more "post-" terms have appeared in a slew of variations across diverse areas, appearing as post-media, post-cinema, post-Internet, post-virtual, post-digital, post-anthropocene, post-feminism, postmaterialism, posthumanities, postracial, post-truth, post-theory and post-algorithmic, just to name a few.[80] There are probably many more in the pipeline; the post-screen clearly has to take a number!

---

77 Jean Baudrillard, "Screened Out," in *Screened Out*, trans. Chris Turner (London; New York: Verso, 2002): 176-180, 176.

78 Michael Polanyi, *Personal Knowledge: Towards a Post-Critical Philosophy* (Chicago, IL: University of Chicago Press, 1958).

79 Rosi Braidotti and Maria Hlavajova, *Posthuman Glossary* (London; New York: Bloomsbury, 2018).

80 It is impractical and unnecessary to list all references featuring these terms; a sample here will hopefully suffice: Roger F. Cook, *Postcinematic Vision: The Coevolution of Moving-Image Media and the Spectator* (Minneapolis, MN: University of Minnesota Press, 2020); David Theo Goldberg, *Are We All Postracial Yet?* (Malden, MA; Cambridge: Polity Press, 2015); Piotr Woycicki, *Post-Cinematic Theatre and Performance* (London: Palgrave, 2014); Vincent Mosco, *Becoming Digital: Toward a Post-Internet Society* (Bingley: Emerald, 2017); Nicos Komninos, *Smart Cities*

In that sense, the "post-" is undoubtedly trendy. However, its popularity also ironically threatens its ontology of distance, and hence critical relevance. Writing a concluding chapter (itself labelled a "post-script") for *Media After Kittler*, Jussi Parikka comments tellingly on the disappearing distance between the object and its "after": "Just when you thought (new) media studies got started it seemed already over."[81] Insert "(new) media studies" with any discipline of choice – including, for that matter, (new) screen studies – and chances are its "post-" is already on the horizon. Yet, proclaiming the closure of an era to justify its "post-" requires genuine consideration, an exercise which entails honest and sometimes agonizing self-reflection. Miriam De Rosa and Vinzenz Hediger's introduction of their edited issue, "Thinking Moving Images Beyond the Post-medium/Post-cinema Condition" in the *Cinéma & Cie* journal, is one example which reflects with candid frankness the "twists, negotiations, or even jolts" of the provocations posed by their choice of issue title. De Rosa writes: "Yet, after all that has been said and written, I am still not quite sure what post-cinema is" – by that honest disclosure, the ensuing examination also doubles up as a contemplation on "what cinema is" (or more accurately, perhaps, what pre- post-cinema is.)[82] By reflecting on the "post-" in its acknowledgement of the ambiguities surrounding the cessation of the "pre-", the discussion becomes a fruitful re-visiting of the latter, while not losing the critical insights of progress and change via the former.

The *post-screen* thus not only jostles for space in a crowded forum, but also needs to justify its terms of discontinuation and bear its share of honest contemplation about the "previous" era. When did *the screen* end, and what is it that the post-screen is "post-" of? At risk of presenting a red herring, this book pursues neither of those arguments. "Post-" here is thus not employed in the sense of the "after" or "later," per its literal meaning; it is not intended to denote any sort of stage in chronology. Indeed, the media examples deployed to argue for the post-screen in the following chapters will range across different eras from the analogue to the electronic to the computational. The

in the Post-Algorithmic Era: Integrating Technologies, Platforms and Governance (Cheltenham: Edward Elgar Publishing, 2019); and a classic: David Bordwell, *Post-Theory: Reconstructing Film Studies* (Madison, WI: University of Wisconsin Press, 1996).

81   Jussi Parikka, "Postscript: Of Disappearances and the Ontology of Media (Studies)," in *Media After Kittler*, eds. Eleni Ikoniadou and Scott Wilson (London, New York: Rowman & Littlefield, 2015): 177-190, 177.

82   Miriam De Rosa and Vinzenz Hediger, "Post-what? Post-when? A Conversation on the 'Posts' of Post-media and Post-cinema" of "Post-what? Post-when? Thinking Moving Images Beyond the Post-medium/Post-cinema Condition," *Cinéma&Cie: International Film Studies Journal*, XCI:26/27 (Spring/Fall 2016): 9-20, 10.

screen was and is still here, at least for now. The intention is not to declare its disappearance and/or account for its putative futures.

Rather, "post-" in terms of *the post-screen* is leveraged in two ways. The first is to hark to the *critical* sense of the "post-", specifically that of the *posthuman* (and its associated tenets, particularly post-anthropocentrism). The post- of the posthuman does not so much define an "after the human" as much as it points to a critique of an ideal in terms of larger politics of entanglement, assemblage, intertwining and networking which today colours our understanding of our world, such as between human and non/inhuman entities, objects and non-anthropomorphic elements, subjects and objects. In parallel thinking, the screen does not warrant a "post-" so much in terms of its demise, but revised thinking of screen-based relations in similar expressions of entanglement, entwinement and new visions arising from them. Just as posthumanism re-oriented the relations of humans and their world, the post-screen colours another imagination of reality across the entanglements that contemporary eroded screen boundaries present in replacing and re-placing virtual and actual realities, viewers and images.

Such entanglement and enfoldment of actual and virtual is also, of course, not new. In the advent of digital imaging technologies at the turn of the twenty-first century, for instance, Lev Manovich ascribes to digital images new powers of connecting across distance between virtual and actual realities. In his essay, "To Lie and to Act," Manovich identifies two functions that representational technologies serve: to deceive, and to enable action. On deception, Manovich discusses the role of cinema, particularly stylistic techniques of film positioning, editing and montage, in what he calls "creating fake realities."[83] More pertinently, on enabling action, he highlights images of "telepresence," such as those in virtual environments, against images of "teleaction," such as those which enable "real-time remote control" – to "drive a toy vehicle, repair a space station, do underwater excavation, operate on a patient or kill – all from a distance."[84] Images of "teleaction" are thus not mere representations of objects (or even representations of fake objects), but *enablers* of a new relation between image and object across the screen's boundaries, whereby objects are not only *"turned into signs, but also the reverse process – manipulation of objects through these signs."*

---

83   Lev Manovich, "To Lie and to Act: Cinema and Telepresence," in *Cinema Futures: Cain, Abel or Cable? The Screen Arts in the Digital Age*, eds. Thomas Elsaesser and Kay Hoffmann (Amsterdam: Amsterdam University Press, 1998): 189-99, 191.
84   Manovich, "To Lie and to Act," 198.

(emphasis in original)[85] Harun Farocki, in a 2003 lecture, echoes this idea as applied in the military context of the 1991 allied war against Iraq, where he calls such images, such as those recorded by cameras fixed on a missile warhead, "operative images" – "images that do not represent an object, but instead are part of an operation [of war]."[86] In the context of drone warfare, the results of embroilment between image and object are actually lethal. As remote sensing technologies become more common, virtuality and actuality thus connect in increasingly intense ways between image and action or consequence. They are not distinct realities separated by their boundaries, but operate across them in complex imbrications and relations.

Other scholars observe similar overlaps between the virtual and actual, if in different contexts. Sherry Turkle, for instance, wrote in 1995 about "life on the screen" as fluid intersections between onscreen and offscreen lives, specifically in terms of identity construction in the context of "eroding boundaries between the real and the virtual, the animate and the inanimate, the unitary and the multiple self."[87] She quotes from "Doug," a player of multi-user dimension (MUD) games: "RL [real-life] is just one more window."[88] If those intersections over MUDs were fluid in their multiple and multi-variegated natures, the smartphone, circa 2007 a decade later, arguably annexed those boundaries as it "brought the internet into everyday life."[89] Previously, "'the internet' and 'real life' were still separate domains, people had to 'get online' to move from one to the other.... A decade later, smartphones in hands, the distinction had evaporated."[90]

The most interesting arguments, though, are those which more than exemplify imbricated virtual and actual relations across screen boundaries. Rather, they *re-characterize* this actual/virtual entwinement by shifting it

85   Manovich, "To Lie and to Act," 199. As elaborated via Farocki, this relationship becomes perhaps most intense in relation to military warfare, where violence-at-a-distance via images becomes almost routine as paradigm and tactic, from the use of aerial photography in the First World War identifying bombing targets to images from cameras on missile warheads to the employment of drones today. For more on aerial photography, including its use in warfare, see Paula Amad, "From God's Eye to Camera-eye: Aerial Photography's Post-humanist and Neo-humanist Visions of the World," *History of Photography* 36:1 (2012): 66-86. The drone, in view of its centrality in recent US remote warfare policies, must also surely be one of the most prominent signature objects of the contemporary moment: see Chamayou, *Drone Theory*.

86   Harun Farocki, "Phantom Images," *Public* 29 (2004): 13-22, 17.

87   Sherry Turkle, *Life On the Screen: Identity in the Age of the Internet* (New York: Simon & Schuster, 1995), 10.

88   Turkle, *Life on the Screen*, 13.

89   "Books and Arts: Histories of the Web: Paradise Lost," *The Economist*, February 22, 2020.

90   Ibid.

into another critical space. One such argument is Jean Baudrillard's thesis on simulation and simulacra, which not only demonstrates how the virtual copies and encroaches on the actual across its boundaries, but escalates that replacement into a different relational reality: the hyperreal. Baudrillard proposes the idea of a simulation of the world so perfect that it becomes a defining component of lived reality: "what was projected psychologically and mentally, what used to be lived out on earth as metaphor, as mental or metaphorical scene, is henceforth projected into reality, without any metaphor at all, into an absolute space which is also that of simulation."[91] Baudrillard demonstrates the simulacra with diverse examples, referencing theme parks (particularly Disneyland),[92] video recorders, virtual cameras, television, talk shows and reality shows.

However, it is Baudrillard's references to screen media, and the exchanges across the screen's boundaries between audience and image, which most viscerally capture the dystopia of the hyperreal. In turn, across various essays, Baudrillard paints this dystopia as a hunger that is not only insatiable, but borne precisely out of an apotheosis of media. In the face of "the collapse of the real and its double" as instantiated by media products such as reality television, Baudrillard charges that "the mediatic class" "is starving on the other side of the screen."[93] His solution? Transfer the viewer "not in front of the screen where he is staying anyway, passively escaping his responsibility as citizen, but into the screen, on the other side of the screen."[94] Or, in other words, initiate "the last phase," "where everybody is invited to present himself as he is, key in hand, and to play his live show on the screen."[95] The virtual here is not so much entwined with as it gobbles up the actual – "we have swallowed our microphones and headsets...we have interiorized our own prosthetic image and become the professional showmen of our own

---

91   Jean Baudrillard, "The Ecstasy of Communication," in *The Anti-Aesthetic: Essays on Postmodern Culture*, ed. Hal Foster (New York: The New Press, 2002): 145-153, 148. Also see generally Jean Baudrillard, *Simulacra and Simulation* (Ann Arbor, MI: The University of Michigan Press, 1994).
92   See Baudrillard, *Simulacra and Simulation*, 12-14. Also see Baudrillard, "Disneyworld Company," in *Screened Out*, 150-154.
93   Jean Baudrillard, "The Virtual Illusion: Or the Automatic Writing of the World," *Theory, Culture & Society*, 12 (1995): 97-107, 100.
94   Baudrillard, "The Virtual Illusion," 100.
95   Baudrillard, "The Virtual Illusion," 99. *The Truman Show*, directed by Peter Weir (1998; Los Angeles, CA: Paramount Home Entertainment, 2019), DVD, is a prime fictional work which co-opts this premise, where its main character, Truman Burbank, literally – if unknowingly – lives and presents his life, as live, on the screen. Chapter 2 elaborates further on the connections between screens and the film's denouement.

lives."[96] To this greed, this eclipsing of the actual by the virtual, this all-consuming reality of the hyperreal, Baudrillard gives a name: "the ecstasy of communication."[97] But there are consequences to this ecstasy, namely, the suspension of moral judgement, as per the opening quotation of this section. This suspension, then, is the final critique of the hyperreal. It is not simply an observation of how far the virtual encroaches onto the actual for "the mediatic class," to the point of a strangulation where "[t]here is no 'Other' out there and no final destination."[98] The critique is about re-cognizing and re-characterizing that space of engagement, and identifying its perils and seductions, with or without any solution in the offering.

Encountering the world is thus as much about enquiring the meaning of its content as it is about bumping up against its expressive relations, with their associated analyses of critique, caution and assessment. Marshall McLuhan nailed this idea sixty years ago with the unbeatably catchy phrase, "the medium is the message," where the study of any media object lies not, or at least not only, with the contents or operations of the object itself: "it mattered not in the least whether [the machine] turned out cornflakes or Cadillacs."[99] What also mattered was the medium's relations to the world which, in turn, shape our understanding of the political, cultural and social consequences and meanings of our actions.[100] Baudrillard echoes this approach, not only explicitly connecting screen media to its relational values across its boundaries, but also underscoring the transfigurations of ourselves and our societies as the true message of media:

> The 'message' of the railways is not the coal or the passengers it carries, but a vision of the world, the new status of urban areas, etc. The 'message' of TV is not in the images it transmits, but the new modes of relating and

96 Baudrillard, "The Virtual Illusion," 97.

97 See Baudrillard's essay as titled "The Ecstasy of Communication." See also Baudrillard, *Screened Out*, where he repeats the phrase in relation to the virtuality of cyberspace: "Both coder and decoder — in fact your own terminal, your own correspondent. That is the ecstasy of communication." (179)

98 Baudrillard, *Screened Out*, 179.

99 Marshall McLuhan, "The Medium is the Massage," in Marshall McLuhan, *Understanding Media: The Extensions of Man*, 3rd ed. (Berkeley, CA: Gingko Press, 2013), 7-8.

100 As with Foucault's regard on discourse, it is not about an expression or representation, but about "ways of constituting knowledge, together with the social practices, forms of subjectivity and power relations which inhere in such knowledges and relations between them." As cited in Chris Weedon, *Feminist Practice and Poststructuralist Theory* (Oxford: Blackwell, 1987), 108.

perceiving it imposes, the alterations to traditional family and group structures.[101]

This book thus treads (and threads) across these ideas in thinking about the contemporary screen to encompass these meanings of relations, complexity and entwinement in its first sense of the "post-": to study the screened world in a framework of expressive relations *as situated across the screen boundary*; to understand the relational constellations of images, viewers and imagination which arise out of contemporary media; to characterize, carve out and name an alternative critical space which may accommodate them. This sense of the "post-" would also be one that re-understands its relational complexities and differences in a way which generates possibilities, rather than spiralling copies and replacements of the actual by the virtual which only sound ominous warnings and admonitions of lost reality.

The second way of leveraging the "post-" would be in terms of its basic sense of the "after" – not by way of heralding the next stage in a chronology, but instead in a spatial sense by reaching for the richness of a critical space that is in some way *beyond* the object. Hence, while the screen is indubitably present in contemporary media, this book's examination of eroding screen boundaries arches towards defining an if still un-defined space of being *past* the screen as an object. Put another way, this is also a gesture towards *no more screen*, a phrase adapted, if freely, from André Bazin's proclamation of "no more cinema" in his 1971 reading of Vittoria De Sica's "perfect aesthetic illusion of reality" as shown in De Sica's 1948 film, *Ladri di Biciclette*.[102] Celebrating the eschewal of spectacle as part of the film's defining neo-realist style, Bazin observes how *Biciclette*'s "'integral' of reality" presents "pure cinema": "no more actors, no more story, no more sets."[103] Bazin's declaration of the purity of medium is inspiring here in how he not only deconstructs a new aesthetic via that recognition, but also embeds a core of truth in the erasure of cinema, where his concern in going *beyond* spectacle, or *beyond* event, nevertheless always retains a steadfast affinity with the real. In thinking about the erasure of screen boundaries to that point or space of its beyond – and

101 Jean Baudrillard, *The Consumer Society: Myths and Structures* (London; Thousand Oaks, CA; New Delhi: Sage Publications, 1998), 142.

102 André Bazin, "Bicycle Thief," trans. Hugh Gray, in André Bazin, *What Is Cinema, Vol. 2*, 2nd ed. (Berkeley, CA: University of California Press, [1971] 2005): 47-60, 60.

103 Bazin, "Bicycle Thief," 60.

hence the sense of the "post-" – this book thus takes Bazin's approach of purity and truth as an inspiration, namely, to see where we can get to on a *tabula rasa*: no more edges, no more perimeters, no more borders, no more screen.

Yet "no more screen" is not a point of finality. Rather, in the void it portends springs a deeper examination of its stakes and meanings. Baudrillard, in the opening quotation of this section, points out the merging of all that "was separated in the past" and, more importantly, the ensuing absence of moral judgement as its result. Attention thus also has to be paid to the moral meaning of separations, demarcations, ontological "cuts" and boundaries. This includes not only thinking about the value of the in-between, but also the losses from and thereby any possible redemptions for that world which now no longer *has difference*, or no longer contains any discernible differentiation between image and object. The message of media is thus also one that should contain space which safeguards the possibility of making moral judgement; losing that space – rather than the real – is the true peril.

The conceptualization, terminology and representation of the *post-screen* thus converge out of these two vectors of thought and against these motivational contexts for discerning meaning, relations and conceptual space. As with Bazin and cinematic realism, as with Magritte's *La Condition Humaine*, as with Baudrillard's dire warning in the opening quotation of this section, the critical argument of the post-screen ultimately rounds back to the human – specifically, what is gained and lost in our understanding of ourselves from the erosion of screen boundaries and the absence of differentiation. In the main, this book is an observation about the screen boundaries in the current screen-based era. Its armature for these observations is three media technologies – Virtual Reality; holograms and holographic projections; and light projections – chosen for their unique manifestations of screen boundaries, and the complications they present on separations and demarcations. However, like religion and art, media is ultimately about the fulfilment of inner human longings, even as it folds and enfolds complex assemblages of materialist concerns, ideological politicking, ethical responsibilities, aesthetic interest and so on. In media lie mysterious appeals by the soul out of which people acquire a more mystical happiness beyond the brute needs for food, water, shelter and so on. Here I bring up, again, John Durham Peters's work in his book, *The Marvellous Clouds*, as another core inspiration. Explaining the premise of what he means by the meaning of media (which is that it does not "mean"; it "is"), Peters illustrates his point via a description of a seemingly banal family call on the pay phone, long-distance from Jerusalem to Tel Aviv, fraught with danger and laden with emotion: "The import of the call was existential, not informational"; the "medium" is about "disclosure of being rather than clarity of signal."[104] In terms of *what media is*,[105] the post-screen – namely, the thinking of screen boundaries in its "post-" space – also rests on its theorization in terms of the more abstract, existential aspects of our being and in how we are in it as humans.

Interestingly, though, while of a higher order, these longings incarnate to a corporeal level as a hunger, with media as the food to satiate it. Here, then, is where screens also become the interface *par excellence*, transforming into channels of nourishment and gratification. Baudrillard, as mentioned above, describes the "starving" "mediatic class" on the other side of the screen, who cross screen boundaries to devour the mediated versions of themselves, microphones and all. In 1984, Sherry Turkle wrote of another hunger – one for intimacy and emotional connection – which got fed by the computer and the mediated connections it provided:

> Terrified of being alone, yet afraid of intimacy, we experience widespread feelings of emptiness, of disconnection, of the unreality of self. And here the computer, a companion without emotional demands, offers a compromise. You can be a loner, but never alone. You can interact, but need never feel vulnerable to another person.[106]

However, as contemporary users are discovering today, media not only feeds the hunger, but perpetuates a vicious circle around it. Media's only dogma is its constant consumption so that users continue paying their account subscription fees to fill the coffers of media and technology companies and generating data and online footprints for them to monetize… so that they may create more media. From Candy Crush addictions to non-stop Google searches to Netflix binge-watching, contemporary media users ceaselessly offer up at that church. Take, for instance, the syndrome of FOMO, an acronym for "fear of missing out," characterized as "the desire to stay continually connected with what others are doing,"

---

104 Peters, *The Marvellous Clouds*, 14.

105 A deliberate play of words here referencing André Bazin's famous cinema book title, *What is Cinema,* as well as Dudley Andrew's reply via his own book title, *What Cinema Is! Bazin's* Quest *and its Charge* (Chicester: Wiley-Blackwell, 2010).

106 Sherry Turkle, *The Second Self* (New York: Simon and Schuster, 1984), 307.

particularly vis-à-vis social media,[107] and also often associated with elements of mental ill health such as anxiety and compulsion.[108] While there are many personal and social complexes which drive FOMO, at its heart is essentially a hunger to devour more media content. Its pushback, then, is to fast via a complete cessation from media consumption such as by "unplugging from technology," per advice from the Sabbath Manifesto, which also advocates a "National Day of Unplugging" (1-2 March 2019), encouraging pledges to "unplug from technology regularly" and "carve a weekly timeout into our lives."[109] This feast-and-famish cycle of FOMO accentuates the nature of media as a real yet chimerical fuel: it sates something deep within the human psyche, but also produces misery out of ever more profound kinds of hunger. Its nature is changeless, but takes various forms. As Turkle writes: "Today we suffer not less but differently."[110]

This book, in its "post-screen" ethos of *no more screen*, thus also attempts to explain our human condition as a quest for *another kind of space to feed the hunger*. Its argument is not a social science ethnography of media consumers to identify their hunger or otherwise. Its argument is to assert and give a name to a mediascape of eroding or erased screen boundaries and to re-think the signified meanings of that world. But boundaries also relate to the existential conditions of humans' inner lives, because they are powerful. Boundaries represent change. They usher in different states, spaces and places. They promise a new way of being. From mirrors to windows to door frames and, of course, to screens, humankind's myths, fairy tales and classic stories contain multiple boundaries which are portals to transformation of selves, worlds and destinies: think Alice's looking glass; Snow White's mirror; Coraline's secret door; the wardrobe door to Narnia; the role of Portunus as the ancient Roman god of keys, doors and ports, just to name a few examples. Even the most prosaic makeover shows on daytime television reveal the made over (and ostensibly better looking) participants through suitably dramatized opening doors.

107 A.K. Przybylski, K. Murayama, C.R. DeHaan and V. Gladwell, "Motivational, emotional, and behavioral correlates of Fear of Missing Out," *Computers in Human Behavior* 29 (2013): 1841-1848, 1841.

108 See Michael Shea, "Living with FoMO," *The Skinny* online, July 27, 2015, https://www.theskinny.co.uk/students/lifestyle/living-with-fomo.

109 As quoted from the cover page of http://www.sabbathmanifesto.org/.

110 Turkle, *The Second Self*, 307.

It is thus possible to read change and transformation across screen boundaries as solutions to the hunger, namely, as an escape. Escape across screens is a familiar connection, made most notably vis-à-vis video games and virtual worlds, whose virtual realities through the screen, allowing for freer expression of self, identity and so on, provide a welcome refuge from the more grounded world of the flesh.[111] This idea takes further root in relation to computers, as established by Apple's famous 1984 television commercial for its first Macintosh computer broadcast at the Super Bowl. A veritable classic today, if perhaps only in the histories of advertising and Silicon Valley lore, a young woman bursts into a large Orwellian screening room, complete with an audience who sits in rows like grey-clad worker automatons in Fritz Lang's 1927 film, *Metropolis*. She runs down the aisle, pursued by riot police, swings a sledgehammer in both hands and lets it fly towards the giant screen. The screen shatters, "and liberates the enslaved audience from the tyranny of command line interfaces and c//: prompts with the power of Mac's GUI (graphical user interface)."[112] On one level, the smashing of the screen heralds its literal visual transformation as an interface; on another level, it is also a nod to how the screen is the frontline to transformations of worlds, ideology and ways of being. It is an escape route, and its boundaries are its threshold.

However, the erosion of screen boundaries melds reality between the virtual and the actual, and seals this escape route. As the distinction between the two disappears, one can no longer become a getaway from the other. The familiar desperation of inescapable simulacra beckons. But we might thus also read this phenomenon as a different space – the post-screen not as a straightforward escape, but a *transcendence*. Or, evoking the sense of Mircea Eliade's oft-used term, a "hierophany," which refers to "something of a wholly different order," of "a reality that does not belong to our world, in objects that are an integral part of our natural 'profane' world."[113] In this reading, the actual and the virtual, still integral in themselves, combine in the post-screen space *beyond* for something *else*, for that "something of a wholly different order." Hunger – and as a theme which threads through

---

111   See, in particular, Sherry Turkle's work in *The Second Self*; also, Henry Jenkins, "'Complete freedom of movement': Video Games as Gendered Play Spaces," in *From Barbie to Mortal Kombat: Gender and Computer Games*, eds. Justine Cassell and Henry Jenkins (Cambridge, MA: MIT Press, 1998): 262-297.

112   Peter Lunenfeld, "The Myths of Interactive Cinema," in *Narrative Across Media: The Languages of Storytelling*, ed. Marie-Laure Ryan (Lincoln; London: University of Nebraska Press, 2004): 377-390, 378.

113   Mircea Eliade, *The Sacred and the Profane* (New York: Harper, 1961), 11.

this book – is thus no longer a void gripped in an overwhelming need to be filled with illusory satiation. It becomes a statement for a different order of things and for what our media histories have become. The hope, then, is that it becomes something else in turn, returning to the higher order of what media has always meant to being human – a mode of spirituality. The fear is its failure to do so.

1   Screen Boundaries as Movement

Abstract

This chapter lays down the starting point for conceptualizing the post-screen by articulating the screen in terms of its boundaries, rather than a surface filled with light. It clarifies screen boundaries as not merely the formal material edges around a lit surface, but an active interplay of movement between actual and virtual boundaries. The essence of the screen, then, is not so much in what it displays but its relational paradoxes between what is shown and not shown, hidden and revealed. A transformative space arises out of the paradoxical interplay, resulting in affective engagements with love, pleasure and information. However, interactivity diminishes the authority of screen boundaries, paving the way for the post-screen of eroded screen boundaries.

Keywords: screens; screenology; boundaries; frame; window; actual; virtual; interactive media

Re-placing the Screen: Play and Display, Appearance and Dis-appearance

We must abandon received definitions and categorizations of what constitutes a screen.

~ Erkki Huhtamo[1]

The villain in the 2019 Spider-Man film, *Spider-Man: Far from Home*,[2] is unusual. Created by Stan Lee and Steve Ditko in 1964 for the thirteenth

---

1   Erkki Huhtamo, "Screen Tests: Why Do We Need an Archaeology of the Screen?" *Cinema Journal* 51(2) (Winter 2012): 144-148, 148.
2   *Spider-Man: Far From Home*, directed by Jon Watts (2019; Los Angeles, CA: Columbia Pictures, Marvel Studios, Pascal Pictures), release.

Ng, J., *The Post-Screen Through Virtual Reality, Holograms and Light Projections. Where Screen Boundaries Lie*. Amsterdam: Amsterdam University Press 2021
DOI: 10.5117/9789463723541_CH01

volume of *The Amazing Spider-Man* comic series,[3] the antagonist character of Mysterio does not have an obvious superhuman power. He is unlike other villains in the series, such as Dr. Octopus, who had mechanical tentacles fused to the spine, or Vulture, who could fly (albeit via an electromagnetic harness).

Instead, Mysterio (or Quentin Beck, as he is also known) is a special effects whiz. In *Far from Home*, Mysterio (played by Jake Gyllenhaal) fools all the characters in the film, including Spider-Man (played by Tom Holland), with large-scale holographic illusions created from Augmented Reality (AR) projectors mounted on flying drones. These illusions appear as realistic multi-sensorial simulations which surround Spider-Man, even following him as he moves. Most importantly, they do not seem to be contained within any kind of screen or framing boundaries.

The twist in the film, then, is that *simulation itself becomes an antagonist.* The real threat in this superhero world is not the usual death and destruction wrought by the villain, but the bewilderment and disorientation in being unable to distinguish between the actual and the virtual. This indistinguishability is not so much – or not only – because the virtual is extremely life-like, but because it is not within a recognizable frame that separates it from the actual. Spider-Man's traditional vanquishing of the villain thus takes on an untraditional route, albeit one in complete keeping with contemporary times: to defeat Mysterio, Spidey has to first break his illusions. Our superhero finally achieves this at the film's climax not by his more celebrated superpower of web-slinging, but his "spider sense," or, in *Far from Home*, the drolly named "Peter Tingle" – a kind of sixth sense normally used in perceiving imminent danger, and in this case becomes Spider-Man's guide to identifying what is reality and what is not. Only by mastering the discernment between illusion and reality could Spider-Man finally triumph over Mysterio, whereby he slays the villain and restores order by re-asserting the diegesis' "reality."

Notwithstanding its unabashed status as a conventional superhero movie geared to be maximum blockbuster entertainment fare, *Far from Home* presents a curiously timely instantiation of the post-screen. As mentioned, there does not seem to be any sort of screen for Mysterio's illusions. If a screen did exist, it was imperceptible and seemingly immaterial, or made

---

3   "*The Amazing Spider-Man* #13 saw [Stan] Lee and [Steve] Ditko return to the creation of new super villains. This issue marked the debut of Mysterio, a former special effects expert named Quentin Beck": see Matthew K. Manning, "1960s," in *Spider-Man Chronicle Celebrating 50 Years of Web-Slinging*, ed. Laura Gilbert (London: Dorling Kindersley, 2012): 14-43, 25.

out (or perhaps made up) of thin air[4] with neither visible physical form nor boundaries. Mysterio's projected illusions were able to take over reality with almost facile efficiency – and so began Spider-Man's problems. *Far from Home* thus moves the old battle of good versus evil, or superhero versus villain, into a new battlefield of contemporary media's conflation between image and reality. Or, we can say, the new battlefield of the post-screen, which emerges not from the perfection of the Baudrillardian hyperreal but the perfection of the *seamlessness in the boundary*. Across this gap, then, order and disorder balance between what is revealed and concealed, what is displayed and in play, what appears and disappears.

***

Screen scholarship constantly revises the concept of the screen. In 2004, Erkki Huhtamo called for what he named "screenology," or "a branch within media studies focusing on screens as 'information surfaces.'"[5] Specifically, this enquiry entails understanding screens through historical antecedents in what Huhtamo calls a "media-archaeological" approach, or one at least "toward an archaeology of the screen."[6] As he puts it, "screenology would be a way of relating different types of screens to each other and assessing their significance within changing cultural, social and ideological frames of reference."[7] His concerns for a historically inflected understanding of screens as "information surfaces" are valuable as they open up the scholarship to a wider ambit of what is a screen. For example, Huhtamo critiques Charles Musser's study of "screen practice" as mapped, from the perspective of early cinema, across a continuum involving only images projected on

---

4   There are many examples in the movies of "light illusions projected into thin air," the most famous one being the Princess Leia hologram in *Star Wars: A New Hope*, directed by George Lucas (1977; Los Angeles, CA: 20th Century Fox, 2004), DVD. Scientifically, such projections "into thin air" are impossible, at least for now. This will be discussed further in chapter 4A's discussion on true holograms.

5   Erkki Huhtamo, "Elements of Screenology: Toward an Archaeology of the Screen," *Iconics: International Studies of the Modern Image*, 7 (2004): 31-82, 32.

6   Per the subtitle of Huhtamo's article, "Elements of Screenology," he describes this methodology as to "excavate manifestations of the screen as they appear in visual representations." (33) Also see Siegfried Zielinski, "Media Archaeology," in *Digital Delirium*, eds. Arthur and Marilouise Kroker (New York: St. Martin's Press, 1997): 272-283; and Jussi Parikka, *What is Media Archaeology?* (London: Polity Press, 2012).

7   Huhtamo, "Elements of Screenology," 32.

a screen, such as the magic lantern traditions.[8] He points out how this approach ignores moving image practices that are not projection-based but are as much part of "screen practice" as early cinema: "Looking back from the vantage point of early cinema leads Musser to omit traditions and forms that cannot be directly linked with the lineage of projecting of films as a public spectacle."[9]

Rather, Huhtamo expands the notion of the screen to include diverse *surfaces* for display, such as those of moving panoramas, shadow theatre and peep show boxes (including their nineteenth-century successors such as the zoetrope and stereoscopes), or what he calls "proto-screens."[10] For instance, Huhtamo argues for moving panoramas as such a proto-screen because they were often presented in the same venues as magic lantern shows. Moreover, as he points out, one "was usually 'framed', either by the proscenium or by pieces of canvas masking the front part of the hall," with "a lecturer" next to it, "explaining it to the audience sitting opposite the painting in the auditorium."[11] Huhtamo's argument thus opens up the thinking on screens beyond a taxonomy of media technologies, providing refreshed conceptualizations which encompass broader engagement with media display and exhibition culture. Other scholars have advanced the same approach of this wider thinking – Mauro Carbone, for instance, beautifully traces a trajectory of proto-screens and arche-screens across what he calls "the enigma of the surface that is invested with such a celebration and therefore delimited from the surrounding space."[12] Carbone draws such surfaces from windows and painting canvases to more intriguing and more ancient delimited areas for contemplation – "the rectangle that the Roman haruspex used to draw with their staff in the sky to wait and see how the eagles would cross it"; "the curtain that, in the sixth century BCE, Pythagoras inherits from the sacerdotal tradition to separate those having the right

---

8   Huhtamo, "Elements of Screenology," 36. He repeats this critique in "Screen Tests: Why Do We Need an Archaeology of the Screen," *Cinema Journal*, 51(2) (Winter 2012): 144-148.

9   Huhtamo, "Elements of Screenology," 38. Musser is not entirely wrong here, though; his categorization echoes McLuhan's distinction of screen technologies between those on which light is projected versus those which emit light – see Marshall McLuhan, *Understanding Media: The Extensions of Man* (New York: McGraw-Hill, 1964), 313. Charles Acland extrapolates McLuhan's idea to argue how even the audience themselves can be screens, namely, as "surfaces on which both projections and emissions settle": Charles R. Acland, "The Crack in the Electric Window," *Cinema Journal*, 15:2 (2012): 167-171, 167.

10   Huhtamo, "Elements of Screenology," 33.

11   Huhtamo, "Elements of Screenology," 38.

12   Mauro Carbone, *Philosophy-Screens: From Cinema to the Digital Revolution*, trans. Marta Nijhuis (Albany: State University of New York Press, 2016), 65.

to see him from those who are only allowed to listen to him"; the walls of the Chauvet Cave in France, approximated to be around 30,000 years old.[13] Such trajectories are valuable in throwing open our imaginations about what screens are and thus expand the thinking into wider waters of philosophical and historical implications.

One such philosophical implication on the role of screens and reality is Introna and Ilharco's reading of screens via Heidegger. In this reading, they present "a Heideggerian phenomenological analysis of screens" for what they call "the *screenness* of screens," or "the necessary meanings that enables [sic] us to identify each and all appearances of screens *as 'screens'* in the first place." (emphases in original)[14] Their critical argument is how these meanings of *"the screen-in-the-world"* thus condition human behaviours and the comports of humans to a particular surface as a screen. This, then, is a different thinking and imagining of the screen "as that which it is, in and only in its *world*," where it is in and relevant to the flow of human involvement "in-the-world in which [humans] dwell." Screens take on meaning – their *"screenness"* – in terms of the phenomenon of their existence, "presenting, displaying, relevant content for our involvement and action in the world." (emphases in original)[15]

Thinking about screens in terms of their boundaries, then, takes on the approach of these wider outlooks not so much to revise but to *re-place* the screen philosophically and conceptually, or to locate it elsewhere besides its visibility as *the media apparatus that displays*. As *Far from Home* demonstrates, if in slick blockbuster fashion, the importance of display in mediatized environments today does not lie in the mere fact of having something seen, since all that was shown to Spider-Man were fakeries and illusions. Rather, display – and the seeing of something, or having it seen – is to grasp at the *gap* between what is seen and not seen, as Spider-Man learns. Hence, across that gap, *display* is also about *play* – play in the slippage of boundaries; in where and how boundaries lie to signify something that is displayed; between what is revealed and what is hidden. The screen is thus discerned not directly through what it shows but indirectly through its boundaries. In this sense, the screen *dis-appears* – not disappear as in vanish into nothingness, but dis-appear as in an appearance that is yet undone in some sense (per the meaning of the "dis-" prefix), or thwarted into some other

13   Carbone, *Philosophy-Screens*, 63-65.
14   Lucas D. Introna and Fernando M. Ilharco, "On the Meaning of Screens: Towards a Phenomenological Account of *Screenness*," *Human Studies*, 29 (2006): 57-76, 62.
15   All quotations from Introna and Ilharco in this paragraph are from "On the Meaning of Screens," 64-66.

form or dimension – into its contours, edges and frame, increasingly fluid and malleable. In turn, its new manifestations pave the way for thinking about the post-screen of diminishing or eroded screen boundaries.

The rest of this chapter thus conceptualizes the screen by way of interrogating the meaning of its boundaries as re-placed via these ideas of play and display, appearance and dis-appearance. As the forms and shapes of screens morph with speed and innovation, re-viewing the screen becomes important in understanding contemporary media displays. Huhtamo's exhortation, per the section's opening quotation, rings as a true directional headlight: "we must abandon received definitions and categorizations of what constitutes a screen."

Screen Boundaries: Physical and Virtual, and of the Movement Betwixt

On one level, the screen seems to be an obvious thing. To quote Charles Acland (mentioned earlier in the Introduction), a screen appears as a straightforward object, intuitively grasped: "we just seem to know it reflexively: a thing that glows and attracts attention with changing images, sounds, and information."[16] Or per Lev Manovich's succinct description of the screen familiar to most twenty-first century media users: "a flat, rectangular surface positioned at some distance from the eyes."[17] The boundaries of the screen thus seem correspondingly clear as the material frame which encloses the surface area that presents a certain amount of visual information.

However, screens also present *virtual* boundaries which, while not as discernible as their physical counterparts, nonetheless introduce important implications for thinking about the screen and the nature of the virtual reality it contains. I use the word "virtual" here in line with Anne Friedberg's employment of the term, i.e.,

> ...to distinguish between any representation or appearance (whether optically, technologically or artisanally produced) that appears 'functionally or effectively *but not formally*' of the same materiality as what it represents [emphasis in original].[18]

---

16   Acland, "The Crack in the Electric Window," 168.

17   Lev Manovich, *The Language of New Media* (Cambridge, MA: MIT Press 2002), 94.

18   Anne Friedberg, *The Virtual Window: From Alberti to Microsoft* (Cambridge, MA: MIT Press, 2009), 11, and particularly the discussion between pages 7-12.

Virtual boundaries thus arise from the screen out of the *not formal* (or "second-order")[19] materiality of the objects that appear onscreen. This is a critical point. As with the formal materiality of actual boundaries such as fences or walls, this *not formal* materiality produces for the screen similar unstated boundary-related directives, such as not to cross, to keep out, even not to touch. On the last, media scholar Malte Hagener notably once described how a little girl ran up to a screen-wall in an auditorium and touched it after a showing of Martin Scorsese's film, *Hugo* (2011), in 3D.[20] This gesture was unusual enough to warrant his comment, for people generally do not touch cinema screens. As Wanda Strauven writes in relation to his account, "[w]hy would you indeed want to do so? How many film specta-tors have ever touched a theatrical film screen in their life?"[21] Strauven contemplates such behaviour as the "return of the rube" – a re-purposed naïvete and wonder of contemporary screen assemblages from the modern viewer. Nevertheless, the child's move, and the commentary it attracted, acknowledges the implicit dictates of the screen's virtual boundaries. What is shown onscreen delimits a separate spatial and haptic space, where the touching of it is understood as a gesture of non-compliance.[22]

Hence, screens are bordered and defined by both *physical* and *virtual* boundaries – the former materializing as the formal physical frame around the display; the latter appearing as normative in nature, and marked by practices, functions and considerations connected to the awareness of the image and the space represented by it. The two do not necessarily overlap, for the virtual space of the image need not occupy the precise area of the physical screen. Their boundaries may thus also differ. Having said that, exhibition practices do seek to overlay one across the other: for instance, a

---

19 Friedman, *The Virtual Window*, 11: "Virtual images have a materiality and a reality but of a different kind, a second-order materiality, liminally immaterial. The terms 'original' and 'copy' will not apply here, because the virtuality of the image does not imply direct mimesis, but a transfer – more like metaphor – from one plane of meaning and appearance to another."

20 As recounted by Wanda Strauven in her article, "Early cinema's touch(able) screens: From Uncle Josh to Ali Barbouyou," *NECSUS* (November 22, 2012), https://necsus-ejms.org/early-cinemas-touchable-screens-from-uncle-josh-to-ali-barbouyou/.

21 Strauven, "Early cinema's touch(able) screens," np.

22 As is the nature of rules, these directives of the screen's boundaries also prove to be change-able with modified function and design. For example, interactive screens, such as those on tablets and smartphones, present different dictates as screens which actively invite touch with fingers or a stylus pen by implicit and explicit instructions, such as via text directives to "swipe," "write" or "tap."

cinema projectionist will try to display an image by fitting its dimensions[23] into the largest possible area on the screen, and then concealing all excess screen space, such as by blocking with curtains, so that the screen appears "filled" with the image. Or, conversely, if the image is "too large" and parts of it fall outside the physical boundaries of the screen, the projectionist may apply an aperture plate – a piece of metal with a cut rectangular hole in the middle which fits the relevant aspect ratio – onto the projector to mask off any light hitting outside of the screen. The result of these practices is that the virtual image thus appears to fit precisely within the screen's physical framework, so that both effectively share the same boundaries, *even if they may not*. Different viewing formats will also affect the interplay between virtual and physical boundaries. For example, a "Fullscreen" version from a DVD will fill the screen with the image (though the sides of the image may be "chopped off" in the process). On the other hand, a "Widescreen" version will display more of the originally filmed image but with unfilled screen areas at the top and bottom.[24]

The distinction between the screen's physical and virtual boundaries is thus important because it clarifies thinking about screen boundaries as more than the rigid material edges of "a thing that glows" in its most obvious sense, for that "glow" captures most attention. Lev Manovich, for instance, notes the forcefulness of the screen's material boundaries: "[T]he screen is aggressive. It functions to filter, to screen out, to take over, rendering non-existent *whatever is outside its frame*." (emphasis added)[25] He thus highlights the viewer's complete concentration on what is seen onscreen as "the singular image completely fills the screen."[26]

---

23   This is also known as the image's aspect ratio, which is the proportional relationship between an image's length and height. The issue of screen formats is beyond the scope of this book, but as a brief guide: images are shot in various dimensions of length and height, or aspect ratios, where the image height is standardized as a single unit in the ratio, and its length is typically longer. Examples include 1.37:1, which is the most common length-to-height aspect ratio for shooting films today, or 2.35:1, which is the widescreen CinemaScope format. As films are shot in various aspect ratios, they are likewise displayed on screens in those different dimensions of length and height, which may not be the same dimensions as the screen.

24   The virtual boundaries between the image and the unfilled screen areas may be creatively co-opted into the story of the film. For example, in the 2016 *Ghostbusters* film in 3D, directed by Paul Feig (2016; Los Angeles, CA: Sony Pictures), the ghosts in the film were shown to breach the letterbox bars as a show of their paranormal power: see Kyle Buchanan, "Why Ghostbusters Looks So Unusual in 3-D," *Vulture*, July 20, 2016, https://www.vulture.com/2016/07/why-ghostbusters-looks-so-unusual-in-3-d.html.

25   Manovich, *Language of New Media*, 100.

26   Ibid.

However, such a reading of the screen takes into account only the obvious message of the screen – of what it shows and is seen, even what it "thinks"[27] – and ignores the more complex senses of *play* in the screen's display that is engaged across its physical and virtual boundaries. *Far from Home* comes to mind once more, and we can recall how Spider-Man's battle hinged as much on Spidey both seeing and seeing *through* the illusion as it did on his apprehending the gap between illusion and reality. The screen as a display surface is thus likewise as much about the *virtuality* of what is shown within its formal *physical* edges as it is also about the *gap* between the display and what is outside it.

Moreover, it is important to note that this sense of play across boundaries is by no means limited to specific "screen media" in the modern sense of the phrase (cinema; television; photography; the perspectival frame), even as much of the discussion in this and the next chapter concentrates on them. The historical exposition of the (dis-)play of images across boundaries is a long one, coming through also, for example, in the optical games and illusionist tricks common in the sixteenth century. Jean Pena, for instance, describes in "De usu optics" the use of mirrors inside a *camera obscura*, so that "the image of the [object], placed outside the room, will be observed inside the room hanging in the air."[28] Frequently deployed as visual effects in acts of conjuring ghosts, demons or evil spirits, such uses of catoptrics to reflect images "in the air" are as much for the appearances of darker magic as also for games to amuse and entertain. Above all, though, they are exemplars of play – via mirror reflection; hidden objects; optical tricks; mirror lighting and other devices and techniques – across physical and virtual boundaries between image and object. They are image displays which, even if not laid out across a screen in the formal sense, rely on the muddying of physical and virtual boundaries not only for their key effects, but for the paradoxical delimitation of the image itself: it is separated from its surroundings even as it is "in the air." Other optical tricks, such as illusionistic ceiling paintings, phantasmagoria and panoramas – all of which will also be discussed in chapter 3 in relation to VR – similarly reflect such play across boundaries. Display – and their paradoxes – across actual and virtual boundaries draw

---

27   On the notion of the film as a "quasi-subject," see Vivian Sobchack, *The Address of the Eye: A Phenomenology of Film Experience* (Princeton: Princeton University Press, 1992), 142.

28   Jean Pena, 1969, "De usu optics Praefatio," in Petrus Ramus – Audomarus Talaeus: Collectanae praefationaes, epistolae, orations, edited by Walter J. Ong. Hildesheim, 140-158, 157, as quoted from Sven Dupré, "Playing with Images in a Dark Room: Kepler's *Ludi* inside the Camera Obscura," in *Inside the Camera Obscura – Optics and Art under the Spell of the Projected Image*, ed. Wolfgang Lefèvre (Berlin: Max-Planck-Institut, 2007), 65.

from a long visual tradition, and one which indubitably feeds into thinking about both screens and, now, the post-screen.

Screen boundaries thus articulate screens in this alternative conceptualization:[29] not as a fixed artefact, but as a more radical set of paradoxes which relate constantly between what is within the boundaries and what is not. These paradoxes multiply, arbitrating between what is shown and not shown; revelation and concealment; the actual and the virtual; and, most importantly, in the grasping of their gap betwixt for the true meaning of what is being seen.[30] This, too, is the central paradox of cinema and, indeed, seeing itself. In Deleuzian terms, seeing one image as it is inevitably relates to *another* image, so that perception is always a *movement* between the objective and the subjective, where the latter subtracts anything that is not of interest.[31] In this sense of the subtractive, seeing something thus also always involves *not* seeing it, the same way a scotoma, even if a spot of blindness, is necessarily part of human vision.[32] The key, then, to genuine sight is the affective movement between them – and in that sense, Henri Bergson's words resonate on movement which bring about action and knowledge, and then creation itself: "To movement, then, everything will be restored, and into movement everything will be resolved."[33] Hence,

29   Or, as Mauro Carbone calls it, an "arche-screen," "understood as a transhistorical whole gathering the fundamental conditions of the possibility of 'showing' (*monstration*) and concealing images on whatever surface": *Philosophy-Screen*, 66. Yet, Carbone is more concerned about expanding imaginations of screens across the prehistoric and historic – hence the idea of an "arche-screen" to also include bodies, curtains, the *templum*, walls, tents, mirrors, veils and so on – than about fundamentally shifting the conception of the screen itself to the interplay between its boundaries. The two ideas are not unrelated, however – as Carbone writes, the notion of the arche-screen across the expanse of its "theme" is also in its "conditions of the possibility of showing and concealing" (ibid), which in turn is connected to positive or negative powers of vision, or what Carbone calls "delimiting to exceed," see 66-72.

30   In this respect, the interplay between the seen and the unseen also contrasts powerfully against sound, which, being omnidirectional, does not have the same relationship with screen boundaries like the image. Sound emanates from all directions around the viewer, particularly given contemporary surround sound systems in the theatre. Sound transcends the boundaries of the screen.

31   Gilles Deleuze, *Cinema 1: The Movement-Image*, trans. Hugh Tomlinson and Barbara Habberjam (Minneapolis, MI: University of Minnesota Press, [1986] 2003), 63-64. Also see Colin Gardner, *Beckett, Deleuze and the Televisual Event: Peephole Art* (London: Palgrave Macmillan, 2012) on seeing in Deleuze's terms: "the material moment of subjectivity is always subtractive," 23.

32   See also James Elkins, *The Object Stares Back: On the Nature of Seeing* (New York: Harvest, 1996).

33   Henri Bergson, *Creative Evolution*, trans. Arthur Mitchell (New York: The Modern Library, 1983), 273.

distinguishing between the physical and virtual boundaries of the screen is important in order to understand how screens also relate to *movement*, as the next sections will discuss. Its material edges are only one half of what determines the screen; its virtual boundaries, in defining the area of light, is the other half. In short, a screen is more than its common appearance as a static thing that glows. Understood in combination with its virtual boundaries, the screen is a *dis-appearance*: it is, rather, a constant play of movement across its boundaries in terms of how and what it displays, dynamic between what it hides and what it literally *brings to light*.[34]

Metaphors for the Screen

Metaphors used by scholars to theorize the screen are critical tools in conceptualizing these movements across screen boundaries. In particular, they also clarify the fluid dialogue that such movement engenders. There are several metaphors at large for the screen, but the principal ones of *frame*; *window*; and *mirror* emerge most coherently from discussions as primarily established by film scholars. This is for good reason as much of classical film theory is constructed from thinking about the role and operation of the frame around the cinematic image – what is shown and what is closed off – if anything to understand and assert the medium's purpose. Or, as Dudley Andrew puts it, "by which we seek to understand (and control) the cinema complex."[35]

Taking from Charles F. Altman's account,[36] Andrew summarizes three positions:

> Bluntly put, [Sergei] Eisenstein and [Rudolf] Arnheim conceived of the spectator as being before *a framed image* (as a painting); [André] Bazin claimed he sat before a *window*; and [Jean] Mitry intertwined the notions

---

34   One might also recall here Jean Baudrillard's writing of photography, or photo-graphy, as "the writing of light" ("l'écriture de la lumière") or from light which emanates out of two sources, one from the object and the other from the gaze: Jean Baudrillard, "La Photographie ou l'Écriture de la Lumière: Litteralite de l'Image," in *L'Echange Impossible* (The Impossible Exchange), trans. Francois Debrix (Paris: Galilee, 1999), 175-184. The implicit paradox of revelation (object) and the hidden (gaze) out of Baudrillardian photography mirrors the same contradictions here across screen boundaries.

35   Dudley Andrew, *Concepts in Film Theory* (Oxford: Oxford University Press, 1984), 273.

36   Charles F. Altman, "Psychoanalysis and Cinema, the Imaginary Discourse," *Quarterly Review of Film Studies* 2, no. 3 (Summer 1977): 260-64.

finding that cinema's specificity lay precisely in the *oscillation between window and frame* [emphasis added].[37]

These metaphors also reflect their respective theorists' philosophies of image and reality.[38] For Eisenstein, Arnheim and other film formalists, the boundaries of the screen signify a blocked-off world within a *frame*: as with a painting, the frame encloses the image to privilege the contents and their composition in its delimited space. Here, the formal material boundary of the screen entirely dictates the delimitation of what is shown. The spectator's attention is fixed on what is inside the boundary, closed off from what is outside it. The expressionist aesthetics of cinema are what mattered to these theorists, and the screen understood as such a frame was the idealized vehicle for displaying them.

On the other hand, for Bazin and others of the realist school, the purpose of cinema lay in its reflection of reality, where "cinema attains its fullness in being the art of the real."[39] To these scholars, the cinema screen is a *window* whose boundaries are not a frame that closes off, but displays a view onto a putatively limitless world.[40] In other words, beyond the screen's boundaries is implied an abundance of space and innumerable other objects. More importantly, there is a *continuity* of reality, namely, "[w]hen a character walks out of the camera's field of vision, we know that he has left the visual field, but he continues to exist in an identical state somewhere else in a hidden part of the setting."[41] In this metaphor of the window, the screen is an aperture. Bazin offers another metaphor along this line – that of the mask: "the screen is not a frame like the frame of a painting, but a *mask* that reveals only part of an event." (emphasis in original)[42] The screen as a window is thus understood not so much in its formality of physical boundaries that rigidly separates what is inside or outside them (as it is for the formalists),

---

37   Andrew, *Concepts in Film Theory*, 12-13.

38   See Dudley Andrew, *The Major Film Theories: An Introduction* (Oxford: Oxford University Press, 1976) for a now-classic discussion of these film theories and philosophies.

39   Bazin, "La Strada," *Crosscurrents* Vol. VI, no. 3 (1956), 20, trans. J.E. Cunneed from *Esprit* XXIII, no. 226 (1955), 487-51, as quoted in Andrew, *Major Film Theories*, 137.

40   Due to cinema's peculiar characteristics as a medium which entails its editing, composition and other filmic processes of representation, Andrew expands this metaphor beyond a window to a *prism*: "We can do this [i.e. suppress the filmic process of representation in favour of the artistic process of expression] because the cinematic process has its own peculiarities. It is not so much a window as a prism." See his discussion at Andrew, *The Major Film Theories*, 30-33.

41   Ibid.

42   André Bazin, *What Is Cinema?*, ed. and trans. Timothy Barnard (Montreal: Caboose, 2009), 193.

but through ongoing and fluid *movement* between physical and virtual boundaries, between what is shown and not shown, both of which, being part of the same whole, continue with and extend from each other. The audience gains information by what they see *through* the aperture of the screen, *and* what they cannot see *beyond* it.

The metaphor of screen-as-window is powerful because of these continuities across its boundaries between seen and unseen, light and darkness, inside and outside. It is simultaneously an enclosed view and an opening to an external world. These contradictions in the metaphor also connect the screen to the *architectural window* which itself makes significant connections and movements, such as between interior and exterior that, in those movements, open up the darkness of the former into the luminosity of the latter. In this sense, we may also recall Leon Battista Alberti's famous and oft-quoted instruction to a painter "to 'regard' the rectangular frame of the painting as an open window (*aperta finestra*)"[43] as the first formal connection crystalized between the simultaneous openness and closedness of the architectural window and the boundaries of an image on display.[44] Taking on Alberti's metaphor, Anne Friedberg further capitalizes on the window's properties of revelation and permeability (to light and "ventilation") to argue how "the cinematic, television, and computer screens have become substitutes for the architectural window."[45] Writing in 2009 and in the wake of Microsoft Windows platforms' dominance in the 1990s through to the noughties (which would be Windows 3.0 through to their 95, 98 and XP versions), Friedberg conceptualizes the screen both as a window and also as Windows. The interplay between the actual and the virtual across the boundaries of the twentieth-century computer screen thus becomes *augmented*, as its content plumbs greater depths of virtuality from computational worlds, including computer applications, video games and the World Wide Web, against the actual. Discussing the computer screen a few

43   Friedberg, *The Virtual Window*, 1. See also Leon Battista Alberti, *On Painting and On Sculpture: The Latin Texts of De pictura and De statua,* trans. Cecil Grayson (London: Phaidon, 1972), 55.
44   Playing on the same ideas of concealment and revelation, the screen-as-a-window also balances between paradoxical opacity and transparency, where the screen, besides showing and not showing, is also a one-way window with "a surface opacity" which allows the viewer to see into the diegetic screen world, but not the other way around. Out of this opacity thus arises the awkwardness when a character "looks into" the camera at the viewer: see generally Tom Brown, *Breaking the Fourth Wall: Direct Address in the Cinema* (Edinburgh: Edinburgh University Press, 2012), and also the text in chapter 2 accompanying footnote 50. Also see Thomas Elsaesser and Malte Hagener, *Film Theory: An Introduction through the Senses* (New York: Routledge, 2010) for readings of cinema and specific films as window, frame, door and threshold, 13-54.
45   Friedberg, *The Virtual Window*, 11.

years before Friedberg, writing in 2002 Lev Manovich also points out the phenomenon of "coexisting windows" on a computer screen (more notable on Windows 3.0 and 3.1 than the later versions) that clamour for the viewer's attention, albeit observing this as more akin to "modern graphic design, which treats a page as a collection of different but equally important blocks of data such as text, images, and graphic elements."[46]

The screen as a window also epitomizes its role as a vital threshold for translation or dialogue. In this exchange, the screen's virtual boundaries in terms of the image it presents become its crux. Vilém Flusser's 1977 discussion of television best demonstrates this correspondence. On television as a window, Flusser writes:

> The TV was projected to be a new type of window. It was meant to provide men with maps of the world to be used in subsequent commitments. This is what the word 'television' means: a better vision that is provided by conventional windows.[47]

Flusser wanted to imbue television with an ideological purpose, where it is "to be looked through and to provide a view and a vision," to be not merely a window to the world, but *"an improved window*, a medium for understanding the world and dialoguing with others." (emphasis added)[48] His positing of television as a window is thus about creating a dialogic society to connect people together; to "recognize" each other "in the sense of perceiving and conceiving his [or her] message," and "allow[ing] the other person to recognize us in the same way"; to "form a true 'polis'"[49] so as to dispel our inmost loneliness.

This position echoes uncannily with how, as discussed above, Bazin and the realists also view cinema – a different medium from TV, for sure, but in this case a superficial detail. What is of more striking note is how the

---

46 Manovich, *Language of New Media*, 101.

47 Vilém Flusser, "Two Approaches to the Phenomenon, Television," in *The New Television: A Public/Private Art*, eds. Douglas Davis and Allison Simmons, trans. Ursula Beiter (Cambridge, MA: MIT Press, 1977): 234-47, 240. Flusser's description of television as a window, *cf* Bazin's deployment of the same metaphor for cinema, is ironic in that Flusser was doing so precisely to distinguish television from cinema, which, unlike Bazin, Flusser regarded as "a late development of wall painting." Or, more accurately, to Flusser films "are a synthesis of paintings and books of fiction, and therefore represent events 'better' than do either." (ibid) Cinema is too engaged as an art form to reflect the world; this perspective similarly reflects Eisenstein's own ideas of cinema.

48 Ibid.

49 Ibid.

ideologies behind the metaphor of the window converge. In the realists'
view, cinema as a window should "give the spectator as perfect an illusion
of reality as possible"[50] via the promise of a continuous world beyond what
the screen reveals. For Bazin, writing in the 1950s still in the wake of the
Second World War's horrors, cinema as a window into reality is to enable
a viewer to look through it to find humanism, to obtain some redemption
for the abhorrence of the war – to have, in short, "an opportunity to savor,
before the time finally runs out on us, a revolutionary flavour in which terror
has yet no part."[51] The screen boundaries *qua* the edges of a "window" are
thus more than a view into an exciting world and the promise of a larger
one beyond its borders. In the act of looking into the specific frame of a
window, the screen delivers redemption and hope for society torn asunder
from war and loneliness. In this sense, the metaphor of the window, placed
as such by Flusser and Bazin, is particularly compelling because it points
to the criticality of its boundaries *as* a window which, far beyond the view
it presents, serves *a more meaningful human connection*.

Across the two metaphors, film theorist Jean Mitry synthesizes the screen
"as *both frame and window*" (emphasis added).[52] He argues that cinema is
indeed, on one hand, an analogue of reality, where cinema's "raw material"
is "the image which gives us an *immediate* (unmediated, untransformed)
perception of the world." (emphasis in original)[53] At the same time, it is also
ordered, presented, "taken, developed, and screened" by someone else: "In
the film theatre somebody else is telling us to look at this or that part of
the world, telling us in addition that we should give it a 'significant' look."[54]
Mitry writes: "The framed image begins to strike us as an ordered image
which we must look at purposefully and in relation to other framed images;
but all the while it never ceases pointing to the world it represents."[55] As
such, the screen is "both frame and window," with editing and sequenc-
ing constituting the former, and the illusionary continuity of the diegetic
world for the latter. The boundaries of the screen here are dual-purposed
(if dubiously trying to get the best of both worlds): to enclose for order, yet
also to point outwards into the world.

50   André Bazin, "An Aesthetic of Reality: Neorealism," in *What is Cinema? Vol. II*, ed. and trans.
Hugh Gray (Berkeley; Los Angeles, CA: University of California Press, [1971]; 2005), 26.
51   Bazin, "An Aesthetic of Reality," 21-22.
52   Andrew, *The Major Film Theories*, 191.
53   Andrew, *The Major Film Theories*, 189-190.
54   Andrew, *The Major Film Theories*, 190.
55   Andrew, *The Major Film Theories*, 191.

With the popular application of psychoanalytic theory to film studies in the 70s and 80s, Andrew highlights yet another "inspired metaphor" – the *mirror* – with which to understand the screen, and to break the apparent impasse between Bazin (realism) and Eisenstein (formalism).[56] Here, cinema is not held in relation to the world in terms of revealing or ordering it, but to the psychoanalytic processes – "the fact and the force of desire"[57] – of both filmmakers and spectators. The boundaries of the screen thus contain not the revelations of an external world, but an interior one – the audience's psyche as reflected to them out of the images on the screen. The virtual boundaries of the screen come into play once more, as the image itself becomes another kind of threshold: not that as between itself and reality, but between the inside and outside of the viewer themselves. Or, as Elsaesser and Hagener put it in their readings of a corpus of European and American films (and referencing the same passages by Andrew as I do here), the moments in cinema "when we are confronted with an image as if with our own reflected self."[58]

More recent scholarship adds to these classic metaphors. Eyal Peretz, for instance, notes, as with Bazin and the realist school, that the space outside the boundaries of the screen indicates "the continuous world beyond what the screen shows."[59] However, he appends an additional element to that "off" space in terms of a sense of "an anytime/anyplace outside of this continuous world."[60] Hence, to Peretz, the "off" – or "the invisible outside of a fictional space" – also contains a more mysterious dimension: "an 'otherworldly' anytime/anyplace"[61] as "the ghostly medium of an unlocatable 'beyond' that has come to haunt the strange, decontextualized zone of the screen."[62] As such, the reality revealed in cinema "is doubled, becoming ghostly and

---

56   Andrew, *Concepts in Film Theory*, 13.

57   Andrew, *Concepts in Film Theory*, 134.

58   Elsaesser and Hagener, *Film Theory*, 55.

59   Eyal Peretz, *The Off-Screen: An Investigation of the Cinematic Frame* (Stanford, CA: Stanford University Press, 2017), 36.

60   Peretz, *The Off-Screen*, 37. See in particular his thorough analysis of the floating leaf in the opening scene of Andrei Tarkovsky's *Solaris* (1972), whose languid movements on and off-screen in the scene shows its "double quality" of being there yet also emerging from another world, ultimately serving as an effective message about the world of Solaris: 35-40.

61   Peretz, *The Off-Screen*, 36. The sense of "anytime/anyplace" also recalls Gilles Deleuze's idea of "any-space-whatever," which echoes the mysterious-yet-recognizable otherworldliness in Peretz's reference of the "off": the "any-space-whatever" is not an unrecognizable space, but one where all the coordinates of understanding that space have changed. See Deleuze, *Cinema 1*, 102-122.

62   Peretz, *The Off-Screen*, 36.

attaining the status of fiction," where any screen reality appears to be both actually there as well as "with the ghostly aura of the fictional."[63] I read his theory of the screen thus as *both window and portal*, a cut-out of reality from a larger world but which also opens out and continues into the "off" as a fictional, mysterious, ghostly, invisible space. In these ways, the imaginings of the screen continue to play off the problematics of concealment and revelation, play and display, abstract and concrete, virtual and actual – all the movements betwixt which colour the paradoxes of screen boundaries and constitute the screen's nature of display, delimitations and demarcations.

Crossing Screen Boundaries: Love, Pleasure, Information, Transformation

In turn, the interplay of these movements between the screen's paradoxes fires up its boundaries as a *transformative* space. This sense of transformation in the boundary once again echoes Branden Hookway's description of the interface (mentioned in the Introduction as applicable to screen boundaries), specifically where he notes it as

> ...a liminal or threshold condition that both delimits the space for a kind of inhibition and opens up otherwise unavailable phenomena, conditions, situations, and territories for *exploration, use, participation, and exploitation* [emphasis added].[64]

But what might be the "exploration, use, participation, and exploitation" which open up from these movements of screen boundaries? Again, film scholarship, given the intensity of the screen image for cinema and its concern with audience reception of the medium, contributes richly to this discussion. Susan Sontag, for instance, writes of *cinéphilia* – "the name of the very specific kind of love that cinema inspired": "the experience of surrender to, of being transported by, what was on the screen. You wanted to be kidnapped by the movie...to be overwhelmed by the physical presence of the image."[65] How the screen opens up its world thus, in turn, invokes

---

63 Peretz, *The Off-Screen*, 37.
64 Branden Hookway, *Interface* (Cambridge, MA: MIT Press, 2014), 5.
65 Susan Sontag, "The Decay of Cinema," *The New York Times*, February 25, 1996, https://www. nytimes.com/1996/02/25/magazine/the-decay-of-cinema.html.

the virtual and emotional portage of the audience into the film's world, transforming their own world in their intensity of *cinéphilia*.

Another answer, again core to cinema spectatorship, to the transformative space out of the viewer's gaze crossing screen boundaries is that of *pleasure*; and, moreover, in different forms. One form is sheer sensorial pleasure, such as an audience's enjoyment in the image's visual intensity.[66] Roger Cardinal, for instance, writes of cinema vivid in what he calls "peripheral detail," namely, "something which does not fit in with the intended meaning of a work."[67] It is through this level of detail that the viewer is "coax[ed]... into a fresh relationship to the image, one in which the whole screen is acknowledged as a surface which is, so to speak *detailed all over*, like a mosaic, available to the gaze as an even field of rippling potency and plenitude." (emphasis in original)[68] Here, in Cardinal's interpretation, the virtuality of the screen's boundary is writ large, whereby the spectator's gaze in crossing that boundary and spanning the image receives the sensorial pleasure of the scene's visual richness and detail. Tellingly, in explaining this visual experience, Cardinal also invokes the metaphor of the window: "If what lies at the periphery of the frame possess what I have called phenomenal density, it will stimulate the viewer to *take the frame as a window* onto a reality which now extends undiminished beyond the limits of the screen." (emphasis added)[69] Written in 1986, Cardinal does not mention Bazin, Flusser or even Alberti, but his "pausing over peripheral detail"[70] certainly converges with their concern for human connection through the screened image. The detail in the corner for the viewer's pause is not just a fetish for quirky interest. As with Bazin's film theory, to Cardinal that detail is a profound promise, *held back yet delivered*

---

66   Such pleasure most commonly refers to visual pleasure, but also includes haptic and tactile pleasure. There is particularly substantial literature on the latter, see, for instance: Jennifer M. Barker, *The Tactile Eye: Touch and the Cinematic Experience* (Berkeley, CA: University of California Press, 2009); or Laura U. Marks, *The Skin of the Film: Intercultural Cinema, Embodiment and the Senses* (Durham, NC: Duke University Press, 2000).

67   Roger Cardinal, "Pausing over Peripheral Detail," *Framework* 30-31 (1986): 112-130, 113.

68   Cardinal, "Pausing over Peripheral Detail," 126. The sense of "detailed all over" from Cardinal also invokes the idea of pixellation, the exploration of which comes to the fore in Michaelangelo Antonioni's *Blow-Up* (1966), discussed later in this chapter: see text accompanying footnote 99.

69   Cardinal, "Pausing over Peripheral Detail," 127.

70   Per the title of his article, but also the thrust of his argument that the peripheral detail – the incidentals – is cause for not only pausing over, but also pause for thought, despite its seeming triviality.

by screen boundaries, of an unbroken and redemptive reality – "a true slice of 'suchness', a giving-back of refreshed contact with the tacit naturalness of things."[71] Cardinal's pleasure across screen boundaries becomes, as with Bazin and Flusser's engagements with the image, something more profoundly connected to the human condition:

> Intellectually I concede that filmic experience is fantasmatic, for the cinema screen is flat and fixed. And, yet, once the illusion of tactile depth quivers before me, I stretch towards it in a concentrated act of participation *which involves my whole being and transcends intellectual reserve* [emphasis added].[72]

If one pleasure of crossing screen boundaries lies in the revelry and revelation of accessing the image, a certain converse also applies in terms of their transgression and trespass, where the encroachment of space across the screen's boundaries becomes inextricably linked to arousal out of peeping, voyeurism and scopophilia. Here, again, manifests critical interplay between the material and virtual boundaries of the screen, where the screen, as an audience's peephole into an event, works both through what it shows and, just as crucially, the limitations of what it does not. Once more, film scholarship offers extensive discussion on the visual pleasures of such spectatorship, conventionally male,[73] which drives film language itself. Numerous films are also self-reflexive of this pleasure, where the audience's gaze through the frame of the screen mirrors the protagonist's voyeurism as they correspondingly spy on other characters through windows (*Rear*

---

71   Ibid.

72   Cardinal, "Pausing over Peripheral Detail," 129.

73   Such visual pleasure is historically attributed to male pleasure: see Laura Mulvey's oft-quoted essay, "Visual Pleasure and Narrative Cinema," *Screen* 16(3) (Autumn 1975): 6-18, as the classic exposition of this stance. However, subsequent accounts of spectator theory have updated these ideas of pleasure: see, for example, Manthia Diawara on black spectatorship ("Black Spectatorship: Problems of Identification and Resistance," in *Black American Cinema*, ed. Manthia Diawara (New York: Routledge, 1993): 211-220). See also, as a sampling, Teresa de Lauretis, Andrea Weiss or Patricia White on lesbian spectatorship, respectively, The *Practice of Love: Lesbian Sexuality and Perverse Desire* (Bloomington: Indiana University Press, 1994); "A Queer Feeling When I Look at You: Hollywood Stars and Lesbian Spectatorship in the 1930s," in *Multiple Voices in Feminist Film Criticism*, eds. Diane Carson et al. (Minneapolis, MI: University of Minnesota Press, 1994): 343-357; and *Uninvited: Classical Hollywood Cinema and Lesbian Representability* (Bloomington: Indiana University Press, 1994).

Window;[74] *American Beauty*);[75] holes in walls or doors (*Psycho*;[76] *By the Sea*;[77] *The Handmaiden*);[78] or even a diegetic camera's viewfinder (*Peeping Tom*).[79] In these cases, voyeurism across the screen's boundaries transforms into yet another kind of movement as between actor and spectator, where one becomes the other as the peeping subject, embroiled in a paradox of activity and passivity.[80]

Finally, screens, articulated as boundaries, also give and conversely withhold *information*. As Anne Friedberg writes, "we know the world by what we see: through a window, in a frame, on a screen."[81] Carbone, again, incisively cuts through this paradox of interception and concealment with his argument of the arche-screen's oppositions of positive and negative overdetermination, whereby the function of the former is that of delimitation (to know what is shown) and the latter as prohibition (to know what is shown as an *excess*).[82] On one level, the screen may disclose more by simply *showing more things* within its frame, such as by the visual richness of detail in the image as described by Cardinal above. It may also do so through cinematic techniques such as different depths of field: one prominent example, out of many, is the use of deep focus in *Citizen Kane* to show the key scene of

74   *Rear Window*, directed by Alfred Hitchcock (1954; London: Universal Pictures UK, 2007), DVD.

75   *American Beauty*, directed by Sam Mendes (1999; University City, CA: DreamWorks Home Entertainment, 2006), DVD.

76   *Psycho*, directed by Alfred Hitchcock (1960; London: Universal Pictures UK, 2006), DVD.

77   *By the Sea*, directed by Angelina Jolie (2015; London: Universal Pictures UK, 2016), DVD.

78   *The Handmaiden*, directed by Park Chan-Wook (2017; London: Curzon Artificial Eye, 2017), DVD.

79   *Peeping Tom*, directed by Michael Powell (1960; London: Optimum Home Entertainment, 2011), DVD.

80   See also Deleuze on "the mental image according to Hitchcock," which Deleuze argues is the image where relation itself, such as the passivity of the spectator and the activity of the actor, becomes the object of the image: "Each image in its frame, by its frame, must exhibit a mental relation": Gilles Deleuze, *Cinema 1: The Movement-Image*, trans. Hugh Tomlinson and Barbara Habberjam (Minneapolis, MI: University of Minnesota Press), 201.

81   Anne Friedberg, *The Virtual Window: From Alberti to Microsoft* (Cambridge, MA: MIT Press, 2009), 1. The fundamental premise here is that visuality is equated to knowledge: see, for instance, Jacques Aumont, "The Variable Eye, or the Mobilization of the Gaze," trans. Charles O'Brien and Sally Shafto, in *The Image in Dispute: Art and Cinema in the Age of Photography*, ed. Dudley Andrew (Austin: University of Texas Press, 1997): 231-252, who traces "seeing as an instrument of knowledge, even of science" from the nineteenth century painting of the étude to panoramas to cinema (234). See also generally Jonathan Crary's classic text, *Techniques of the Observer: On Vision and Modernity in the Nineteenth Century* (Cambridge, MA: MIT Press, 1991).

82   Carbone, *Philosophy-Screens*, 68-72.

the young Charles Foster Kane playing with his Rosebud sled.[83] Moreover, the mobility of the camera, in terms of the distances and heights it reaches, presents ever greater amounts of visual information. In this respect, one might recall Jacques Aumont's writing in 1989 of "the unleashed camera" ("the famous *entfesselte Kamera*") – "a camera enabled not only to see all, but also to see from anywhere."[84] He described the movements of this camera in relation to the then-astonishing shots it produced in 1920s German films such as *Variety* (1925) and *The Last Laugh* (1924), which were "filmed from the great Ferris wheel of an amusement park or, even more strikingly, from a flying trapeze."[85] Today, virtual cameras substantively augment the provision of information in the image not only with their ever more unfettered and rapid movements through virtual space, but also their ability to take on humanly impossible perspectives.[86] Again, cinematic examples abound; a random few choices will suffice here as illustration: *Russian Ark*,[87] for instance, whose digital camera snakes through the labyrinthine halls of the Hermitage in a single 90-minute shot to reveal much of the vast museum to the audience; Alfonso Cuarón's swirling CGI shots of the emptiness of outer space in *Gravity* (2013);[88] the swooping shots across the vast vistas of Middle-earth in the *Lord of the Rings* trilogy films;[89] the rush through the

83   *Citizen Kane*, directed by Orson Welles (1941; Burbank, CA: Warner Home Video, 2016), DVD. Interestingly, this scene is set as viewed across the boundaries of *a window*. Its significance of the young Kane, innocently playing with his sled beyond the window's frame in this scene, is compounded by its deeply hidden revelation of Rosebud as the name of the sled. The utterance of the name of "Rosebud," in turn, is the mystery which drives the film's narrative from its first scene.

84   Jacques Aumont, "The Variable Eye," 247. This notion of visual omnipotence is also echoed in other more recent technologies, such as the technology of iPIX, widely used in the early 2000s, whose camera lens technology that creates 360-degree pictures, as with a panorama, promises viewers the ability to "see everything!... Anywhere. Anytime. In any direction," as quoted from Alison Griffiths, "'The Largest Picture Ever Executed by Man': Panoramas and the Emergence of Large-Screen and 360 Degree Internet Technologies," in *Screen Culture: History and Textuality*, ed. John Fullerton (Eastleigh: John Libbey, 2004): 199-220, 199.

85   Ibid.

86   For a persuasive argument of a cinema shaped by such visuality, see William Brown, "Man Without a Movie Camera – Movies Without Men: Towards a posthumanist cinema?", in *Film Theory and Contemporary Hollywood Movies*, ed. Warren Buckland (Abingdon; New York: Routledge, 2009): 66-85.

87   *Russian Ark*, directed by Alexander Sokurov (2002; London: Artificial Eye, 2003), DVD.

88   *Gravity*, directed by Alfonso Cuarón (2013; Burbank, CA: Warner Home Video, 2014), DVD.

89   *The Lord of the Rings* trilogy, directed by Peter Jackson (2001-2003; Burbank, CA: Warner Home Video, 2016), DVD.

streets of Victorian London in *A Christmas Carol* (2009);[90] and many others. This unrestrained mobility of the camera in cinema carries over as well to interactive media as connected to a third-person perspective, rather than that of a character. Will Brooker, for instance, writes of "a literal flying-eye camera, independent of the Mario avatar" in *Super Mario 64*;[91] he notes how "the mode evolved through *Tomb Raider* (Eidos, 1996) and became the dominant POV [point of view] in Rockstar's *Grand Theft Auto* when the series went 3D in 2001."[92] Referencing *Lara Croft: Tomb Raider*,[93] Steven Poole also describes how the camera's point of view is "constantly moving, swooping, creeping up behind [Lara Croft] and giddily soaring above, even diving below the putative floor level."[94] Across different media, camera mobility thus radically increases the amount of information an audience may obtain out of the screen's fixed boundaries.

However, on another level, the physical boundaries of the screen also delimit the visual information, where the viewer, while in a most basic sense can literally only see what is shown, also sees *through* what the screen withholds. Or, rather, they are able to know its excess in the negative outside space of the screen. Carbone's reading of the *teikhíon* – a low wall by a road – of Plato's Cave as a screen illustrates not only this paradox between showing and concealment, but also the screen's selective role in such displaying and withholding of information. Where the screen in Plato's "Allegory of Cave" is commonly read as the cave wall which depicts the shadows of objects that its prisoners then understand as their reality, Carbone also reads another screen in the *teikhíon* as that which "conceals the men who are part of the machinery of the Cave": via the *teikhíon*, these men both produce shadows (of their objects) and conceal shadows (of themselves). As Carbone puts it: "indeed, the *teikhíon* picks out what has to be displayed on the opposite wall and screens off what, instead, has to remain hidden to the prisoners' eyes."[95] The wall, as a screen, thus both displays and conceals, selects and withholds.

Moreover, the limits of information from the screen are due not only to the paradoxes of its boundaries, but also other parts of the filmmaking

---

90   *A Christmas Carol*, directed by Robert Zemeckis (2009; Burbank, CA: Walt Disney Home Entertainment, 2010), DVD.

91   Nintendo, *Super Mario 64*, 1996.

92   Will Brooker, "Camera-Eye, CG-Eye: Videogames and the 'Cinematic,'" *Cinema Journal* 48(3) (Spring 2009): 122-128, 127.

93   Core Design, *Lara Croft: Tomb Raider*, 2001.

94   Steven Poole, *Trigger Happy: The Inner Life of Videogames* (London: Fourth Estate, 1999), 145-146.

95   All quotations from Carbone in this paragraph are from *Philosophy-Screens*, 66-69.

process. Like Bazin, Italian filmmaker Michelangelo Antonioni writes of the screen as a limited aperture through which an audience peers into an ostensibly unlimited world beyond it. However, his metaphor is not of the eye-mask but the *keyhole*, where filming entails spying from behind it: "the movie camera hidden behind the keyhole is a gossipy eye which records what it can."[96] Unlike Bazin, Antonioni does not think that the perimeters of the "keyhole" limit what is shown. To him, that is an easy obstacle to overcome, per his blithe suggestion: "So one hole is not enough. You make ten, one hundred, two hundred holes, place that many cameras behind them, and let several miles of film roll through."[97] Access to more information is simply a mitigation of making multiple holes in the visual range until there are enough to capture the whole event. The tension – and the ensuing interplay – between the physical and virtual boundaries of the screen thus collapses by this approach to the multitude of the image. Antonioni's perspective on the facileness of acquiring a boundless image through "two hundred holes" seems a little glib, but the hypothetical point remains: the withholding and provision of visual information via the boundaries of the screen is really a facetious issue, where boundaries can always be expanded, with its limitations therefore also overcome. In this perspective thus also lies the seeds of the post-screen, where the image is endlessly expandable, and the boundary is diminished or inconsequential.

Rather, to Antonioni, the issue is what happens to those "miles of film." Thinking as a filmmaker whose job is to create a story or some kind of coherence out of raw footage, as with Mitry, he focuses on the "ordered" version of what is finally shown on the screen. The limitations for information onscreen are thus not through the movements between what is seen or not seen out of the keyhole (or window), but *the editing process* – and, by extension, the filmmaker's interpretation – after the recording:

> What will you have [on making multiple holes]? A mountain of material in which not only will there be the essential aspects of an event, but also the marginal, absurd, and ridiculous aspects. Or the less interesting aspects,

96   Michaelangelo Antonioni, "Let's Talk About Zabrieski Point," *Esquire* 74 (August 1970), https://antonioni9.wordpress.com/2011/10/24/lets-talk-about-zabriskie-point-august-1970/.
97   Antonioni, "Let's Talk About Zabriski Point," np. His idea of multiple keyholes also brings to mind Italian filmmaker Pier Paolo Pasolini's views on the long take (as compared to the edited cut) which asked the viewer to imagine footage shot from every possible point of view for a multitude of subjectivities as an interrogation of the subjectivity of the long take: "Observations on the Long Take," trans. Norman MacAfee and Craig Owens, *October*, Vol. 13 (Summer, 1980): 3-6.

I mean to say. Then your job is to reduce and select. But the actual event really did include all these aspects. In selection, you will be falsifying. So you say, I am interpreting.[98]

Moreover, to Antonioni, the limits of the screen's information via the filmmaker's editing go even further – ultimately, anything an audience sees onscreen is futile in conveying any meaningful information, an idea Antonioni explored extensively in his acclaimed 1966 film, *Blow-Up*.[99] In the film, a photographer, Thomas (played by David Hemmings), is convinced that a murdered body lay beyond some bushes in a park which he had photographed. In one scene, Thomas "blows up" or enlarges the photographs he had taken of the bushes and stares at their grainy surfaces in search for the truth of the murder. But the images are impenetrable and Thomas's questions remain unanswered. The film, as Asbjørn Grønstad puts it, "reveals not only the materiality of film but, more importantly, the attendant opacity of its images," where "the diegetic world envisioned by a filmmaker is neither less inscrutable nor more mimetic than that of the painter or the writer."[100] In *Blow-Up*, it is not the boundaries of the Antonionian screen which limit its information, nor even the subjectivity of the filmmaker's interpretation, but the ontology of film itself, whose indexical imprint of light seals in the truths of its reality. This is a medium which gives away nothing; its screen is consequently also a showing of nothing.

Interactivity and the Moveable Window

To break this seal of unfathomability in the film image would require the introduction of an ontologically different medium, one which also, in the process, radically changes the nature of screen boundaries. This medium form, then, is interactive media, where the user is *in control* of what the screen shows. As videogame designer Chris Burke writes with respect to interactive media, "a system [of space] is modeled and the exploration [of that space] is given over to the player."[101] The interactive media user is thus able to explore the virtual space presented to them onscreen by manipulating

98   Antonioni, "Let's Talk About Zabrieski Point," np.

99   *Blow-Up*, directed by Michaelangelo Antonioni (1966; Burbank, CA: Warner Home Video, 2004), DVD.

100   Asbjørn Grønstad, "Anatomy of a Murder: Bazin, Barthes, *Blow-Up*," *Film Journal*, 2004, www.thefilmjournal.com/issue9/blow-up.html (no longer accessible).

101   Chris Burke, "Beyond Bullet Time," in *Understanding Machinima: essays on filmmaking in virtual worlds*, ed. Jenna Ng (London; New York: Bloomsbury, 2013): 23-40, 31.

their avatar's first-person perspective or the game's third-person virtual camera. In short, they can simply see *for themselves* what is otherwise concealed beyond the edges of the screen.

Interactive media thus upends two key issues in relation to screen boundaries and information. Firstly, the filmmaker's selection and editing processes for what is to be displayed onscreen become irrelevant, as the "truth" is available and up to the user to discover for themselves. Secondly, the opacity of the photographic image, per Grønstad's comment above, as imprinted in a particular time at a particular angle gives way to the transparency of interactive space examinable at the user's will to obtain information as needed. As Burke further comments, leveraging on Thomas's predicament in *Blow-Up*:

> ...[discovering the evidence in *Blow-Up* in a video game] would simply be a matter of finding the park and looking in the right place for the body. Additionally, because video games are topological by nature, one can imagine shifting into replay or 'theater mode' after the moment when the photographer takes the incriminating pictures. One could then explore the bushes, for example, to see whether or not a murderer was hiding there.[102]

Interactivity thus undermines the boundaries of its screen, which then no longer define or delimit the relational movements of revelation and conceal-ment, but become a *moveable window of reference* to serve the interactive user's agency.[103] Here, too, lay preliminary gestures to the post-screen in references to the diminishing screen boundary, if in this case not by way of the physical boundaries disappearing or ceasing to exist, but by the blunting of its definitional force. Where interactivity relegates authorial control of the virtual camera, the boundary no longer delimits what is seen and not seen. As the relevance of the boundary ebbs, the post-screen emerges.

Screen Boundaries Across Dimensions

Screen boundaries thus present screens in physical and virtual delimitations. More importantly, they evince screens not as rigid displays, but as fluid

102 Ibid.
103 Correspondingly, this moveability augurs the post-screen through Virtual Reality (VR), where the boundaries of the VR screen similarly give way to the user's agency in exploring its VR space. See chapter 3 for a fuller discussion of the post-screen and VR.

borders of movements between multiple oppositions: revelation against concealment; limited against potentially limitless vision; in appearance against dis-appearance; object against image; play against display. The result is the articulation of screens as movement in constant reciprocity between these oppositional poles. The *post-screen*, then, radicalizes the manifestation of these exchanges, where the dis-appearance of screen boundaries augments the complexity of the screen's movements of reciprocity, not least between the actual and the virtual, and destabilizes both regimes. The results are revised realities which fool and confuse, as with *Spider-Man: Far from Home*, where illusion takes over as the disorder the superhero must make right. This shift becomes the new peril, jeopardizing, as it did with Spidey, our sense of reality and orientation of truth. But it can also become the basis for other kinds of interesting and multi-faceted realities to emerge, as this book will also argue. These are implications of the post-screen that will be discussed further along this book.

For now, though, is a concluding exemplar on the paradoxical subtleties of screen boundaries with which to close this chapter, as a symmetrical bookend to the use of *Spider-Man: Far from Home* for its opening. As a deliberate counter to *Far from Home*'s blockbuster spectacle, this exemplar is an independent short film titled *The Facts in the Case of Mister Hollow*.[104] Just under five minutes long, the near-entirety of *Mister Hollow* focuses on a single photograph at which the audience is instructed to "look closely."[105] These "close looks," in turn, are achieved entirely by the virtual camera's movements. First, the camera pans and tilts around parts of the photograph, zooming into various close-ups to reveal more information to the audience. As the music intensifies, the virtual camera moves into the *depths* of the photographic image (in the process augmenting the photograph's two-dimensionality into three) to divulge further secrets and, indeed, "the facts in the case of Mister Hollow."

If *Far from Home* erases screen boundaries in its diegesis via Mysterio's illusions, *Mister Hollow* affirms in spades the paradoxes of screens *and* screen boundaries. The film's full title, *The Facts in the Case of Mister Hollow*, unwittingly or otherwise echoes the title of Edgar Allan Poe's short story, *The Facts in the Case of M. Valdemar*.[106] In Poe's story, the narrator is a mesmerist

104 *The Facts in the Case of Mister Hollow*, directed by Vincent Marcone and Rodrigo Gudiño (2008), online at https://www.youtube.com/watch?v=bzw8qdXCep8.

105 Per the text presented to the audience which appears at 1"07'.

106 Edgar Allen Poe, "The Facts in the Case of M. Valdemar," *Broadway Journal*, December 20, 1845.

who puts M. Valdemar in suspended hypnosis at the very moment of the latter's passing. Thus lies the story's masterful exploration of both movement and suspension (namely, absence of movement) between the boundaries of life, death, afterlife and "afterdeath."[107]

In resonance, *Mister Hollow* is also entirely about the movements and suspensions between the screen's boundaries, particularly as between surfaces, space and dimensions. First of all, other than some text in the opening seconds to set up its narrative, the entirety of the film's visual focus is the photograph. The photograph fills the screen, emphasizing the screen's boundaries by corresponding them to the borders of the photograph, beyond which the camera does not move. Eventually, this fixity becomes ironic for the screen, rather than being a window to a limitless outside world in the usual meaning of the metaphor, opens out instead to the confined space of a still and composed image.

Yet, the mobility of the film's virtual camera deftly defies those limitations of the screen and, by extension, those of the photograph as well *qua* a two-dimensional image. The camera's uses of pans, tilts, zooms and depth of field subvert the screen's material boundaries to reveal a great deal more information beyond the surface intelligibility of the two-dimensional photograph. If Roger Cardinal was tantalized by "the *illusion* of tactile depth quiver[ing]" in the image before him, *Mister Hollow*'s mission is to present the *truth* lying in those tactile depths. The film thus demonstrates fluent and critical interplay between revelation and concealment across its virtual boundaries, moving adroitly between two-dimensional surfaces and three-dimensional spaces.[108] This remediation between surface and space, of course, follows a long media tradition, such as perspective painting by artists from the Renaissance or trompe-l'oeil art which employed – indeed, pioneered – linear perspective to create illusions of depth. However, the virtual camera's unexpectedly fluid and omnidirectional movements in *Mister Hollow* through the virtual three-dimensional space of its photograph are entirely contemporary in articulating the fluid nature of the screen and its boundaries today – on one hand, an acknowledgement of the limitations of the screen; on the other,

---

107 Both the short story and these ideas of life, death and afterlife will be further discussed in chapter 4 on the post-screen through holographic projections.

108 Their shift across two- and three-dimensions also arguably represents the break between the analogue and the digital, where the two-dimensionality of the print photograph, as an analogue medium of light imprinted on a flat light-sensitive surface per *Blow-Up*, expands into the potentially infinite depths of three-dimensional virtual space inscribed by the digital logic of the mobile virtual camera and the computational.

counterpointing their very rigidities.[109] It is a proto-film of the post-screen, whose virtual boundaries dis-appear as movement between two and three dimensions, working the same contradictions as those of the Möbius strip imposing two-dimensional surface onto three-dimensional object. In turn, as this book will argue, these movements become the key contestations in contemporary screen culture as not only changing conceptualizations of screen boundaries in service of the post-screen, but also the unravelling of various theoretical threads in how images relate to reality across disappearing boundaries, dimensions and borders. More importantly, these unravelling relations of images and reality constitute a politics of the post-screen for post-democratic media and communications: the boundaries between information and misinformation that are increasingly difficult to discern; the exponentially uncertain grounds on which to premise values of truth, such as those once held to the indexicality of the photographic image and now annihilated in the wake of algorithmically generated deep fakery. The next chapters will continue to work out this unravelling.

109 The short film's creative antecedents are thus both *Blow-Up* and *Citizen Kane*: the opacity of the two-dimensional photograph per the former is resolved by the use of field depth per the latter. The camera moves constantly to give us knowledge, not in terms of moving *elsewhere* in the world, but, as in *Citizen Kane*, of moving *in the same space*, only to different depths.

2   Leaking at the Edges

Abstract

This chapter continues thinking about the screen via its boundaries, where, despite strong associations of protection and partition, screens are also shown to be read as subject to rupture, breach and violation, whose violence, moreover, may be presented in gendered terms. Breaches of the screen also imbricate the virtual and the actual in (i) exposing the factual realness of the actual by the fictional realness of the virtual; and (ii) spillover, or "leakage," of the virtual into the actual. With mobile and interactive media, screen boundaries diminish further as parties become virtually co-located and, as virtual worlds grow deeper and more complex, share them with algorithms. As demarcations between actual and virtual realities, screen boundaries grow increasingly unstable and fragile.

Keywords: protection; screen boundaries; found-footage; Covid-19; virtual co-location; algorithm

Protections and Partitions

> *Say something, goddammit! You're on television!*
> ~ Christof[1]

Everybody watches Truman Burbanks' life onscreen. A character in Peter Weir's *The Truman Show*,[2] Truman (played by Jim Carrey) is the unwitting star of a reality television programme. Unknown to Truman until near the film's end, the programme was created by svengali producer Christof (played by Ed Harris) to broadcast every aspect of Truman's life, supported by a cast of actors. Truman eventually cottons onto his fabricated life and escapes by

---

1   Dialogue line from *The Truman Show*, directed by Peter Weir (1998; Los Angeles, CA: Paramount Home Entertainment, 2000), DVD, as spoken by its character of Christof, played by Ed Harris.
2   Ibid.

Ng, J., *The Post-Screen Through Virtual Reality, Holograms and Light Projections. Where Screen Boundaries Lie.* Amsterdam: Amsterdam University Press 2021
DOI: 10.5117/9789463723541_CH02

literally sailing away from his "island" home, determined to leave behind his fake reality. The film's denouement lies in Truman's boat puncturing the wall of the dome which encloses the giant studio space in which he had lived his whole life, painted ironically to look like a cloud-dotted horizon. Truman gets off the boat and steps into the shallow water around it. He finds the door on the wall marked "Exit," and opens it. In this climactic moment, Truman prepares to step across the boundary from a directed, created-for-television reality to ostensibly "true" reality.

In this moment, too, the wall becomes a double metaphor. It represents all the obstacles Truman had to overcome in order to escape – his ignorance of the circumstances of his life; his imagined aquaphobia; his uncertainty of what lay beyond the island he lived on – all of which he conquers by breaching the wall. More pertinently, the wall symbolizes the screen on which Truman has appeared all his life. It contains the world in which Truman lived his screened life, where he is, as Christof declared many times, "safe." It sets apart that screened world as a discrete space which contains its fabricated reality and artificial behaviours, such as actors behaving to Christof's direction. Truman leaves his onscreen life only by literally breaking through the wall, first via his boat's prow, then by walking through the "Exit" door. Only with Truman right by the door at the wall does the fictional world get its ultimate exposure of truth, where Christof, forgetting that Truman is the one actor he cannot direct, barks at him a directorial order: "say something...you're on television!"[3] Like the great and powerful Wizard of Oz exposed (by way of curtains being drawn open, as with a screen!), Christof likewise becomes unmasked as "the man behind the curtain,"[4] revealing the smoke-screen charade of his television show. And only when poised before the wall and about to burst through his screen world was Truman ready to receive this truth.

The wall-as-screen in *The Truman Show* thus marks the boundaries not only with which the screen-world begins and ends, but also as a space in which truth erupts. Here, the screen's boundaries signify yet another cluster of contradictions: it is a protective covering, yet it leads to rupture and exposure; its boundaries separate, yet are prone to breach. Once more,

---

3   The failure of the actors to follow Christof's orders and direction in Truman's world directly correlates with, and at times instigates, the breakdown of its illusory world.

4   Dialogue line from *The Wizard of Oz*, directed by Victor Fleming (1939; Burbank, CA: Warner Home Video, 2006), DVD, in the scene where Toto, Dorothy's little dog, pulls aside the curtain to reveal the "Wizard," who, attempting to cover up the exposure of his theatrical illusions, exhorts Dorothy and her friends to "pay no attention to the man behind the curtain." The phrase has entered popular parlance as a reference to seeing someone for who they really are.

these paradoxes expose the various instabilities of the screen boundary on which, as we will see in further chapters, the post-screen capitalizes for its different regimes of reality and image.

***

Media scholars closely associate ideas of protection – and partition as protection – with the notion of the screen. For starters, these ideas appear in the very etymology of the word "screen" itself, even in translations. Exploring the definitions of the word "screen" in three languages – English (*screen*); German (*bildschirm*) and French (*écran*) – Gunther Kress reflects that in English, the word *screen* gives rise to two meanings: *sheltering* – such as to shelter from "a too intense heat" – and *partitioning*, by way of "something on one side that does not get through, is prevented from getting through, to the other side."[5] Erkki Huhtamo similarly points out how the word in sixteenth century usage, "and probably earlier," "was used to refer to a 'contrivance for warding off the heat of fire or a draught of air.'"[6] Wanda Strauven adds: "The connotation [of screen] is that of a barrier, of an object that is placed in-between, to protect or to separate."[7] She highlights the connection between the traditional fire screen and the hand-held face screen, both having the same purpose of protection from the heat of the fire.[8] Mauro Carbone also reflects on these ideas via the ancient walls of Plato's Cave – not of the more commonly referenced "the opposite wall" (i.e., the surface on which the Cave's prisoners see the shadows they perceive to be reality), but the *teikhíon*, which is "a low wall built along a road." In relation to the Cave, the *teikhíon* conceals the people carrying the objects

---

5   Gunther Kress, "'Screen': metaphors of display, partition, concealment and defence," *Visual Communication* 5(2) (2006): 199-204, 199. Kress cites two what he calls "example-meanings" as illustration: "a screen to set between one and the fire" (200); and "a screen for gravel or corn is a grating which wards off the coarser particles and prevents them from coming through." (201) The *Oxford English Dictionary* (OED) similarly sets out the meaning of a screen as "a contrivance for warding off the heat of a fire or a draught of air."

6   Erkki Huhtamo, "Elements of Screenology: Toward an Archaeology of the Screen," *Iconics: International Studies of the Modern Image*, 7 (2004): 31-82, 35.

7   Wanda Strauven, "Early Cinema's Touch(able) Screens: From Uncle Josh to Ali Barbouyou," *NECSUS* (November 22, 2012), https://necsus-ejms.org/early-cinemas-touchable-screens-from-uncle-josh-to-ali-barbouyou/.

8   Transposed into another and modern context, we can also think of sun*screen*, where the "screen" in the term connotes the same ideas of defence and protection resonant with the item's function as a similar barrier against heat – in this case, harmful ultraviolet rays from the sun. My thanks to James Barrett for this suggestion.

whose shadows then appear on the opposite wall; Carbone thus also reads the *teikhíon* as a screen, "in the sense that it *protects*... and hence conceals the men who are part of the machinery of the Cave." (emphasis in original)[9] In terms of its French version, Kress explains *écran* as "a *protective barrier*, 'to protect from sight or view'"; "at a further metaphorical level, *écran* can also mean to protect someone from, say, criticism." (emphasis in original)[10] In German, the word *Bildschirm* ("picture-shield") denotes the screen for the computer or TV, but the etymology of *Schirm* itself "comes from a much older Germanic *skermi-*, or the animal hide stretched across the shield used in fighting (as a protection for the surface of the shield). The verb *schirmen* means *to (safe)guard, protect, defend*." (emphasis in original)[11] The sense of protection here in the screen thus expands on the idea of the militaristic, of warlike defence against aggression; as Strauven comments, "a trace of the Old German *skirm* is still visible in the English expression *skir*mish."[12] Similarly, Franco Casetti remarks how "in the fourteenth century the Italian word *schermo* denoted something that protects against outside agents."[13]

Indeed, Casetti notes that it is only "at the beginning of the nineteenth century" that "the term [of the "screen"] began to enter into the sphere of entertainment," primarily with the popularity of visual illusions such as phantasmagoria and magic lanterns.[14] These media, then, developed the idea of the screen, more in line with chapter 1's discussion, as "open[ing] our gaze to something hidden."[15] Casetti thus highlights the "slippage of meaning" in our thinking of "screen" as that from "a surface that covers and protects, to one that allows us to glimpse images projected from behind."[16] Kress echoes this disparity between protection/concealment and display in the linguistic meanings of "screen" he examined. In particular, he notes the "one startling commonality" in his "brief excursion into history and cultural

9   All references to Carbone in this paragraph are from Mauro Carbone, *Philosophy-Screens: From Cinema to the Digital Revolution*, trans. Marta Nijhuis (Albany: State University of New York Press, 2016), 67.

10   Kress, "'Screen,'" 201.

11   Ibid. Strauven also traces the Old German *skirm* as the root of the Dutch word *scherm*, or the Middle Dutch *screm*, which itself is "another possible root for the French *écran*": Strauven, "Early cinema's touch(able) screens," 6.

12   Strauven, "Early Cinema's Touch(able) Screens," 6.

13   Francesco Casetti, "What is a Screen Nowadays," in *Public Space, Media Space*, eds. Chris Berry, Janet Harbord and Rachel O'Moore (Basingstoke: Palgrave Macmillan, 2013): 16-40, 17, quoting Salvatore Battaglia, *Grande Dizionario della Lingua Italiana*, Vol. 17, Turin: Utet, 1994.

14   Casetti, "What is a Screen Nowadays," 17-18.

15   Casetti, "What is a Screen Nowadays," 18.

16   Ibid.

variation," whereby "in that history and across languages and cultures, meanings of *defence, protection, shelter, concealment, partition* dominate over that of *display*, though different in the three different languages/ cultures." (emphasis in original)[17]

In this contestation, then, also lies competing articulations of the screen across the current and last chapter. The previous chapter discussed the screen in terms of its boundaries by way of paradoxical movement between different dimensionalities, revelation and concealment, actual and virtual, and argued for a fluidity to screen boundaries in play and display. Here, we continue that expression of screens and their boundaries via their other meanings of protection and partition. Just as screen boundaries play out their paradoxes of movement, within those continuities of movement also lay ostensible *discontinuities* by way of defence and separation, whereby screen boundaries are *stoppages* of movement. Yet, as Truman Burbanks demonstrates with such flourish and as much of this chapter will argue, walls and boundaries can also be broken and shattered. As with the paradoxes of screens as movement, screens as protection likewise present contradictions in being simultaneously defence and breach. One quick instantiation of such contradiction is Huhtamo's observation of how the nineteenth-century feminine "hand-screen," like hand-held fans, were held against a lady's face both as protective covering against intrusive or unwanted gazes, even as it also invited interest and enticement, where "veiling one's face behind a hand-screen incited desire and curiosity, like a mask."[18] In a more literal context, Galit Wellner discusses the cell phone screen as what she calls a "wall-window," a portmanteau word which connotes the mobile phone screen's simultaneous functions as both barrier against the user's immediate environment and opening to information.[19] Protection in both cases is thus paradoxical in closing *off* as well as *dis*closing for further exploration. In turn, as this chapter will show, another more forceful and energetic movement emerges across the screen's boundaries, by which the virtual encroaches or spills over, and actively co-locates with the actual.

In this sense, movement/display and protection/partition are not necessarily in conflict. Of greater value are their various re-conceptualizations of the screen via its boundaries so as to better understand the nature of its display – not, as a now-familiar theme in this book, in terms of what it

17   Kress, "'Screen'," 202.
18   Huhtamo, "Elements of Screenology," 35.
19   Galit Wellner, "Wall-Window-Screen: How the Cell Phone Mediates a Worldview for Us," *Humanities and Technology Review* 30 (Fall 2011): 77-103.

shows, but of its gap betwixt the virtual and the actual. This chapter will specifically discuss how that gap of screen boundaries pulls both against and across these tensions of separation, with results of rupture, encroachment and spillover which further destabilize its boundaries. The gap grows ever more porous and ambiguous; the post-screen beckons.

Rupturing Screen Boundaries

As with the discussion in chapter 1 on screens as display surfaces, the screen as protective defence attracts its own metaphors. One example is Serge Daney's comparison of the screen to "the skin, the transparent," or how

> ...[t]he transparent continuum that clings to the real takes its form, the bandages that preserve for us the mummy of reality, its still living corpse, its eternal presentness: that which allows us to see and protects us from what is seen: *the screen* [emphasis in original].[20]

The notion of skin in relation to the screen is already mentioned above in the etymology of the word in terms of the Old German *skermi-*, referring to animal hide stretched across a shield. Like the screen, skin itself also contains paradoxes, such as that of simultaneous permeability and impermeability. On one hand, skin is porous, absorbing elements such as air and, less benignly, toxins and chemicals. On the other hand, skin is also a barrier, preventing dangerous processes such as water loss from the body as well as entry of harmful microorganisms or irritants into it.[21] Skin is stretchable and resilient, yet delicate and vulnerable to bruising, penetration and other violent force. Alluding to these qualities, Daney takes on André Bazin's idea of reality as the essence of cinema,[22] and writes of the screen as such a barrier against "the fire of the real," paradoxically impermeable and fragile: "The screen, the skin, the celluloid, the surface of the pan, exposed to the fire of the real and on which is going to be inscribed

---

20   Serge Daney, "The Screen of Fantasy (Bazin and Animals)," in *Rites of Realism: Essays on Corporeal Cinema*, ed. Ivone Margulies, trans. Mark A. Cohen (Durham, NC: Duke University Press, 2002): 32-41, 34.

21   Indeed, the skin has been described as a barrier in various models, such as the Elias model of the skin barrier, or "the domain mosaic model of the human skin barrier," as taken from Bo Forslind, "A Domain Mosaic Model of the Skin Barrier," *Acta Derm Venereol (Stockh)* 74 (1994): 1-6.

22   See discussion in chapter 1, particularly on Bazin and the realist school of cinema.

metaphorically and figuratively – everything that could burst them."[23] As skin, the screen according to Daney presents its paradoxes as both defence and vulnerability: it is an outer barrier against reality, yet also a membrane sliver on which reality imprints itself, and so delicate it risks being penetrated at any moment.[24]

Hence, to Daney, the screen is not – like the *skermi-* on the military shield – an all-protective covering to barricade against reality on the other side of the screen.[25] Rather, it is a covering that, even while being a protective defence, is so breakable it can be taken to the point of fetishism akin to a fixation with virginity. Daney's argument, then, follows that the skin of the screen is the hymen, and the breaking of it – the breach of reality through the screen – a violent, almost profane, assault as in rape:

> That tiny difference, the screen: 'Of course,' says Bazin, 'a woman who has been raped is still beautiful but she is no longer the same woman.' *The obscenity perpetrated by the rape of reality* cannot fail to send us back to the rape of the woman *and the screen, the hymen* [emphasis added].[26]

In such characterization, the screen is not only a paradoxically fragile protective surface. Significantly, it becomes ruptured as a *gendered* surface. To that extent, Strauven draws similarly gendered connections between the screen and the female body. Via readings of Georges Méliès's films, Strauven points out how female bodies are often "put on display," such as in *Le Merveilleux éventail vivant* (*The Wonderful Living Fan*, 1904), where display panels, reminiscent of fire screens, are "magicked" into, and thus

---

23   Daney, "The Screen of Fantasy," 34-35. It is interesting that he characterizes the real before the camera as a fire, unwittingly or otherwise invoking the sense of "screen" as protection against it.

24   This tension in the screen as both barrier and penetrable lamina is used to particularly endearing effect in Paul King's *Paddington* (2014) when the eponymous bear, homesick in London, is shown a video of his homeland of "darkest Peru": he steps up to the screen displaying the black-and-white footage and first pauses before it – here the screen is a physical barrier, a hurdle of the vast distance from London to Peru. Paddington then *walks through the screen* and emerges on the other side in his homeland, awash in colour and sights – here the screen is a penetrable lamina, bringing virtual reality within such tangible and sensorial proximity that one need simply reach through for it.

25   Daney contrasts this fragility of skin against Bazin's notions of capturing reality in cinema, which he labels a "fantasy," characterizing in it "a comical vision of the screen as the surface of a Teflon saucepan (in glass), capable of 'sealing' [in the culinary sense] (*saisir*) the signifier." Daney, "The Screen of Fantasy," 34.

26   Daney, "The Screen of Fantasy," 35.

equated to, living women.[27] In his oeuvre of films frequently presented as magic trick shows, women's bodies are also often co-opted into the trickery, where they are "treated as concrete barriers in the execution of magic (and filmic) tricks," "constantly covered and uncovered by Méliès by means of screens, cloths, curtains, and so on, to eventually be turned into a screen itself – that is, a screen *for* and *on* display." (emphasis in original)[28]

While Méliès's presentation of infringement or breach of screens/bodies in his films is not presented in overtly violent terms, a tone of vicious transgression nevertheless sometimes appears. For example, in *L'illusionniste double et la tête vivante* (*The Triple Conjurer and the Living Head*, 1900), a living woman's head was first placed on a small table before being "magicked" into a full body of an upright woman. The two "magicians" standing on either side of her (played by Méliès himself as duplicated) are amazed and triumphant in their conjuring, and they try to kiss and touch her. As it becomes clear that she is a superimposed image, one of them, with registered astonishment, passes his hands several times through the image of her body.[29] If we read the woman's body here – itself a displayed virtual reality – as a screen, this "breach" of the screen-as-female-body is at best a comical play between the virtuality of the image and the corporeality of the body. At worst, it carries the same connotations of rape as alluded to by Daney, unmistakeable in its figurations of male invasiveness of the female body.

On such terms, the boundaries of the screen thus demarcate gendered spaces, the breaching of which inevitably signifies gendered violence. While Daney and Strauven have construed this violence to be assaults against women, a more modern take may be read in *Ringu*,[30] which turns gendered violence across screen boundaries on its head in a remarkably original way. A Japanese horror thriller film which performed to great success at the box office and spawned several follow-ups in a franchise as well as a Hollywood remake, the premise of *Ringu*'s plot is a curse which takes the form of a

27   In ungendered terms, Carbone also describes how the human body is itself a "specular wall," or an "arche-screen." As he writes: "the body can produce images simply by being interposed between a luminous source and a wall (as it happens in the myth of the origin of painting narrated by Pliny the Elder) or by being decorated with inscriptions, drawings, or tattoos." *Philosophy-Screens*, 66.

28   Strauven, "Early cinema's touch(able) screens," np. In relation to screens for display, Strauven describes an instructive text by way of American Mutoscope and Biograph's *A Midnight Fantasy* (1899), "where Rose Sydell appears framed as a (living) billboard among three other life-sized vaudeville posters on the street," np.

29   The woman's position – standing with her hands behind her back, as if pinned to the wall – reinforces this reading of gendered assault.

30   *Ringu,* directed by Hideo Nakata (1998; London: Tartan, 2001), DVD.

videotape that will a week later kill anyone who has watched it.[31] The film follows an investigative reporter, Reiko (played by Nanako Matsushima), who, having inadvertently watched the tape and is now under its curse, sets out to save herself with the help of her ex-husband, Ryuji (played by Hiroyuki Sanada). Reiko traces the curse to its originary location: a well on Izu Oshima Island in which Sadako, the girl-spirit who had created the curse, was ostensibly murdered by her father.

The film's climax arrives just after the audience had been led to believe that Ryuji and Reiko had managed to break the curse. In the quiet after the storm, the curse resumes its malevolence in the form of Ryuji's television set suddenly turning on by itself. Ryuji's television screen shows the image of the well on Izu Oshima Island which he and Reiko had previously visited. After a few moments, a movement stirs at the edge of the well: a figure dressed in white, its face covered entirely with long black hair, starts to climb out of the well. It is Sadako, who, on exiting the well, proceeds to lurch towards the diegetic camera in front of her and which is ostensibly recording the shot. The film cuts often between the television image and Ryuji's face, initially relatively composed: after all, the threat is *on the other side of the screen*. The screen separates the virtual reality of the image of Sadako's horrifying form from Ryuji's own diegetic actual reality; it is a surface of apotropaic magic, protecting him from this menace. Yet where, for instance, a film like *Uncle Josh at the Moving Picture Show* (to be discussed in the next section) leverages screen boundaries for comedy, *Ringu* flips the notion of the screen as defensive partition for its horror. In the most memorable shot of the film,[32] Sadako arrives in front of the presumed camera and, by first pushing with her head, *crawls out of the television screen into Ryuji's diegetic world*. She drags herself on her hands and knees *over* the literal physical boundaries of the television screen, so that Sadako *herself* becomes the manifested transgression of the screen's virtual boundaries as she eventually stands upright in Ryuji's living room. As she walks over, Ryuji stumbles around the room in horror and dies of a heart attack, fulfilling the curse.

---

31 The "get-out" from the curse is to copy the tape and show it to another person who, in turn, can save themselves by repeating the process.

32 This shot of Sadako has become so inseparably and memorably associated with *Ringu* that advertisers re-created it in promoting *Ringu*'s sequel, *Sadako 3D* (2012), by hiring groups of female models to pose on the streets of Tokyo with a television screen around their shoulders and their heads pushed out of it as a nod to the shot as the film's visual trope: see Emily Balistrieri, "Sadako Mob Terrorizes Tokyo to Promote 'Ringu' Sequel," *Crunchyroll*, May 6, 2012, http://www.crunchy-roll.com/anime-news/2012/05/06-1/sadako-mob-terrorizes-shibuya-to-promote-ringu-sequel.

This climactic scene also represents the first time the audience sees the direct effect of the curse, which has so far taken the form of the played videotape showing various incoherent images. Notably, the curse-as-videotape was previously contained behind the boundaries of the screen which in effect act as a defence against its malevolence. When the curse comes to pass in its killing, it is thus appropriate that its horror should be visualized not only in Sadako's terrifying form, but also in terms of *the violation of screen boundaries*. As with Truman Burbanks, the breaching of screen boundaries, which had previously been of protection and defence, is transformational. For Truman, the breach exposes the truth of his world; in *Ringu*, it demonstrates that the true horror of the curse is not Sadako at all. Appropriately, the fulfilment of the curse is the breach of screen boundaries as protective partition; the disintegration of the screen's apotropaic magic; the inexorability of Sadako's encroachment into the side of the screen which, until then, was the safe space from her.

In terms of the gendering of screen boundaries and the violence of their infringement, we can thus read the space behind the demarcation of the screen as that of the female, occupied first by Sadako's evil powers via the videotape emanating the curse and later by her humanized form as she appears out of the well. Taking up Daney again, if the screen is a fragile covering exposed to the fire of the real, in this case there is a literal "bursting" of that "real" of Sadako's malevolence as she crawls across the boundaries of Ryuji's television screen. However, the breach of the screen's boundaries in *Ringu* is not an assault on the female body/virginity on Daney's terms, but the exact converse: this is about Sadako's empowerment, and a very precise occasion for her to turn the tables so as to be the aggressor. If we think back, again, to Daney's metaphor of the hymen for the screen, Sadako's (her name in Japanese meaning "chaste child" (*sada*: chaste and *ko*: child)) reinterpreted transgression becomes even more ironic: the rupturing of the screen, then, is not about the violation of virtue (as in female virginity), but about how the virtuous (the "chaste child") violates her innocence (in killing her victims). By literally pushing through the television screen and crawling slowly across its boundaries on her bloodied hands and knees, Sadako shatters a different kind of glass ceiling.

Interplay between Fictional and Factual Threat

The rupture of the screen boundary is not only of violence and violation. In their contradictions of protection and breach, screen boundaries also demonstrate inherent vulnerability and instability, and ultimately a fluid and confounding engagement between the virtual and the actual. In *Ringu*, the virtual threat or danger crosses screen boundaries to displace actual reality; alternatively, as we will see in this and the following sections, one also *merges* with and into the other. In this sense, the screen boundary is more than merely a defensive wall subject to penetration or dismantlement.[33] It is also a more insidious separation between the virtuality and actuality of threats and which differentiates or disentangles their fiction and factuality. The famous, if anecdotal, reception to the Lumière brothers' 1895 film, *Arrival of a Train at La Ciotat*, comes to mind here, where – oft-discussed as "one of the key myths of cinema's spectatorial origin"[34] – the audience allegedly fled the film theatre in panic from the perceived danger of the oncoming train. The veracity of the account has been much contested, but the point here is what it says, even as an apocryphal story, about the virtual real as threat and, in crossing screen boundaries, its embroilment with the actual real. As Elizabeth Grosz writes, "virtual objects are now capable of generating the same perceptual effects as 'real' objects."[35] In this embroilment, the question of the actual and the virtual is not one of a binary, but connects to a third term of *the real* not in any Lacanian or related sense, but simply by the absoluteness of their palpable effect on the viewer. The key here to screen boundaries as a defence is thus not so much about the breach of harm per se as with the crossing of Sadako's malevolence, but the entanglements across its thresholds of the real-ness of virtuality with the real-ness of actuality. In short, both realities are real, but one is fictional, while the other is factual. More importantly, they show up *against each other*: the factual realness of one (the audience's reactions) exposes the fictional realness of the other (the oncoming train).

It only took a few years for the audience to become savvy with the most basic of such an exchange between virtual and actual realities or threats across the screen's boundaries. This is clear from Edwin Porter's 1902 film,

---

33   This idea also brings to mind David Theo Goldberg's line on walls as "the last line of defense": see "Wallcraft: The Politics of Walling," *Theory, Culture and Society*, February 27, 2015, https://www.theoryculturesociety.org/david-theo-goldberg-on-wallcraft-the-politics-of-walling/, a web piece on the politics of walls, or what Goldberg calls wallcraft, where he traces the erections, falls and forces of walls through a broad swathe of history and geopolitical spheres. As with the paradoxes of screens, Goldberg also notes the paradoxes of walls, such as where they are "force fields" of activity, even as they also contain and restrain.

34   Anne Friedberg, *The Virtual Window*, 155. The audio equivalent would be Orson Welles's *War of the Worlds* radio broadcast, which similarly spilled over into the actual.

35   Elizabeth Grosz, *Architecture from the Outside: Essays on Virtual and Real Space* (Cambridge, MA: MIT Press, 2001), 78.

Uncle Josh at the Moving Picture Show (re-made from Robert W. Paul's 1901 *The Countryman and the Cinematograph (aka The Countryman's First Sight of the Animated Pictures)*), the other oft-discussed early cinema film, particularly regarding the screen's partitions of on-/offscreen and material/virtual space. *Uncle Josh* not only demonstrates the virtual against the actual, but leverages the idea as a comedic element for the audience's appreciation. The short film shows our eponymous hero watching three projected movies in a theatre hall: one of a dancing woman; one of an oncoming train; and one of a courting scene between a man and a woman whom, the Edison catalogue explained, Uncle Josh identified to be his daughter. The film's comedy lies in how Uncle Josh reacts to the movies: he imitates the dancer, "matched in scale as he mirrors her movements in a complex parody of transgendered identification and blatant gender difference";[36] jumps away and cowers offstage from the train; and shadow-punches the man flirting with whom he thought to be his daughter. Finally, he tears down the film screen, only "to reveal its surface, the projectionist, and the apparatus behind it."[37] But this does not faze Uncle Josh, who continues to throw punches at where the man in the image was, and eventually lands one on the projectionist already furious with Uncle Josh's destruction of his equipment. The two fall brawling onto the stage floor and the film ends.

The comedy of the film works because, unlike Uncle Josh, the audience understands the screen boundary as the barrier between two realities, partitioning the actuality of the audience's reality against the virtuality of the onscreen image. Their amusement arises from their superior appreciation of this division against the naiveté of Uncle Josh as the "rube"; they realize there is no need to leap away from the oncoming train, just as there is no point in shadow-boxing the courting paramour. Wanda Strauven argues that Uncle Josh's ignorance lies in his lack of understanding how cinema works, namely, that cinema is only to be viewed, not touched. She writes: "[Uncle Josh] is clearly acting like a rube who does not understand that moving pictures are produced by a light projection onto a screen and that this screen should not be touched, only looked at."[38]

Yet if Uncle Josh had merely touched the screen, the film would not be complete. Its full effect and meaning lie in Uncle Josh *tearing down* the screen, for only in that total destruction of the screen boundary does he rip up the screen's demarcations of actual and virtual realities, and in so

36   Friedberg, *The Virtual Window*, 158.
37   Friedberg, *The Virtual Window*, 160.
38   Strauven, "Early Cinema's Touch(able) Screens," np.

doing reveal their entanglements with each other.[39] As Uncle Josh rushes up to the screen, the image of the courting man superimposes over him; as he tears down the screen, the angry projectionist is revealed in the place of the image. Paramour and projectionist – virtual and actual – thus double up, not to substitute each other (for they remain distinct), but to expose the real-ness of one specifically in relation to the other, and ultimately reveal the true threat (namely, the angry projectionist who brawls with him).

Uncle Josh's ending is thus about screen boundaries which more than carve up different realities and spaces, as most interpretations have it. Rather, it discloses two other messages about the screen. The first is that the screen as defence is a double con: it protects against virtual threats (the paramour) even as it conceals actual ones (the projectionist); it confounds the actuality of one with the virtuality of the other, and vice versa. The trick, then, is to recognize each for what it truly is. Specifically, as with Spider-Man in *Far from Home*, discussed in chapter 1, this trick is to understand that such recognition lies not in seeing the image, but in identifying the boundaries of their respective realities, and grasping the *gap* between each.

The second message is more specifically about the threat against which screens defend and protect: it is, at the end of the day, really a larger point about *the trauma of the real* and the protections from that, for which *Uncle Josh*'s angry projectionist is merely a metaphor. If screens are the interface out of which we experience the world in all its horror, bewilderment and confusion, they also display the multiple truths of our experiences. Screens as barriers against threat are thus, again, paradoxical in their simultaneous functions as opening and defence: on one hand, the facing and confrontation of the world; on the other, protection against the real, whereby T.S. Eliot's words at once come to mind: "human kind / cannot bear very much reality."[40] Screen boundaries thus do not only demarcate the virtual against the actual. Rather, they diagnose the larger problem of all representation,

---

39   *Uncle Josh* is only the start of a line of films that self-reflexively mark the separation between the virtuality of the film world and the actuality of the viewer's world, and in the process remind the viewer of the technological artifice out of which the diegetic world is constructed. Other well-known examples include a sudden sequence of the film stock burning up in Ingmar Bergman's *Persona* (1966), where the image annihilates before the viewer's eyes in literal visual terms; or the failure of the sound equipment in the last sequence of Abbas Kiarostami's *Close-Up* (1991), reminding the viewer of the reality of filmmaking equipment in creating the film illusion, and the boundaries between the two worlds. See also Jean-Louis Baudry and Alan Williams's well-known article, "Ideological Effects of the Basic Cinematographic Apparatus," *Film Quarterly* 28(2) (Winter, 1974-75): 39-47, which argues for the ideological meanings out of cinema's technological and aesthetic artifice.

40   T.S. Eliot, "Burnt Norton," in *Four Quartets* (New York: Houghton Mifflin Harcourt), 2.

namely, the confrontation of reality so stark and unbearable it can only be negotiated through a tightrope act of balances and counter-balances, shadowboxing and shadow boxes. The true nature of screens as protection and partition thus shows up against the porosity and instability of their boundaries between the virtual and the actual. Or, rather, their lies about being protections against confounding dangers, and their confounding of dangers against protection.

Leaking at the Edges: The Merging of the Amalgamated Real

These implications of the virtual across screen boundaries also facilitate another kind of breach of the screen – one which does not so much rupture as it disintegrates its separation. Here, the virtual *spills over* into the actual: they "leak" at the edges of the screen, so that virtual and actual cease to be a binary. Rather, they merge into a kind of amalgamated real, simultaneously virtual and actual. *Ringu*, again, comes to mind, where its critical plot point is that the curse visits all who watch the video. As the film's main characters watch the video, *so does the audience*. Through various shot-reverse-shot sequences, they are *sutured* to the videotape victims and, in tandem, become tainted with the curse from the video they have watched in complicity.[41] Their gaze across *their* screen boundaries thus co-opts them into the film's diegetic world. Similarly, as Ryuji stares, horrified, at his television screen showing Sadako staggering towards him, the director takes care to include a few shots which frame the diegetic screen squarely within them. The result is that the audience, again as with Ryuji, looks directly at the malevolence of Sadako stumbling towards them on their *own* screen. Of course, Sadako does not crawl through the audience's screen. Nevertheless, the film's plot points, its key shots and the audience's own act of gazing implicate them into the diegesis. The horror of the screen's breach in *Ringu* is thus two-fold: the first is in Sadako's encroachment across the television screen's boundaries, as discussed earlier; the second is the "spillage" of that horror into the audience's *own* world through their complicity of looking, so that its virtuality amalgamates into their actuality.

This "leakage" is most prominent in what has been called "found-footage horror," a film genre popularized by the late 1990s box-office successes of *The*

---

41   On suture, see J.-P. Oudart, "Cinema and Suture," *Screen* 18(4) (Winter 1977/8): 35-47.

Last Broadcast[42] and, a little later, the 1999 film of *The Blair Witch Project*.[43] The genre's main characteristic is that a substantive portion of the film, if not its entirety, is presented to the audience, usually via introductory title cards, as "found" from cameras "recovered" from the diegetic world and "recorded" by characters who have either "died" or "disappeared."[44] The effect is to present the film to the audience as an object from a fictional world that is somehow present in theirs, thus dissolving screen boundaries as a partition between the actual realities of their space and the virtual realities of the film's world, and somehow amalgamating the two. As Cecilia Sayad describes:

> The films are not presented to us as 'inspired by' real events [as is the usual premise for fiction films]—they are supposed to constitute the audiovisual documentation of these events. What we see, we are told, are real people, not characters based on them... The horror movie is thus presented not as mere artefact but *as a fragment of the real world, and the implication is that its material might well spill over into it* [emphasis added].[45]

Several cinematic strategies in the "found-footage" genre work hard to promote this porousness of the screen's boundaries, and to solicit the audience's psychological belief of such a "spillover" across them. For instance, the story always involves a plausible premise for the video recording to exist. This premise generally runs along the lines of a pair or group of ordinary people, usually teenagers or young adults, setting out with a camera to record their lives or make a film, and end up recording an adventure which "turns out" badly for them, almost invariably due to paranormal or supernatural reasons. This narrative thus provides reasonable justification for the existence of

---

42   *The Last Broadcast*, directed by Stefan Avalos and Lance Weiler (1998; London: Metrodome, 2003), DVD.

43   *The Blair Witch Project*, directed by Daniel Myrick and Eduardo Sánchez (1999; London: Pathe, 2003), DVD. However, its provenance can be traced from as early as 1980, as seen in the Italian snuff movie, *Cannibal Holocaust*, directed by Ruggero Deodato, a large part of which, as Cecilia Sayad describes, "displayed mock found footage of the tragic deaths of a TV crew shooting a film in the Amazon within the context of a fictional narrative." See Cecilia Sayad, "Found-Footage Horror and the Frame's Undoing," *Cinema Journal* 55:2 (Winter 2016): 43-66, 44.

44   See Sayad, "Found-Footage Horror," 44-45, for a detailed list of films in this genre. More recent films closer to the time of writing include M. Night Shyamalan's *The Visit* (2015); Michael McQuown's *The Dark Tapes* (2017); and Justin Barber's *Phoenix Forgotten* (2017).

45   Sayad, "Found-Footage Horror," 45.

the "footage" appearing before the audience, boosting belief of its veracity as being from another reality now "spilling over" into theirs.[46]

The story is also buttressed by form and aesthetic which, as Sayad contends, are key to this "merging of the diegetic world and that which exists outside of it."[47] Analysing the strategies of framing in her chosen "found-footage" case study, *Paranormal Activity* (2009; directed by Oren Peli), she argues how framing contributes to the instability of the image between on- and off-screen spaces. This instability across boundaries effectively presents the continuity of the screen-world with the audience's and reinforces the "connection between horror and reality": if the horror is framed to enter from a corner of the screen, it is not implausible to conceive of it lurking just *beyond the screen* in the audience's world.[48] Complementing this belief, as Sayad further notes, is the film's aesthetic of relative amateurishness, characteristic as well across the genre, marked by "raw cutting, elliptical narrative, and grainy, shaky, and precariously framed images that mimic the style of amateur filmmaking."[49] This unpolished "home-video" aesthetic, whereby the "footage" appears in "recovered," unedited and "rough-cut" form, thus further supports the film's professed authenticity as "documentation" of its diegetic world.

Finally, the use of direct address – the recording of speech or gestures made specifically to a camera which, normally invisible, thereby becomes

---

46   This plot premise of "documentary" footage posing as fabricated witness accounts is also used in other media forms to solicit user belief in similar slippage of the virtual into the actual. One example is the "paranormal pictures" Photoshop contest launched on the Something Awful (SA) Forums in 2009, which "required participants to turn ordinary photographs into creepy-looking images through digital manipulation and then pass them on as authentic photographs on a number of paranormal forums." As forum users shared their Photoshop creations, they usually supplement their images with fabricated witness accounts for greater conviction. The most notorious of these is Slender Man, posted by SA user Victor Surge (real name Eric Knudsen), as "a mysterious creature who stalked children." The Slender Man myth propagated online in various memes, websites, fan art, vlogs and alternate reality games; it came to international prominence when it was cited in the stabbing of 12-year-old Peyton Leutner in Wisconsin by her two classmates. Information and all quotations taken from "Slender Man" at KnowYourMeme, http://knowyourmeme.com/memes/slender-man#origin.

47   Sayad, "Found-Footage Horror," 52. Specifically, she underpins her argument by identifying the shift of the horror film genre as represented by "the interpenetration of reality and fiction," namely, from traditional allegory to "a new locus: the film's form": 43.

48   Sayad, "Found-Footage Horror," 58-63. Sayad pursues her close reading along two strands: (i) the film's "decentered composition," with the mise-en-scène of elements within the frame as relegated to the image's periphery and background to "subvert the rules of composition" (60); and (ii) the illusion of control in relation to "time, space and narrative perspective." (61)

49   Sayad, "Found-Footage Horror," 43-44.

part of the diegesis – likewise adds to the form of the genre. As Tom Brown points out, direct address in "found-footage" films differs from its usage in other kinds of cinema – it is normally an anomalous disruption or, taking Pascal Bonitzer's words, "a rupture of the fabric of the cinematic fiction."[50] However, direct address in "found-footage" films is non-anomalous in that it is "performed to a diegetic camera – a camera that is part of the story world."[51] Hence, "talking to the camera is explained (explained away) by the *possibility* that other diegetic characters might see this testimony." (emphasis in original)[52] This is sometimes played on the nose: for instance, at the end of the Godzilla-homage film, *Cloverfield*,[53] the protagonists give tearful "if-you-are-watching-this-it-means-I'm-dead" speeches to the camera as the rampaging monster roars above them. A similar speech also appears in the famous apology scene of *The Blair Witch Project*, where the main character weeps and self-consciously apologizes directly to the camera for her part in leading her friends into their misadventure.[54] In these speeches, the characters' addresses demonstrate both their self-consciousness as a recording object *and* their consciousness of a potential audience watching the recording, one which includes the present cinema audience. They disrupt the viewer's "sense of the real place of spectating from a seat in a cinema theatre or from a sofa in a domestic setting."[55] In turn, this disruption co-opts the audience into the story, intermingling actual and virtual realities.

Through these various cinematic strategies and genre characteristics, "found-footage horror" demonstrates the porosity of screen boundaries and the effective "leakage" of the virtual into the actual. But genre is not the only expression of fluidity across screen boundaries; other film strategies similarly destabilize their spatial partitioning. Alfred Hitchcock's uses of plot,

---

50   As quoted in Tom Brown, *Breaking the Fourth Wall: Direct Address in the Cinema* (Edinburgh: Edinburgh University Press, 2012), 23.

51   Brown, *Breaking the Fourth Wall*, xiii.

52   Ibid.

53   *Cloverfield*, directed by Matt Reeves (2008; Los Angeles, CA: Paramount Home Entertainment, 2008), DVD.

54   Such "factual" recordings of "fiction," in turn, imparts a kind of agency to the camera: see Jenna Ng, "The Handheld Digital Camera Aesthetics of *The Blair Witch Project* and *Cloverfield* (via *Strange Days*)," *16:9, A Danish Journal of Film Studies*, vol. 32, http://www.16-9.dk/2009-06/pdf/16-9_juni2009_side11_inenglish.pdf.

55   Constance Balides, "Immersion in the Virtual Ornament: Contemporary 'Movie Ride' Films," in *Rethinking Media Change*, eds. David Thorburn and Henry Jenkins (Cambridge, MA: MIT Press, 2003): 315-336, 317.

camera framing and cuts in *Rear Window*,[56] for example, famously set up a relation of viewing and action which reverses the positions between actor and spectator across different sides of the screen boundaries. The actor in the film is thus a spectator unable to act and who can only spectate across the boundaries of his window ("he is reduced as it were to a pure optical situation").[57] Conversely, the spectator to the film is an actor drawn into the film's action across the boundaries of their screen. The effect of the film, as Deleuze puts it, is "to implicate the spectator in the film," with "the characters themselves... assimilated to spectators."[58] This relation not only articulates the film's suspense, but also further illustrates the fragility of the screen's boundaries in metaphorical exchange between spectatorship and action.[59]

Advancing film technologies also erode screen boundaries, of which 3D cinema, in its aim to dissipate the two-dimensional *surface* of the screen by emphasizing the three-dimensional *volume* of screen objects, forms the clearest example. The earliest forms of 3D cinema exploit that violation of the screen surface as its prime thrill, as seen in the trailer for the 1953 *House of Wax*,[60] the first colour 3-D feature film from a major American studio, which promises things that "come off the screen right at you!"[61] While more recent offerings of 3D cinema have modulated their use of the "shock" effect and nuanced its use of depth,[62] 3D technology remains essentially invasive across screen boundaries. As Simon Lefebvre puts it, if with some

56 *Rear Window*, directed by Alfred Hitchcock (1954; London: Universal Pictures UK, 2007), DVD.

57 Gilles Deleuze, *Cinema 1: The Movement-Image*, trans. Hugh Tomlinson and Barbara Habberjam (Minneapolis, MI: University of Minnesota Press, [1986] 2003), 205.

58 Deleuze, *Cinema 1*, 205.

59 This interplay between screen and audience also brings Truffaut's words to mind: charged by American film critics for favouring *Rear Window* "because, as a stranger to New York, you know nothing about Greenwich Village," Truffaut, revealing his different understanding of the film, retorts, "*Rear Window* is not about Greenwich Village, it is a film about cinema, and I *do* know cinema": François Truffaut, *Hitchcock*, collaborated with Helen G. Scott (New York: Touchstone, 1983), 11; emphasis in original.

60 *House of Wax*, directed by André De Toth (1953; Los Angeles, CA: Warner Bros), release.

61 As cited in Philip Sandifer, "Out of the Screen and into the Theater: 3-D Film as Demo," *Cinema Journal* 50(3) (Spring 2011): 62-78, 63. However, Sandifer proceeds to dismantle that promise by 3D for shock, and argues for 3D cinema as a showcase of technological wonder rather than of immersive experience.

62 See in particular the insightful readings of various 3D films in Miriam Ross, *3D Cinema: Optical Illusions and Tactile Experiences* (Basingstoke; New York: Palgrave Macmillan, 2015); and Owen Weetch, *Expressive Spaces in Digital 3D Cinema* (Basingstoke; New York: Palgrave Macmillan, 2016).

intensity, its purpose is "to pulverize the screen, its edges and its surface, its framing and its flatness,"[63] where screen objects enter the spectator's space with movement and/or volume. The violence in that sense of breach is, again, palpable: "The screen is a wall to break down, the wall that one must shatter."[64] Screen boundaries thus play creative roles in partitioning and protecting against different kinds of spaces, and their transgressions, as with all forms of trespasses, become continual contestations for the audience's increasingly unstable territory.

Finally, cinema's marketing practices also evidence a long history of attempting to expand "the virtual worlds of films"[65] beyond the screen's limits, implying the same sense of "spillover" of fictional space into the actual for promotion and publicity. For instance, Sayad describes William Castle's "extrafilmic stunts" from the 1950s for his films to "surpass the screen," including

> ...the selling of death-by-fright insurance policies to ticket holders for *Macabre* (1958), a skeleton hanging from the cinema's ceiling in screenings of *House on Haunted Hill* (1959), and vibrating motors located under the seats in venues showing *The Tingler* (1959).[66]

Similarly, Sarah Atkinson traces "film marketing techniques which blended cinema and the world of the film into reality" – apparently first known as "ballyhoo" and later dubbed "pseudo-events"[67] – from as early as the 1920s. In what she calls "extended cinema," Atkinson argues for their dominance in contemporary ways that broaden the film's narratives through other media, such as "'in-film' websites which allude to a fictional organization or person" so that "the parallel film and online world imbricated fact with fiction."[68] The virtual

63   Simon Lefebvre, "The Disappearance of the Surface," in *Screens: From Materiality to Spectatorship – A Historical and Theoretical Reassessment*, eds. Dominique Chateau and José Moure (Amsterdam: Amsterdam University Press, 2016): 97-106, 98.

64   Ibid.

65   Daniel Yacavone, *Film Worlds: A Philosophical Aesthetics of Cinema* (New York: Columbia University Press, 2015), xiii.

66   Sayad, "Found-footage horror," 47-48.

67   Phrase as taken from Daniel J. Boorstin's *The Image: A Guide to Pseudo-Events in America* (New York: Vintage, 1961), to which Sarah Atkinson refers in her discussion on such "extended [film] promotion" in *Beyond the Screen: Emerging Cinema and Engaging Audiences* (New York; London: Bloomsbury, 2014), 18.

68   Atkinson, *Beyond the Screen*, 19. See also Kristen Daly, "Cinema 3.0: The Interactive-Image," *Cinema Journal* 50(1) (Fall 2010): 81-98, on a similar analysis of, among other films, *The Blair Witch Project* and, more specifically, how the film becomes "one more artefact, along with the

world of the film thus spills over its screen boundary (of the cinema theatre) into another virtuality across another screen boundary (of the computer). From an urban architectural point of view, Richard Koeck discusses how films' virtual realities "penetrated and gained substance in real architectural space," creating cinema as "a multidimensional practice." For instance, the publicity campaign for *District 9*,[69] a film about a world shared in tension between humans and aliens who look like prawns, included "a viral outdoor campaign on benches and kiosks featuring a silhouette of the film's signature prawn-like aliens and a 1-800 number." The *virtual* reality of the film thus converges with the *actual* space of the city, signifying what Koeck describes as a "redefinition of the boundary where a virtual campaign stopped and the physical space of the city began."[70] Sometimes, the spillover of the virtual gets taken to the point of the zany (which is probably the whole point of its marketing tactic), such as J.G. Ballard's relating of Chinese hunchbacks "recruited by the management of the theatre from every back alley in Shanghai," who, "in medieval costume," formed an honour guard to greet guests at the Shanghai premiere of William Dieterle's *The Hunchback of Notre Dame* (1939).[71]

Film worlds and their onscreen worlds are thus not simply "separated" against the one side of the screen's boundary; they encroach into and merge with the audience's actual reality in creative and self-reflective ways. As Sayad, again, puts it, "the image itself has often threatened to break loose…both to expand the space of representation and to demolish the walls separating art from everyday life."[72] The result is to demonstrate increasing porosity and instability of the screen boundary as the demarcation or protection between art and life, operating not as replacements (as in simulacra) or re-placements (as in realities located elsewhere), but extensions of belief in one world across to another. Actual and virtual realities spill into each other's spheres, merging in myriad and multifariously creative ways and destabilizing each's territories. The viewer, in turn, becomes a fused, yet distinct, subject of these overlapping realities, co-opted into one yet embodied in the other.

Web materials, to use in figuring out 'what really happened,'" and in that sense becomes "more like a project," reflective of contemporary digital pleasures and anxieties of "interactivity, web navigation, and digital communication." (86)

69   *District 9*, directed by Neill Blomkamp (2009; Culver City, CA: Sony Pictures Home Entertainment, 2009), DVD.

70   All references to Koeck in this paragraph are from Richard Koeck, *Cine-scapes: Cinematic Spaces in Architecture and Cities* (Abingdon; New York: Routledge, 2013), 147-149.

71   J.G. Ballard, *Empire of the Sun, Vol. 1* (New York: Simon & Schuster, 1984), 23-24.

72   Sayad, "Found-footage horror," 48.

Virtual Co-location in Real-time... and in the Era of Covid-19

As with the final section of chapter 1, interactive media similarly undoes the rigidity of screen boundaries here by exacerbating and/or accelerating the destabilization of their partition and protection. The very fact of interaction erodes the screen boundary as a partition, as the user is able to participate and actively affect events in the virtual world across it. Acknowledging this agency, Janet Murray distinguishes between *looking through* and *looking at*: in view of users' active participation in a responsive environment, she writes that "at some point we will find ourselves looking *through* the medium instead of *at* it." (emphasis added)[73] Specifically, users acting *through* the screen will affect the events in the story and thereby author, or at least co-author, it. These modes of engagement in interactive media thus further erode screen boundaries, where the user's gaze and actions from their place of spectating or gaming look *through* and participate in the onscreen world, in increasing piecemeal weakening the gap between fictional and audience realities.

In relation to interactive media, and specifically with the advent of embedded webcams and microphones, looking *through* the screen becomes looking *into* the screen – or, more accurately, looking into the camera and speaking into the microphone. The result is a kind of amplified direct address, and a further diminishment of screen boundaries. Direct address in "found-footage" films, as discussed above, "spills" the virtual into the actual in the sense of an encroachment, whose invasiveness from the virtual world's space-time is part of the genre's thrill and attraction as one that is almost always related to horror. Conversely, with respect to interactive media, the acts of "looking and speaking into the screen" involve real-time consent and commonality, whereby all parties onscreen occupy a shared virtual space at the same time; they *co-locate* virtually. The screen boundary thus conjoins parties together virtually in real-time rather than separates them, eroding further as a wall of protection and partition.

Virtual real-time co-location is, of course, not a phenomenon associated only with digital media or mobile cameras; as with so many concepts of "new" media, it emerges from the "old."[74] An established analogue antecedent,

---

73   Janet Murray, *Hamlet on the Holodeck: The Future of Narrative in Cyberspace* (Cambridge, MA: MIT Press, 1997), 271-272.

74   The literature on this argument is substantive: see, as a good starting point, Benjamin Peters, "And Lead Us Not Into Thinking the New is New: A Bibliographic Case for New Media History," *New Media & Society* 11(102) (February 2009): 13-30.

for instance, is the radio addresses of former U.S. President Franklin D. Roosevelt's famous "fireside chats" in the 1930s, which Allucquère Rosanne Stone describes as "an imaginary locus of interaction created by communal agreement."[75] The radio broadcasts' sense of co-location creating the effect of sitting directly with the President was crucial to their success through which Roosevelt influenced public opinion and promoted his policies. Electronic bulletin board systems (BBS) in the 1980s such as the Whole Earth 'Lectronic Link (WELL), the teleconferencing successor to the *Whole Earth Catalog* which indelibly imprinted the counterculture on digital communications,[76] forged the same sense of shared co-location through real-time text-based messaging across their virtual communities. In the 1990s and the advent of the World Wide Web, co-location through virtual messaging became popular in the mainstream via cross-platform systems such as ICQ and web-based chatrooms.

Terrestrial broadcast television introduced the use of direct address into a camera, such as by a newsreader, whose visual element more strongly stitches co-location of the newsreader with the TV viewer across the screen boundary as compared to Roosevelt's radio addresses. Today, with the advent of webcams, the common virtual space transfers from the family-oriented television screen to the screens of personal computers and devices. "Looking and speaking *into* the screen" to communicate in virtualized face-to-face conversation with another party in real-time becomes conventional practice amongst ordinary users, rather than specialized for televisual newsreaders. Today's ubiquitous usage of video conferencing applications such as Skype, WhatsApp, Telegram Messenger and WeChat (just as a few examples) evidence the widespread practice of online video calls. Their variations,

---

75   Allucquère Rosanne Stone, "Will the Real Body Please Stand Up," in *Cyberspace: First Steps*, ed. Michael Benedikt (Cambridge, MA: MIT Press, 1991): 81-118, 82. Note, though, that there were equally several live national addresses by other U.S. Presidents which were deemed failures due to mishandling or being improperly employed: see Michael J. Socolow, "How the fireside chat provided a model for calming the nation that President Trump failed to follow," March 11, 2020, *The Conversation*, https://theconversation.com/how-the-fireside-chat-provided-a-model-for-calming-the-nation-that-president-trump-failed-to-follow-133473. Notably, Donald Trump, the former U.S. President, had also mooted using an FDR-style "fireside chat on live television" in defending his call with the Ukrainian President against various allegations of its impropriety: Rob Crilly, "'Fireside Chat on Live Television': Trump Says He Wants to Read Ukraine Call Transcript to American People," *Washington Examiner*, October 13, 2019, https://www.washingtonexaminer.com/news/fireside-chat-on-live-television-trump-says-he-wants-to-read-transcript-of-ukraine-call-to-american-people.

76   See Fred Turner, *From Counterculture to Cyberculture: Stewart Brand, the Whole Earth Network and the Rise of Digital Utopianism* (Chicago, IL: University of Chicago Press, 2006).

such as social websites which facilitate webcam-based conversations (for example, Chatroulette) or live streaming apps (such as Periscope), are also immensely popular.

As of writing during the height of the Covid-19 pandemic in Europe and, in time, around the world, real-time communication in virtual co-location takes on entirely different quantitative and qualitative change as millions of people under lockdowns take to multi-party video conferencing applications, such as Zoom, Microsoft Teams or Google Meet, to gather virtually for work and to socialize. All manner of human activity, on a scale and in ways unimaginable just months before the summer of 2020, have started to take place online. Music concerts and theatre performances are livestreamed. Dinner dates happen online, each party eating on each's side of the screen. In April 2020 before their Easter recess, UK Members of Parliament joined parliamentary meetings through Zoom-powered video links, each looking and speaking *into* their individual screens to address other members.[77] Even activities which *by definition* require physical proximity, such as professional cuddling (a service of providing physical non-sexual touch for stress and trauma relief), adapted to various kinds of real-time online communications under the pressure of lockdown.[78]

In terms of how acting through the screen diminishes its boundaries, the vast megabytes of communication in virtual shared spaces during the (ongoing) era of Covid-19 are unprecedented in effecting screen boundaries as bridges and connection, rather than protection and partition. Yet under the pandemic's pressure of enforced separation and social distancing, virtual co-location also amplifies the remoteness of physical distance, where the screen connects but also emphasizes how apart we are from each other. Screen boundaries in the time of Covid-19 thus take on an additional level of ambiguity, where the poles of intimacy and connection in virtual co-location against physical distance are unprecedentedly stark. In this sense, we may also recall Huhtamo's double senses of the hand-held screen in terms of both *defence* (as a veil) and *invitation* (enticing curiosity and flirtation). The screen of web-based real-time communication under Covid-19 contains the same contradictions, if profoundly augmented. Where we may think of

77 Tom Warren, "Zoom grows to 300 million meeting participants despite security backlash," *The Verge*, April 23, 2020, https://www.theverge.com/2020/4/23/21232401/zoom-300-million-users-growth-coronavirus-pandemic-security-privacy-concerns-response.
78 Allie Volpe, "Embracing Change: Pandemic Forces Professional Cuddlers to Get Creative," *The Guardian* online, July 22, 2020, https://www.theguardian.com/lifeandstyle/2020/jul/22/cuddle-therapy-coronavirus-social-distancing-virtual.

screen boundaries in relation to the twin poles of protection and rupture, those poles expand here into further oppositions of intimacy and distance.

The Screen Boundary Against the Algorithm

The more radical twist in contemporary interactive media is not the shift in addressing the screen, but in how screen boundaries are increasingly shared and porous between the actuality of human users and the virtuality of the *algorithmic*. As increasing numbers of algorithms run the computational processes which affect correspondingly large portions of our lives, screen boundaries are no longer shared between humans in virtual co-location, or poised for spillover between a human audience and the virtuality of their world as recorded image. Rather, they become the conduit for intelligible human engagement with a *thickened* virtuality – or, perhaps more accurately, a virtuality made *opaque* through a pile-up of recording, computational simulation and "black-box" algorithmic processes.[79]

One particularly creative example of screen boundaries in such algorithmic functionality is the new media project, *Karen*, created by British digital media art group Blast Theory and partnered with National Theatre Wales. *Karen* runs as a smartphone app which interacts with the user through its central character, Karen, presented as a "life coach" for the user. She appears to the user by way of an actress in recorded video clips who "looks and speaks *into*" the screen to directly address the user. So far, so familiar. At the end of each session, "Karen" sets a task for the viewer to complete, and instructs them to only log in again after a specific amount of time had passed, thus mirroring the appointment patterns of real-life therapy. However, at this stage, the algorithmic takes over: at the conclusion of each session, *the app shuts down in real-time and remains inoperable for "contact" with "Karen" until the instructed amount of time has actually passed.* Hence, on one hand, the actress's use of direct address breaks down the partitioning of the screen boundary in sharing a virtual co-located space between the human user and the virtuality of "Karen's" screen-world, its porosity not much different from direct address as used in "found-footage" films. On the other hand, the

79   See, as a representative scattering of the literature on this question, Frank Pasquale, The *Black Box Society: The Secret Algorithms That Control Money and Information* (Cambridge, MA: Harvard University Press, 2015); N. Diakopoulos, "Accountability in Algorithmic Decision Making," *Communications of the ACM* 59(2) (2016): 56–62; and Mike Ananny and Kate Crawford, "Seeing without knowing: Limitations of the transparency ideal and its application to algorithmic accountability," *New Media & Society*, 20(3) (2018): 973-989.

algorithmically driven processes which monitor in real-time the availability of the app signal the sharing of a computational virtuality between the human user and the algorithm. The screen boundary thus becomes more than a partition between the virtuality of "Karen's" world and the actuality of the human user; it is also the conduit to an intelligible visualization of the app's algorithmic operations.

Cinema, always self-reflexive on the cultural "turns" and social evolutions of screens, is quick to capitalize on this increasing friction across screen boundaries between the human and the algorithmic. In a sense, this is unsurprising as contemporary film screens, for better or worse, also converge with computer screens via DVDs, file downloads and, increasingly, livestreaming. If "found-footage" films paved the way in the 2000s for a porosity of screen boundaries across which audiences experienced a "leakage" of virtuality, a new film genre in the 2010s presents its revised nature as the human-computer screen boundary. This genre is the "desktop" film (or "computer screen film"), which frames the entirety of the film through a laptop screen belonging to a character. It first became popular in the mid-2010s via the box office hits of *Unfriended*[80] in 2014 (as well as its sequel, *Unfriended: Dark Web* (2018))[81] and *Searching*[82] in 2018. As with "found-footage" horror, "desktop" films consciously exploit self-referential screen boundaries to re-define the thickened virtuality of the algorithmic, visualized here in two ways. The first is via computational operations, as with the *Karen* app. However, rather than the app's rhythms of availability, the algorithmic world is visualized in the films as disrupted screen refresh rates, delayed computer reactions, malfunctioning disk sectors, buffering, glitches, damaged programs, computer viruses. The second is through the films' range of virtualized humans, whereby all human drama are seen via Skype video windows, live casting websites or recorded videos. Unlike "found footage" films, this virtualized action is not meant for the audience, but across the diegetic screen boundary for the character, themselves now placed in a relationship of distance between their own virtual-actuality and the virtual-virtuality of the drama they see through the laptop – a veritable *mise-en-abyme* multiplication of screen boundaries. Virtual space thus further carves itself up between computerized action and virtualized

80   *Unfriended*, directed by Leo Gabriadze and Stephen Susco (2014; London: Universal Pictures UK, 2015), DVD.

81   *Unfriended: Dark Web*, directed by Stephen Susco (2018; London: Universal Pictures UK, 2018), DVD.

82   *Searching*, directed by Aneesh Chaganty (2018; Culver City, CA: Sony Pictures, 2019), DVD.

humans. The result is that the actual audience, on the other side of the screen, becomes utterly remote. In turn, their screen boundary rescinds into a mere formality; it is not so much a protective partition between the actual and the virtual subject to breach or leakage, for the film does not profess to have any relation to the virtual reality's time-space. Rather, the screen boundary is a doubly distant separation that is merely observational of action taking form and place entirely on a doubly virtualized plane.

However, the *mobile screen boundary* yet again changes this radically. Advancing generations of mobile technologies and sophistication of peripherals in their devices, such as in-built microphones and dual-facing cameras,[83] enable smartphone apps not only to capitalize on real-time porosity between virtual and actual spaces, but also to do so with the affordance of user portability. Hence, as with the *Karen* app, the mobile screen boundary across which the user interacts with the app becomes a conduit to its algorithmic operations. However, unlike "desktop" cinema and as a nod to their portability as apps on mobile devices, mobile screen boundaries are not borders demarcating a distanced observational post to the computational, but *portals* opening to *both* virtual and actual worlds. In Augmented Reality (AR) applications, for instance, users interact with the virtual reality within the app comprising of both a virtualization of their actual surroundings as well as virtual objects supplied by the app. As with virtual worlds, the virtual becomes the site of primary engagement and, to all intents and purposes, the dominant reality, or the "first order," in *contra* to its convention as the "second order" of reality. The mobile screen boundaries to AR apps thus become a gateway to its algorithmically thickened virtuality, with which users act, rather than look at.

Conversely, mobile apps may also open out to the user's actual surroundings. One example is *Happn*, a mobile dating app whose algorithm, like that of most dating apps, processes users' data to find matches of potentially compatible partners as registered on the app's system. The notable feature of *Happn* is its (self-promoted) functionality of using GPS tracking on users' phones to identify registered partners who happen/happn to move into their physical proximities.[84] *Happn* thus shifts dating and partner-finding

---

83   Technology moves fast: first released in September 2019, the iPhone 11 Pro (and Pro Max) has what its producing company Apple calls "a transformative triple-camera system": see https://www.apple.com/uk/iphone-11-pro/.

84   This thus fulfils the app's advertising promise to "find the people you've crossed paths with": https://www.happn.com/en/.

from the wholly virtual space of the Web[85] to the fluidly intersectional space between the actual and the virtual, with mobile screen boundaries as their swinging portal. It interfaces the user between their virtual and actual worlds, from virtual co-location with algorithms and virtual humans to algorithmically driven actual co-location with actual humans.

Screen Boundaries in Flux

The subversion, diminishment and erosion of screen boundaries are not a phenomenon of the contemporary media world. However, with increased depths of our virtualized and computational worlds come increased complexities and fluidities of the screen boundary as a covering of protection and partition. With new entanglements through computational culture, algorithmic processing, mobility and real-time interaction, the nature of the screen boundary changes and warrants different metaphors. Of significance is also the role of the audience/user, shifting from their complicity of belief to their active interactions with and re-negotiations of virtual and computational spaces in relation to their actual spaces.

The shifting screen boundaries of contemporary media thus facilitate alternative virtualizations of our world which, in turn, deliver alternative visualizations, and vice versa. These are the revised territories of representations with which we contend today – images that we need to understand and learn to read not only in terms of their cultural meanings, but also their modes for and conditions of our imaginations regarding our contemporary realities. This, then, is the space of the post-screen – the continual interrogation of the mediated relations between image and reality that is as old as the cave drawings of ancient histories and the storytelling that describe mythic imaginations – which now indubitably colour contemporary politics of information and truths. In the remaining chapters, I address the post-screen with respect to three specific media technologies – virtual reality, holographic projections/holograms and light projections – as demonstrations of screen boundaries in flux, and the new imaginations and politics which arise from their changes.

---

85   Online dating soared in popularity in the 1990s – and mobile dating apps in the 2000s – on the basis of the perceived vaster possibilities in the virtualized spaces of the Internet for finding potential partners. As Eva Illouz puts it, albeit in heavy critique of online romancing, "if the Internet has a spirit, it is that of abundance and interchangeability": *Cold Intimacies: The Making of Emotional Capitalism* (London: Polity Press, 2007), 90.

3 Virtual Reality: Confinement and Engulfment; Replacement and Re-placement

Abstract

This chapter explicates Virtual Reality (VR) as the first instantiation of the post-screen. Specifically, it interrogates VR's sense of immersion via two vectors in the post-screen's "forgetting" of screen boundaries – confinement of a viewer's visual field with restricted viewing devices; and engulfment by being surrounded with large screens. The chapter's key idea is its alternative expression of VR's relations of reality as an immersive media form, which it argues shifts from the critical paradigms of replacement to re-placement. Through theoretical critique and readings of various applications of VR, the chapter argues for re-placement as a more ethical and generative space for thinking through VR's relations of the real. In turn, where and how the actual and the virtual is re-placed informs the very purpose of media itself.

Keywords: virtual reality; replacement; re-placement; immersion; simulacra; totalization

"Multitudes of Amys"

> *Multitudes of Amys*
> *Ev'rywhere I look*
> *Sentences of Amys*
> *Paragraphs of Amys*
> *Filling ev'ry book*
> ~ Stephen Sondheim[1]

---

1   Stephen Sondheim, composer and lyricist, "Multitude of Amys," 1970, sung by Mandy Patinkin on *Experiment* (US: Elektra Nonesuch), 1994, compact disc, track 6.

Ng, J., *The Post-Screen Through Virtual Reality, Holograms and Light Projections. Where Screen Boundaries Lie*. Amsterdam: Amsterdam University Press 2021
DOI: 10.5117/9789463723541_CH03

The lines above are lyrics from a love song. Written by Stephen Sondheim for the acclaimed 1970 Broadway comedy musical, *Company*, "Multitudes of Amys" was to mark the moment the protagonist, a bachelor named Robert who had been unable to commit to a relationship, realizes he is in love with Amy.[2] What is interesting for our purposes is how Robert sings of his love not as an emotion but an *omnipresence* – his being in love is to be surrounded by "*multitudes*," or to see and hear his object of affection everywhere he looks. The grand emotion of love replaces Robert's everyday reality with sights of Amy – "I see them waiting for the lights, running for the bus, milling in the stores" – and sounds of her: "choruses of Amys, symphonies of Amys, ringing in my ear." Robert is not only in love, he is also *in* love, as if sunk into a vat of it. He is immersed.

Conceptually, the sentiment of "Multitudes of Amys" is miles from the experience of Virtual Reality (VR) as the focus of this chapter. Yet it might appear that love and VR bear surprising similarities – VR, too, is an immersion in, we can say, *multitudes of virtual reality*, or reality that is *real yet fabricated*. As Peter Lunenfeld puts it, VR is "immersion in (synthetic) experience."[3] The hallmark of the VR environment is its technological multi-sensorial immersive realism – after all, one of the earliest and most enduring visions of VR is the Holodeck of the *Star Trek* TV series: a fictional voice-controlled room which creates a simulated environment that appears realistically and seamlessly around its user so as to *sink* them into the experience.

The key to VR's immersion thus lies in the totalization of its virtual real whereby, as will be discussed, its entire technological set-up eradicates any perceptible presence of its screen boundaries: hence, *the post-screen emerges*. The previous chapters discussed the porosity, instability and flux of screen boundaries as charted across movements of revelation/concealment, play/display, protection/partition in cinema, television and other screen media. Here, those fluctuations come to a fruition in this first instantiation of the post-screen: the purported seamlessness of the Holodeck, mirrored in the way the VR headset powers up an all-encompassing virtual reality that surrounds the viewer wherever they look. Its totality eliminates the perceptibility of its screen boundaries so that they are "forgotten": a further movement in this *pas de deux* between the regressions and (re-)insertions

2   At least, that was the intention. For various reasons, the song was eventually cut from the musical and so does not appear on its song list.

3   Peter Lunenfeld, "Digital Dialectics: A Hybrid Theory of Computer Media," *Afterimage* (November 1993): 5-7, 5.

of the frame (the latter such as by way of the Rijksmuseum's promotion of Rembrandt's *De Nachtwatcht*, as discussed in the introduction).

Of course, in various ways VR is not a perfect totalization, or at least not yet. By definition of VR being mediated reality, Lunenfeld takes on a Baudry-esque position regarding the ideological nature of media apparatus[4] to point out that virtual space is "just as constructed as the 2-D space of film and video, and therefore subjects its participants to analogous spectatorial and ideological positionings."[5] As we shall later see in this chapter, VR's realism as a "natural" immersive space also lapses to varying degrees due to a range of reasons, from "pixel bleed" (where pixels appear blurred) to the intrusiveness of the VR headset. In many ways, as Janet Murray insists, VR remains far from the desired outcome of being "a magical technology for creating seamless illusions."[6] Réne Magritte's *La Condition Humaine* painting, as discussed in the introduction on the artist's desire to substitute nature with creation, comes to mind again here: that desire is doomed for failure; *the gap always shows*.

Hence, on one level, this chapter re-visits the friction between the actual and the virtual real across VR's screen boundaries. However, it avoids taking any side of the divide – VR's technological affordances for some degree of immersion in virtual reality are clear, as is also the foolishness of "magical thinking" associated with its unfettered credibility as a virtual-for-actual substitution machine. These are both easy positions to adopt, but ultimately irreducible, leading to an unconstructive dialogue.

Instead, the chapter seeks a different and more generative conceptual space out of *VR as post-screen*, where the "forgetting" of its screen boundaries is not so much about the eradication of the real between the virtual against the actual. Rather, it is about the discerning of *a different friction* between them – namely, a movement from *replacement* to *re-placement* as exposed in the screen boundaries' flashpoints of subversion and disruption. The shift itself is paradoxically subtle: the same word with the slightest change, yet whose meanings run in radically different directions. My reading of VR as the post-screen thus turns on this shuffle of meaning to demonstrate a different unmasking of the real across its screen boundaries in terms of unexpected yet generative rupturing. Yet again, we can think here about

---

4   See Jean-Louis Baudry and Alan Williams, "Ideological Effects of the Basic Cinematographic Apparatus," *Film Quarterly* 28(2) (Winter, 1974-75): 39-47.
5   Lunenfeld, "Digital Dialectics," 6.
6   Janet Murray, "Virtual/reality: how to tell the difference," *Journal of Visual Culture* 19(1) (2020): 11-27; phrase as quoted is from the article's abstract.

Robert falling in love as he sees around him "multitudes of Amys," where falling *in* love may also turn completely upside down, on a very small word change, into falling *out* of love. The reversal in direction pivots profoundly on minuscule adjustment, yet with whose difference comes acute insight and knowledge.

***

On Immersion (Briefly)

In its post-screen instantiation, key to the purported erasure of screen boundaries in VR is to bring about an experience of *immersion* – in the sense of being steeped in or surrounded by the virtual real. This sense also connects to the word's Latin root – *immergere* – meaning to plunge or dip into.

More typically, though, immersion in screen media refers to *a mental state or process* – an intense state of belief in what is shown onscreen, where the viewer becomes so occupied with it that they temporarily forget about their own reality.[7] According to François Dominic Laramée, this mental state is a "suspension of disbelief, a state in which the player's mind forgets that it is being subjected to entertainment and instead accepts what it perceives as reality."[8] Similarly, Oliver Grau describes immersion in visual media as "mentally absorbing and a process, a change, a passage from one mental state to another...characterized by diminishing critical distance to what is shown and increasing emotional involvement in what is happening."[9]

In that sense, immersion is about a user's *own* mental engagement, in varying degrees of activity, with the media. Janet Murray, flipping Coleridge's suspension of disbelief, writes of how screen media users "actively *create belief*," (emphasis in original) where "we focus our attention on the enveloping world and we use our intelligence to reinforce rather than to question the

---

7   *Cf* other kinds of immersive experiences that do not involve a screen, e.g. reading as an immersive experience. See Marie-Laure Ryan, *Narrative as Virtual Reality 2: Revisiting Immersion and Interactivity in Literature and Electronic Media* (Baltimore: Johns Hopkins University Press, 2001, 2015), especially 85-116.

8   François Dominic Laramée, "Immersion," in *Game Design Perspectives*, ed. François Dominic Laramée (Hingham, MA: Charles River Media, 2002): 61.

9   Oliver Grau, *Virtual Art: From Illusion to Immersion*, trans. Gloria Custance (Cambridge, MA: MIT Press, 2003), 13.

reality of the experience" to undergo immersion.[10] Manovich highlights a viewer's "psychological processes of filling-in, hypothesis forming, recall and identification" as critical components of interacting with any text or image.[11] Geoff Dyer's book-length contemplation-cum-panegyric of Andrei Tarkovsky's 2005 film, *Stalker*, describes his own immersion in Tarkovsky's cinema by way of how he *imagines* persistent images of Tarkovsky's cinema in the films of fellow Russian director Andrei Zvyagintsev. On viewing Zvyagintsev's *The Banishment* (2008), Dyer writes: "Three of the first half dozen shots evoke, in turn, *Nostalghia...*, *Stalker...* and *Solaris* [all films by Tarkovsky]. Thereafter it's impossible not to succumb to spotting Tarkovsky allusions and references."[12] So *immersed* – so mentally absorbed – is Dyer in Tarkovsky's onscreen worlds that, like Robert seeing Amys all around him, Dyer sees the cinematic world of Tarkovskian colours, images and sounds everywhere he looks. His imagination is so captured by Tarkovsky's films that they affect his perceptions of his surroundings in an almost existential way. Dyer writes of

> ...being so absorbed by *Stalker* that I can see nothing but Tarkovsky, so steeped in his view of the world that I mistake it for the world itself... Like all the greatest filmmakers [Tarkovsky] immerses you so completely in his world that it never occurs to you... that the world on-screen ceases to exist at the edges of screen... No, the world beyond the screen is just a continuation of the world we are seeing. To either side and behind there is more of the same. We are not even in a cinema; we are in a world.[13]

In this respect, immersion is not so much about an apogee of sensorial realism in illusory environments such that one is able to naively replace the other – a position which various scholars take or object against. Salen and Zimmerman, for instance, situate their critique of immersion in this sense of what they call "the immersive fallacy," in which they posit immersion as being in an illusory reality so complete that "the frame falls away so that

10   Janet H. Murray, *Hamlet on the Holodeck: The Future of Narrative in Cyberspace* (Cambridge, MA: MIT Press, 1997), 110.

11   Lev Manovich, *The Language of New Media* (Cambridge, MA: MIT Press 2002), 71-72. These acts which he identifies as interaction also form the basis of Manovich's "Myth of Interactivity," where he argues that interactivity is not a feature of digital media; that "all classical, and even more so modern art, was already 'interactive' in a number of ways": 71.

12   Geoff Dyer, *Zona: A Book about a Film about a Journey to a Room* (Edinburgh: Canongate Books, 2012), 194.

13   Dyer, *Zona*, 195-6.

the player truly believes that he or she is part of an imaginary world."[14] While they have some grounds for their scepticism, nonetheless immersion in media environments in this sense is not so much about totally believable simulation as it is about storytelling, description and creating sensorial vividness to engage the viewer's imagination, as well as the viewer's own mental work in creating and reinforcing their belief in the reality onscreen, or suspending disbelief and diminishing their critical distance.

Yet filmmakers, curators, artists, designers and engineers continue to push on sunk-into-a-vat style of immersive experiences, where, as Salen and Zimmerman put it, the frame does indeed "fall away." Like Robert immersed in Amys, in this immersive environment the viewer is ostensibly surrounded by "multitudes" of represented objects, seeing them wherever they turn. It is important to note that such environments of multitudes are not necessarily always advanced computational apparatuses, such as Holodeck constructions or VR. Fred Turner, for instance, writes compellingly of multimedia systems he calls "surrounds" – "multi-image, multi-sound-source media environments" spurred by the ideas of American anthropologists, psychologists and sociologists of the 1930s and 1940s, and built by the Bauhaus refugee artists and designers in America. These "surrounds" took embryonic form as exhibition shows in Europe in the early 1930s, whose blueprints featured, among others, screens which surrounded a viewer on all sides – "some [screens] arrayed at eye level, but others angled down from the ceiling and still others angled up from the ground."[15] This display style served their larger idea of an "extended field of vision," or what the Bauhaus faculty theorized to be "the new vision" as part of "the New Man" in industrial society, rejuvenated as a new humanity from the horrors of World War I. Integrating art and technology, "the New Man" was enabled by technological advances of the era, such as photography and cinema, to become more conscious of their place in space and time.

In the late 1930s and 1940s, as prominent Bauhaus members fled Nazi Germany for America, they brought these ideas with them – so that "the new vision comes to America"[16] – which they turned to the political ends of promoting democracy and public morale in the war effort. One way in which the Bauhaus manifested their ideas was through their designs of

14 Katie Salen and Eric Zimmerman, *Rules of Play: Game Design Fundamentals* (Cambridge, MA: MIT Press, 2004), 451.

15 Fred Turner, *The Democratic Surround: Multimedia and American Liberalism from World War II to the Psychedelic Sixties* (Chicago, IL: University of Chicago Press, 2013), 87.

16 Turner, *The Democratic Surround*, 92.

various art shows at the Museum of Modern Art in New York, from the 1938 exhibition of Bauhaus's work at the Rockefeller Center to the *Road to Victory* show in 1942 and *Airways to Peace* in 1943 at the Museum itself. The latter two, featuring exhibitions of American involvement in the Second World War, were particularly important, acclaimed and well-attended. Of note is the way key Bauhaus figures, such as Herbert Bayer and Walter Gropius, curated the shows' artefacts, which always included, among other formats, enormous, floor-to-ceiling, panoramic photographs, portraits, photograph montages and collages, and multiscreen environments. Turner reads these exhibitions as "democratic surrounds" – large-scale exhibitions which are not just media environments whose scale sought to impress and surround the viewer, but are encoded with dreams of democratic liberalism channelled through theorizations of new visual perceptions and new ways of seeing that aspired to bring about a freer and more humanistic world. In keeping with their ethos of democratic ideals, the shows emphasized individual effort by viewers to come to their own conclusions as they viewed the displays: they "presented [its visitors] with an array of visual materials and three-dimensional environments within which they could mold themselves"; they "offered visitors a chance to experience a democratic degree of agency."[17] In particular, Turner compares these large-scale surrounds to those put up by the Fascists, particularly the Italian Fascists who "were equally fascinated by the power of surrounding viewers with images to influence their ideals" and who, like Bayer, "arrayed images from floor to ceiling" and displayed variations of huge photomontages. Conversely, Turner reads their exhibitions as demonstrative of Fascist ideology, and in particular Fascist attempts at de-individuation and voidance of individual agency:

> But far from offering the eye a set of images for the viewer to bring together in his own independent mind, as Bayer recommended, [Giuseppe] Terragni's wall [of a huge photomontage] *did the work of integration on behalf of the viewer.* As the viewer moved from right to left along the wall, he saw crowds of individuals slowly morph first into turbines and then into an abstract field of hands raised in the fascist salute [emphasis added].[18]

Turner's readings of the "democratic surround" thus weave an incisive thread through the intellectual movements of the war years in the United States to illuminate a historical period of significant audio-visual curations.

17   Turner, *The Democratic Surround*, 113.
18   Turner, *The Democratic Surround*, 90.

More importantly, they also highlight the shows' political meanings and impact which helped shape a then-future global power's consciousness of both its wartime spirit and identity, as well as its post-war victory and destiny. The sense of boundaries between realities is not pronounced in these displays, but neither is it their point. For our purposes, these works demonstrated the effective immersion of the spectator in all-encompassing media displays, with compelling settings of sound and images constructed to surround – even overwhelm – the spectator. As the next section will show, this aspect of totalizing media environments[19] substantively constitutes the historical and cultural footing for contemporary VR in terms of its key effect of "forgetting" its screen boundaries. More than that, particularly in terms of their post-screen discourse of managing and controlling boundaries in the frictions between the real of the actual and the virtual, the totalizing effects of VR also drive dialectics of reality and simulacra for renewed contestations of their respective legitimacies, as we shall see.

The Affective Surround: The Two Vectors of Immersion

Branching off Turner's analysis of the "democratic surround," we may also think of another brand of totalizing media environments which similarly surrounds the spectator with sounds and images, whose politics is not for propaganda or the ideals of person-shaping, but for *affect*. These end-points are not mutually exclusive – much of propaganda, after all, relies on affect, and affect is not without politics. Rather, they are distinguished by their relative focus on each outcome. In that respect, the *affective surround* aims

---

19   I use the term "media environments" here simply to mean environments *constituted* by media, principally of image and sound, not to be confused with the thinking of environments *of* media, specifically where environments are read *as* media. Insightful examples of scholarship of the latter include Erica Robles-Anderson's reading of the megachurch of the Crystal Cathedral in California in the 1980s, arguing that the church's architecture of glass, layout and numerous cameras communicated its religiosity and evangelical spirit ("The Crystal Cathedral: Architecture for Mediated Congregation," *Public Culture* 24:3 (2012): 577-599); or Chandra Mukerji's *Territorial Ambitions and the Gardens of Versailles* (Cambridge: Cambridge University Press, 2010), which read the layout, engineering and garden design of the gardens of Versailles as media that communicated French politics and power. The spirit of their scholarship brings to mind as well John Durham Peters's well-taken wider questioning, if "in media res" (1), of what exactly is media (*cf* "message-bearing institutions such as newspapers, radio, television" (2)), and his argument of media as "containers of possibility that anchor our existence and make what we are doing possible." (2) See *The Marvelous Clouds: Toward a Philosophy of Elemental Media* (Chicago, IL; London: The University of Chicago Press, 2015).

not only to envelope – even engulf – the viewer with "multitudes" of media;[20] the more important objective is to enable the viewer to consign to oblivion differentiations between actual and virtual reality – to "forget" the frame and boundaries of the screen. In so doing, they, like Robert in love, intensely experience a *state of being*. In the case of VR, this state would almost always be about being an or in an Other, such as another person or in another place.

This affective totalizing media environment generally exhibits two characteristics. The first is a high, and ever increasing, degree of realism or, as game designer Warren Spector puts it, "ever more faithful approximations of reality."[21] This realism is achieved primarily through relatively basic features such as the incorporation of colour, illusion of depth and multiple sensorial information, including sound. Where the media environment is digitally generated, this also means a greater density of pixilation which increases the fineness and amount of detail perceived, and higher computational power to update the screen quickly in response to user interaction, such as head movements.

The second characteristic is the totalization of the senses, such as by presenting the viewer with a 360° environment that, as the name suggests, surrounds the body in a full rotation, or at least a large part of their visual field (usually around 120° of arc). The surround effect further involves the user perceiving themselves to be at all times in the centre of the environment, whereby the relevant scene of the environment changes in correspondence with their movements, such as walking or head-turning. The apparatus or setting also shuts out all other external stimuli so that the viewer receives sensorial input only from the media environment.

In the confluence of both characteristics, then, lies the post-screen's critical friction of the virtual real and the actual real by way of the viewer's *"forgetting" of screen boundaries, cf* the fluxes of movement, protection and partition per the discussion in the previous two chapters. Its analyses turn on that critical concealment or purported erasure of boundaries between the virtual and the actual, whereby one ostensibly becomes as real as the

20   *Cf* Rebecca Schneider's application of time as another kind of affective dimension that envelopes the spectator: "a negotiated future that is never simply *in front of us* (like a past that is never simply behind is) but in a kind of vicious, affective surround [emphasis in original]": *Performing Remains: Art and War in Times of Theatrical Reenactment* (London; New York: Routledge, 2011), 37. Ditto fear, in terms of how it can "self-cause," becoming "uncontainable, so much so that it 'possesses' the subject," turns into "experience's affective surround": see Brian Massumi, *Ontopower: War, Powers, and the State of Perception* (Durham, NC: Duke University Press, 2015), 181.

21   As quoted from Salen and Zimmerman, *Rules of Play*, 451.

other. In this instantiation of totalizing media environments as the post-screen, the spectator is immersed as if "sunk in the vat," though there are other phrasings: they are put "in the picture"; they believe they are "there,"[22] or sometimes also termed "present,"[23] in the image; they are "inside the experience."[24] Chris Milk, self-styled VR visionary and artist, specifically draws on VR having its user placed "in the picture," or, as he puts it, "through the window":

> [A] frame is just a window. I mean, all the media that we watch – television, cinema – they're these windows into these other worlds.... But I don't want you in the frame, I don't want you in the window, I want you through the window, I want you on the other side in the world, inhabiting the world.[25]

Given its affective power, this brand of immersion unsurprisingly dogs the ambitions of the screen media industry. As Steven Spielberg claimed at a USC School of Cinematic Arts panel in 2013, "[we] need to get rid of the proscenium. We're never going to be totally immersive as long as we're looking at a square, whether it's a movie screen or whether it's a computer screen."[26] To Spielberg, the holy grail for the experience of media is to "put the player inside the experience, where no matter where you look you're surrounded by a three-dimensional experience."[27]

To analyse the post-screen through VR, then, is to deconstruct VR's framework of "forgetting" screen boundaries. We may think of this framework as held up by two oppositional vectors – *confinement* and *engulfment* – where both are indubitably accompanied by long histories of "older" media that we may see as "prototypes" of post-screen media; in due course, these vectors also become key constituents of VR per the post-screen. The first vector of confinement is the restriction of a viewer's visual field, usually with a device held to the face that features an opening for the viewer to see the

22   Chris Milk, "How virtual reality can create the ultimate empathy machine," TED talk video, 10:00, March 2015, https://www.ted.com/talks/chris_milk_how_virtual_reality_can_create_the_ultimate_empathy_machine?language=en#t-494267, at 2'28".

23   Thomas B. Sheridan, "Musings on Telepresence and Virtual Presence," *Presence* 1(1) (Winter 1992): 120-126.

24   Frank Rose, "Movies of the Future," *The New York Times*, June 22, 2013, http://www.nytimes.com/2013/06/23/opinion/sunday/moviesof-the-future.html?_r=0.

25   Milk, "How virtual reality," at 4'48".

26   As reported in David S. Cohen, "George Lucas & Steven Spielberg: Studios Will Implode; VOD Is the Future," *Variety*, June 12, 2013, https://variety.com/2013/digital/news/lucas-spielberg-on-future-of-entertainment-1200496241/.

27   Ibid.

visual imagery inside it. As the aperture is pressed close to one or both eyes, the viewer's field of vision becomes effectively confined to the scene "inside" the box; at the same time, thus "blinkered," they are also deprived of visual stimuli "outside" it. The edges of the image – or the boundaries of the display area – thus get "forgotten." Jonathan Crary, charting an incisive trajectory across "sites of reality in the early nineteenth century" (via, no less, Théodore Géricault's career-launching 1818-19 painting, *The Raft of the Medusa*),[28] notes this "forgetting" of boundaries as a sense of "privatization of vision" in viewing *The Raft of the Medusa* to Géricault's "portraits of the insane" to peep shows to the panorama. For each of these visual stops, Crary explains how vision becomes narrowed into increasing solipsism that threatens an extreme separation from the outer world.[29] In turn, this separation illuminates two important propositions of this vector of visual confinement: the first is the gap between the self and the other in this visual regime, and the fundamental tension of subjective truth or reality against the need for experiences – and expression of experiences – in common with others. The second proposition, if more pertinent here though related to the first, is the "forgetting" of boundaries in that – and here Crary takes from Bakhtin's words – "private chamber character of experience," "where the peep-show model of looking describes both an intensification of visuality and also an isolation of the subject from a lived embeddedness in a given social milieu."[30] In that isolation thus also lies the elimination of realities' boundaries, so that the subject subsists in their enclosed and privatized isolation with their viewed reality.

In this sense, early media, via Erkki Huhtamo's tracing of what he calls "peep practice," effectively connects this narrowing of vision into solipsism with the elimination of boundaries between the realities of self and the outside world to serve its confinement. Early examples of such "peep practice" include "perspective machines" or "perspective boxes."[31] Developed by Dutch painters, these boxes contained "illusionistic interiors" painted on

28   Théodore Géricault, *The Raft of the Medusa (Le Radeau de la Méduse)*, 1818-1819, oil on canvas, 490 cm x 716 cm, Louvre Museum, Paris.

29   Jonathan Crary, "Géricault, the Panorama, and Sites of Reality in the Early Nineteenth Century," *Grey Room*, No. 9 (Autumn, 2002): 5-25, 15. This is also a politically charged separation, whereby Crary also connects the isolation of vision to what he calls "social docility," thus implying a dynamic of governance and control of the viewing subject through these nineteenth-century regimes of the visual.

30   Ibid.

31   Huhtamo, "Toward a History of Peep Practice," in *A Companion to Early Cinema*, eds. André Gaudreault, Nicolas Dulac and Santiago Hidalgo (Malden, MA; Oxford: John Wiley & Sons, 2012): 32-51, 32.

their inner walls and were viewed through "a carefully positioned hole" for "a perfect spatial illusion."[32] In the seventeenth century, the Jesuit polymath Athanasius Kircher "described a device called the parastatic microscope, a handheld viewer for peeping at images painted on a rotary glass disc."[33] In the eighteenth century, peep shows became itinerant public attractions,[34] where customers paid to peer – peep – through the restricted aperture of large peep boxes to view the scene painted inside it. At the same time, myriad small devices and toys similarly constitute "peep media," including "peep eggs," "stanhopes" and kaleidoscopes.[35] In the nineteenth century, peeping continued through stereoscopic devices of all kinds, including Edison's Kinetoscope, and the novelty picture gallery of the Cosmorama (which featured paintings to be peeped at through a series of magnifying lenses in the walls of the salon). The viewer is not only isolated in their viewed reality out of the peep medium, as with the singular and non-reciprocal viewing of *The Raft of Medusa*, but their visual world is also confined within it and with boundaries forgotten.

This vector of visual confinement is not often, if at all, discussed in relation to VR,[36] yet it is a critical aspect of VR and VR videos (also known as "VR cinema" or "360° videos"), where in their case a more technologically sophisticated computerized headset substitutes a box of painted interiors. Moreover, there are sundry intermediate post-screen variations, such as the Omni-Directional Video (ODV), a video system developed by a media research group at the University of Hasselt and used for *CRASH*, a 2004-2005 "immersive performance" by CREW, a performance group based in Brussels.[37] The ODV system includes a head-mounted display which provides an immobile spectator with a panoramic video-captured image. Unlike VR,

32   Ibid.

33   Ibid.

34   They were also known as "the Raree Show" (for "rarity show"): see Erkki Huhtamo, "Elements of Screenology: Toward an Archaeology of the Screen," *Iconics: International Studies of the Modern Image*, 7 (2004): 31-82, 42.

35   See Jason Farman, "The Forgotten Kaleidoscope Craze in Victorian England," *Atlas Obscura*, November 9, 2015, https://www.atlasobscura.com/articles/the-forgotten-kaleidoscope-craze-in-victorian-england.

36   See, as one example out of many, Marie-Laure Ryan, *Narrative as Virtual Reality* 2, 39-44, whose explication of "earlier technologies" for VR covered only the panorama, the cyclorama, Cinerama movies and the Sensorama, but nothing on peep media and this important vector of confining visual field (stereoscopes were mentioned but only in relation to sense of depth).

37   Kurt Vanhoutte and Nele Wynants, "Instance: The Work of CREW with Eric Joris," in *Mapping Intermediality in Performance*, eds. Sarah Bay-Cheng, Chiel Kattenbelt, Andy Lavender and Robin Nelson (Amsterdam: Amsterdam University Press, 2010): 69-74, 69.

the image is not generated computationally but by recorded video; however, similar to VR, the headset contains an in-built orientation tracker which matches the image to whichever view direction the user takes. As with "older" forms of peep media, the restricted view of the headset mounted close to their eyes enables the spectator to be "blinkered" into viewing only what is shown to them on its display screen, thereby "forgetting" the screen boundaries that fall outside their visual field.

If a scene is *confined* to fill a viewer's visual field for "forgetting" its screen boundaries, it can also be expanded to *engulf* that visual field. The second vector of engulfment to "forgetting" boundaries in immersive media environments, then, is to display images on a screen of such largeness around the spectator that it becomes physically impossible or near-impossible for them to perceive the edges of the image. This sense of largeness of immersive media display is the more commonly discussed precedent for VR's all-encompassing visual environment. Unlike the Bauhaus shows whose screens and images were designed as large-scale visual experiences to correspond with an intellectual journey, the primary effect of these expanded displays is to give the spectator an impression of *a continuous coherent space* that surrounds them, with their body placed at its centre. They are, literally and bodily, "in the image."[38]

As with "peep media," a long history of media similarly demonstrates this vector of the affective surround. The most prominent example is the panorama, a visual attraction popular in the late eighteenth and much of the nineteenth century. Patented in 1787, the panorama features large-scale realistic paintings, usually of exotic settings or scenes, housed in specially constructed rotunda buildings.[39] The "all-encompassing realism" of the panorama and its myriad strategies to achieve its totalizing effect, such as the use of false terrain and clever shifts in lighting, has been much discussed and is unnecessary to repeat.[40] It suffices here to emphasize the panorama's largeness, realism and central perspective as primary features that enable

---

38   Indeed, the patent for the eighteenth century panorama specifically states that the aim of all its features is "to make the observers, on whatever situations he [the artist] may wish they should imagine themselves, *feel as if really on the spot* [of the painted environment] [emphasis added.]": Scott B. Wilcox, "Unlimiting the Bounds of Painting," in *Panoramania! The Art and Entertainment of the 'All-Embracing View'*, ed. Ralph Hyde (London: Trefoil Publications in association with Barbican Art Gallery, 1988): 13-44, 17.

39   *Cf* Jihoon Kim, "Remediating Panorama on the Small Screen: Scale, Movement and Spectatorship in Software-Driven Panoramic Photography," *Animation* 9(2) (2014): 159-176.

40   For a comprehensive description, see Hyde, *Panoramania!*, 7. In brief, besides faux terrain and lighting, an umbrella-shaped roof (velum) constructed above the observation platform conceals the upper edge of the unframed canvas stretched in full circle around the room, as

the spectator to be surrounded by an all-encompassing and coherent view. More importantly, they enable the spectator to "forget" the boundaries of the scene, whereby its edges effectively disappear, earning the panorama its epithets such as "the frameless painting,"[41] or one that displays "a vast, seemingly unbounded vista."[42] As Fred Leeman writes, "by its very nature, this subject lacks the demarcation setting off the picture's border."[43] Dominique Dufourny affirms: "The panorama was to produce an uninterrupted area of representation with no edges, no break, no exterior."[44] Denise Blake Oleksijczuk comments that "the function of the 'frame' is greatly diminished when the spectator is situated at the painting's center."[45] Similarly, Eadward Muybridge's ground-breaking 360° image of San Francisco in 1877, which enabled the city to be seen in continuity from a single central point, was described as "an extended linear image," where "the left- and right-hand ends are no longer the limits of vision, but are in fact contiguous with one another."[46] As Ton Rombout sums up:

> Quite simply, the secret of the panorama lies in the elimination of the possibility to compare the work of art with the reality outside, by *taking away all boundaries which remind the spectator that he is observing a separate object within his total visual field.* Not without reason the panorama used to be called the 'all-view' or 'the picture without boundaries' [emphasis added].[47]

Besides the panorama, numerous other media through the ages made similar use of large-scale screens in the sense of this vector of engulfment enabling the viewer to "forget" the screen's boundaries. A rapid walkthrough of a

does a railing which limits viewers' vision vertically, so they do not peer too far upwards to see the panorama's "screen" boundaries.

41   The Metropolitan Museum of Art, *John Vanderlyn's Panoramic View of the Palace and Gardens of Versailles* (New York: Kevin J., Avery, Peter L. Fodera, 1988), 11.

42   Description as attributed to Hector Berlioz, quoted from Inge Van Rij, *The Other Worlds of Hector Berlioz: Travels with the Orchestra* (Cambridge: Cambridge University Press, 2015), 137.

43   Anonymous, *The Magical Panorama: The Mesdag Panorama, an experience in space and time* (The Hague: B.V. Panorama Mesdag, 1996), 48.

44   Bernard Comment, *The Painted Panorama* (New York: Harry N. Abrams Inc, 2000), 101.

45   Denise Blake Oleksijczuk, *The First Panoramas: Visions of British Imperialism* (Minneapolis, MI: University of Minnesota Press, 2011), 43.

46   David Harris, *Eadward Muybridge and the Photographic Panorama of San Francisco* (Cambridge, MA: MIT Press, 1993), 51.

47   Ton Rombout, *The Panorama Phenomenon: The World Round!* (The Hague: Uitgeverij, Panorama Mesdag, IPC, 2006), 18.

few notable examples should suffice to illustrate the point. For instance, the georama, contemporaneous with and a variation of the panorama, is another large-scale display designed to surround and inundate the viewer. Erected as an enormous hollow sphere ranging from forty to 180 feet in diameter, its inner surface is painted with a topographical map of the Earth's outer surface. As Beslisle writes in her description of the Wyld's Great Globe, a prominent georama displayed in London's Leicester Square in 1851: "Stairways led to multiple viewing platforms from which visitors could examine, *wrapped around them*, all the world's oceans, continents, rivers, and mountain ranges." (emphasis added)[48] Again, the scale, surround effect and distance of the display to the viewer render its edges imperceptible in order for the viewer to "forget" about them.

Nor were such boundary-effacing media limited to paid outdoor attractions. "Panoramic wallpaper" moved the phenomenon indoors, where multiple strips of wallpaper in the reception rooms of the burgeoning nineteenth-century middle classes turned room interiors "into a simulated exterior complete with trees, houses and painted birds frozen in mid-air."[49] While the image indoors is not as large as the panorama or the georama, their immersive effect is similar. Assuming the viewer is in a reasonably sized room and standing at some distance from the wall, they would have been able to experience the same sense of inundation by the surrounding image and disappearance of boundaries as with the panorama. The desired result is that the viewer is under the illusion that they are "in" an outdoor scene, even as they are situated indoors. As Huhtamo comments: "the issue of the screen temporarily retreated to the background – the inhabitants were as if permanently living in a virtual environment; the sense of the frame had disappeared."[50]

As cinema became the prime visual attraction at the start of the twentieth century, attention turned to its technologies of movement, sound and colour for upping the technological ante of realism and immersion. However, interest in the largeness of scale for the "total surround" sense of

---

48   Brooke Belisle, "Nature at a Glance: Immersive Maps from Panoramic to Digital," *Early Popular Visual Culture* 13(4): 313-335, 318.

49   Huhtamo, "Elements of Screenology," 41. Oliver Grau also describes "a 'room of illusion'" painted by Paul Sandby in 1793 "for Sir Nigel Bowyer Gresley at his seat of Drakelowe Hall near Burton-on-Trent in Derbyshire" which "covered three walls with a wild and romantic landscape without framing elements." With scale and faux terrain, the "room of illusion" likewise demonstrates the blurring of boundaries "between the real space and the space of the illusion." Grau, *Virtual Art*, 54.

50   Huhtamo, "Elements of Screenology," 41.

immersion returned in the mid-twentieth century with displays on screens more familiar to the contemporary user. In the 1950s, cinema re-introduced wider aspect ratio forms, such as "widescreen" films (i.e. films projected at any width-to-height aspect ratio greater than the standard 1.37:1 of 35mm film) and Cinerama (which simultaneously projected images from three synchronized 35mm projectors onto a huge, curved screen). Both formats projected images on a larger than usual scale in a bid to create cinema which engulfed the spectator's visual field and, coupled with surround sound, inundated their senses. Around this time, too, Walt Disney debuted Circarama, with eleven 16mm projectors displaying films on big screens arranged in a circle around the audience; in the 1960s, this changed to using 35mm film and became known as Circle-Vision 360°.[51]

Though these formats never really took off, greater mainstream popularity in large-scale projection was achieved in the 1970s and 80s with the introduction of IMAX theatres around the world. With screens reaching more than twenty metres high, IMAX remediates the nineteenth-century panorama with a single massive screen that surrounds the viewer (this time sitting rather than perambulatory), usually with a gentle curvature or hemispheric dome geometry to cohere the viewer's perspective for a more complete illusion. As with the panorama, the engulfing scale of the IMAX screen likewise enables its boundaries to fall beyond the viewer's visual field and be "forgotten," so that it appears as a "frameless visual space."[52] Or, as self-advertised, an experience in which "[t]he theatre disappears as images float through the air, enveloping the viewer in film."[53] As Charles Acland writes, "IMAX is unambiguously a film technology and form designed to create the experience of being there, or getting there, for spectators."[54]

Finally, nineteenth-century panoramic wallpaper as a totalizing media environment remediates into computational versions in the early 1990s by way of CAVEs, the acronym for Cave Automatic Virtual Environment. These are rooms with visual displays projected onto surrounding walls, and sometimes also the floor and ceiling, thus described as "themselves

51   Sam Gennawey, *The Disneyland Story: The Unofficial Guide to the Evolution of Walt Disney's Dream* (Birmingham, AL: Keen Communications, 2014): 108-109.

52   Alison Griffiths, "'The Largest Picture Ever Executed by Man': Panoramas and the Emergence of Large-Screen and 360-Degree Technologies," in *Screen Culture: History and Textuality*, ed. John Fullerton (Eastleigh: John Libbey Publishing, 2004): 199-220, 205.

53   "Technology Description," IMAX 3D Fact Sheet, IMAX Corporation, 1999, np, as cited in Griffiths, "The Largest Picture," 199.

54   Charles R. Acland, "IMAX Technology and Tourist Gaze," *Cultural Studies* 12:3 (1998): 429-445, 431.

giant computer screens."[55] Similar to its preceding formats, media displays surround the CAVE viewer, providing the "multitudes" of images for the spectator's "sunk-into-a-vat" immersion in the multimedia illusion. Appearing near the end of the twentieth century, the CAVE thus forms the latest instalment in a long history of large-scale media displays designed for viewers to "forget" the screen's boundaries; to be immersed in "multitudes" of reality, as did Robert when he fell *in* love.

Thinking about the post-screen in terms of the interfacing work of screen boundaries thus necessitates re-aligning "older" media into these two vectors of confinement and engulfment in the "forgetting" of their boundaries. In that sense, while most, if not all, of these media forms may be considered *proto-screens* in how their images are displayed,[56] the ironic argument here is that they are also prototypical of the *post-screen* in terms of how, across the two vectors as described, they shift and subvert the viewer's perceptions of screen boundaries. The totalization of the media environment that is at the motivating centre of these media across the two vectors is likewise the heart of the post-screen in thinking through the ethos and philosophies of reality at stake in the erosion and disappearance of screen boundaries. As we will see, the instantiation *par excellence* of this thinking is VR, to which we now turn.

The Post-Screen Through VR (1): Confinement and Engulfment

VR technology, in its form as a head-mounted display of immersive simulations, is usually traced to Ivan Sutherland's efforts in 1968.[57] However, the contemporary VR scene buzzes with a renaissance driven by multiple new players, such as Oculus Rift, HTC Vive, Samsung Gear, Playstation VR and Valve Index, among others.[58] Fresh excitement arises from cutting-edge technological advancement, such as greater density of pixels which increases

55   Jay David Bolter and Richard Grusin, *Remediation: Understanding New Media* (Cambridge, MA: MIT Press, 1999), 22.

56   Huhtamo, "Elements of Screenology," 33 and 42.

57   Ivan E. Sutherland, "A Head-mounted Three Dimensional Display," *Proceedings of AFIPS* (1968), 757-764.

58   See Nick Pino, "Best VR headset 2020: which headset offers the best virtual reality experience?," *Techradar*, March 28, 2020, https://www.techradar.com/uk/news/the-best-vr-headset. Moreover, there are also lo-tech options for the VR experience today such as Google Cardboard, which is literally a cardboard template that can be folded into a box-like set to be held against the user's face, while a mobile phone slotted into a pocket of the box provides the image and sound via a relevant VR app.

the sharpness of detail in the image. Accelerated computing power also cuts the lag and latency of the virtual scene, so that it can be re-drawn quickly enough to cohere the user's visual input with the body's balance system.

These developments renew VR's promise of a genuinely totalizing virtual environment. More importantly, they converge the two vectors of "forgetting" screen boundaries to constitute VR as a positive first instantiation of the post-screen. Firstly, in relation to the first vector of *confinement*, the VR user in essence "peeps" into its restrictive device – here, a goggles-like VR headset containing an in-built screen that displays the image to the viewer. Like a paradoxical blindfold, it is placed over the viewer's eyes to block off all visual perception of their surroundings, usually with aids such as foam cushioning against the face to prevent additional light seeping in, *even as* its screen supplies an all-encompassing image of the virtual environment. Other features, such as stereoscopic lenses which provide the illusion of depth, boost the realism of the visual experience, as does the incorporation of multiple stimuli, such as sound piped to the user through headphones or in-built speakers in the VR headset;[59] this also cancels out external noise from the user's surroundings. Via controller sticks grasped in their hands or by physically moving within a sensor-mapped zone, the user is also able to experience haptic sensations such as vibrations and resistance, as well as interact bodily with the virtual environment to influence its elements.[60]

59 Sensorial diversity has always been a part of the experience of screen media – a basic example is the incorporation of sound in cinema, whether by way of live music (for "silent" films in the early 1900s) or recorded synchronized sound from the 1920s. In the 1960s, smell featured as well: devices with olfactorily evocative names such as Aroma-Rama and Smell-o-Vision piped in atmospheric smells to cinemagoers at strategic points of the film to heighten their experience of being in the fictional world. In 1956, Morton Heilig introduced the Sensorama Simulator, a mechanical machine which surrounded its viewer in a 180° arc and, as Ken Hillis puts it, "offered the sensation of real experience through multimediated use of 3-D images, binaural sound, and scent": see *Digital Sensations: Space, Identity, and Embodiment in Virtual Reality* (Minneapolis, MI: University of Minnesota Press, 1999): 7. A short film of a motorcycle ride through New York, for example, would be accompanied not only by the sights and sounds of a street in Manhattan, but also fan-generated wind, handlebar vibrations, body tilting and chemically created smells of the city, such as exhaust fumes and pizza cooking aromas. Heilig's efforts ended shortly due to lack of funding, but contemporary VR clearly incorporates that legacy of multiple stimuli as part of its totalizing efforts.

60 As current VR narratives stand to be mostly experiences built for thrill, there is not very much in the VR world which the user can meaningfully influence. There are a few exceptions; one might be "Home – A VR Spacewalk," where the user is "tasked with making a repair on the outside of the International Space Station, before being confronted with a terrifying emergency situation": see BBC News, "Walk in space with new virtual reality experience from the BBC," November 30, 2017, http://www.bbc.co.uk/mediacentre/latestnews/2017/vr-spacewalk. But they are still few in number.

Brenda Laurel describes this "tight linkage" between multi-modalities as a virtual-reality system's "key to the sense of immersion."[61] In turn, these linkages become an effective confinement of the user's senses in terms of "forgetting" VR's screen boundaries.

At the same time, the VR headset also evokes the second vector of *engulfment* by enabling the user to command a 360° view of the virtual scene. As Oliver Grau puts it, the image space of VR is "a totality or at least fills the observer's entire field of vision."[62] Moreover, the "surround" effect is computational, as the software of VR adjusts the image in correspondence to the user's head or body movements. As Sutherland, earlier mentioned as responsible for the earliest introduction of VR, expressed in 1968: "The fundamental idea behind the three-dimensional display is to present the user with a perspective image which changes as [they] moves."[63] Tracked to the user's movements, usually manifested as a swivelling of their heads while in a fixed spot or taking a few steps in the space around them, the software re-draws the virtual environment in correspondence so that the VR world changes and appears in visual coherence wherever the user looks or moves. Mark Hansen, writing in 2004, specifically notes this image-movement correspondence as "the dynamic coupling of body and image," and emphasizes it as "the defining aesthetic feature of VR."[64] More recently, stand-alone, or sometimes called untethered, VR have expanded sensor capacities whose software is able to track movement across comparatively larger spaces. Previously limited to movements in an area of around a square meter, the untethered VR user, headset strapped over their faces, is able to move through a larger space akin to the perambulatory spectators of the panorama, all the while maintaining a coherent view of the virtual scene.[65]

VR as the post-screen thus emerges from the cross-hairs of these two vectors: it becomes an affective space which confines *and* expands to surround

---

61   Brenda Laurel, *Computers as Theater* (Reading, MA: AddisonWesley, 1991), 161.

62   Grau, *Virtual Art*, 13.

63   Sutherland, "A Head-mounted Three Dimensional Display," 757.

64   Mark B.N. Hansen, *New Philosophy for New Media* (Cambridge, MA: The MIT Press, 2004), 166.

65   In the "untethered" version of VR, the user's headset is not connected by a cable to the system which limits the distance they may move from it. Instead, by donning a "VR backpack," the user is able to move through a relatively large space, thus expanding the breadth of their physical interactivity with the virtual environment. See BBC News, "CES 2018: Hands-on with HTC's untethered VR headset," *BBC News*, January 9, 2018, https://www.bbc.co.uk/news/av/technology-42619808/ces-2018-hands-on-with-htc-s-untethered-vr-headset.

the viewer's sensorial field with "multitudes" of the virtual scene appearing wherever they look and changing coherently however they move. Hence, the transparency or "forgetting" of screen boundaries is not merely in the sense of largeness which engulfs, per the panorama, IMAX etc; nor is it only in the confinement which restricts, as with "peep" media. Specifically, the "forgotten" boundaries of VR as the post-screen take place in the convergence of *both* largeness and smallness: they are about the totalization of a media environment that appears both at scale in surrounding the viewer, but also at the most minute level, so small that it appears everywhere. They relate more to a viewing *on different scales and dimensions* than to any specific sense of big or small screen dimension. The indiscernibility of screen boundaries in the post-screen is thus, in that sense, not even a question of not seeing the boundaries, but, more importantly, one of visual shifts in scales and dimensions.

In this convergence of largeness and smallness, VR becomes a space designed in every way for its screen boundaries to fall beyond the viewer's visual field, and thus be "forgotten." It is also this sense of VR's "forgotten boundaries" in terms of the post-screen to which its transparency of medium is most often referenced, described variously as an "invisibility [of the computer]";[66] with "transparent, perceptual immediacy";[67] with "no veneer of symbolic 'interface'";[68] or, simply, "a medium whose purpose is to disappear."[69] Sometimes, the purpose is elided altogether, so that the medium somehow ends up ceasing to exist. For instance, Jaron Lanier, considered one of the founders of the VR field, remarks on how VR eliminates the abstractness of the computer in favour of the sheer physicality of the virtual experience: "when you use a computer, you tend to start to think of yourself as being like a computer... With a virtual reality system, you don't see the computer any more – it's gone."[70] Lev Manovich, as well, notes repeatedly how

> ...with VR, *the screen disappears altogether.* ... Or, more precisely, we can say that the two spaces, the real, physical space and the virtual simulated

---

66 Brenda Laurel, *Computers as Theatre*, 143.

67 Bolter and Grusin, *Remediation*, 22.

68 Derrick de Kerckhove, "Virtual Reality for Collective Processing," in *Ars Electronica: Facing the Future*, ed. Timothy Druckrey (Cambridge, MA: MIT Press, 1999), 237, citing engineer Eric Gullichson.

69 Bolter and Grusin, *Remediation*, 21.

70 Jaron Lanier and Frank Biocca, "An Insider's View of the Future of Virtual Reality," *Journal of Communication* 42(4) (Autumn 1992): 150-72, 166.

space, coincide. The virtual space, previously confined to a painting or a movie screen, now completely encompasses the real space. Frontality, rectangular surface, difference in scale are all gone. *The screen has vanished* [emphasis added].[71]

The vanishing of the screen – or, in a wider sense, the apparatus of the computer – is, of course, only in a rhetorical sense. The screen is still there, whether in-built within the headset, or via a compatible mobile phone slotted into lower-cost apparatus such as Google Cardboard. The screen has "vanished" only in the sense that the *differentiations* between the actual and the virtual which give rise to its formalization – as Manovich notes, "frontality, rectangular surface, difference in scale" – have in essence become imperceptible to the user and thereby "forgotten." As we shall see in the following sections, these boundaries and differentiations do not so much disappear as they *dis-appear* – in flashes, in paradoxes and in re-placement, to which we now turn.

The Post-Screen Through VR (2): Replacement and Re-placement

In its unique "forgetting" of screen boundaries, then, VR signals its key *frictions* between actual and virtual realities as the new criticalities of the post-screen. Given the totalization of media in VR through the twin vectors of confinement and expansion, the dominant manifestation of this friction tends to be in the critical framework of *replacement*: putting on the VR headset whereby VR's screen boundaries become imperceptible so that the actual *disappears* and the virtual – as *a different object* – *replaces* it. In this sense of replacement, one takes the other's place as *another* physical or material structure experienced as reality. In such replacement thus lies VR's much-vaunted sense of seamlessness or perfect substitution, signalled, for instance, by this blithe summation by Jeremy Bailenson from the Stanford Virtual Human Interaction Lab:

VR systems block out the perceptual input from the real world and *replace* it with perceptual input from a virtual environment that surrounds

71   Lev Manovich, "An Archaeology of a Computer Screen," *Kunstforum International*, Germany, 1995, http://manovich.net/content/04-projects/010-archeology-of-a-computer-screen/09_article_1995.pdf.

the user, is fully responsive to the user's actions, and elicits feelings of presence. Because of these affordances, VR allows users to vividly and viscerally experience any situation *as if it were happening to them from any perspective* [emphases added].[72]

This paradigm of replacement is echoed by many other scholars and from multiple perspectives. Bolter and Grusin, for instance, likewise discuss VR's effect of replacement in terms of the media logic of transparency and immediacy, whereby VR's three-dimensional immersion and capacity for interaction can be understood "as the next step in the quest for a transparent medium."[73] This transparency thus places VR in the same conceptual framework of replacement, whereby the immediacy of media, by which the viewer forgets the presence of the medium, replaces the actual with the virtual, or perception with experiential effect. Mark Hansen, again, writing about VR in terms of a "dynamic coupling of body and image" similarly refers to its effects by way of a simulation of bodily affect, namely, "a process of construction or data-rendering that takes place in the body-brain... and not an inscription or registering of an outside object or reality."[74] As such, VR is a copy which, via its elimination of screen boundaries, *replaces* the perceived image with a body-brain simulation: one takes the place of the other, whereby visual perceptions of physical space are supplanted by the effects of bodily affects.

It is probably unsurprising that VR's framework of replacement in relation to realistic re-presentation has received heavy criticism in how it indicates "magical" or wishful thinking.[75] To anyone who has donned a VR headset, the replacement of reality in VR is certainly by no means perfect – the images are not realistic enough; there is still lag, pixel bleed and so on. VR as yet does not "feel like life," and probably never will. But, as already mentioned, the bigger argument here is not to deconstruct these claims of replacement. Rather, it is to think through the *models* for the relations of reality in VR as the post-screen via its eroded screen boundaries. As Andrew Murphie puts it, his interest, as is mine here, in thinking about VR is not so much "as a form of representation of reality as an *expression* of it;" (emphasis in

---

72 F. Herrera, J. Bailenson, E. Weisz, E. Ogle, J. Zaki, "Building long-term empathy: A large-scale comparison of traditional and virtual reality perspective-taking," *PLoS ONE* 13(10), 2018: e0204494, 4.

73 Bolter and Grusin, *Remediation*, 162.

74 Hansen, *New Philosophy for New Media*, 166.

75 See generally Murray's argument in her article, "Virtual/Reality."

original) or as a *conceptual model* for the relations of realities between object and image:

> ... [T]here is no doubt that VR, as yet, provides a very poor representation of reality and may, in the foreseeable future at least, not reach the degree of high-fidelity reproduction of reality that we already associate with older media such as television. Nevertheless, the high-fidelity reproduction of the world is not necessary to an expression of it, and there can be no doubt that VR, as with everything else in the world, expresses the world in a particular fashion.[76]

Hence, in this vein of VR as an expression of interrelational realities, I want to think through an alternative model for the realities of VR in its post-screen space. The relation of VR as replacement is unsatisfactory and flawed for the reasons outlined above and as already discussed by many. But that is a technical critique. Far more important to my concerns here is the *sterility* of replacement as a conceptual framework for VR, as seen above in various instantiations. For, once replaced, an impasse ensues: what does one do with a replacement, or with a perfect copy?

Articulated as such, Jean Baudrillard's thesis of "simulacra and simulation" inevitably comes to mind as arguably the theoretical mothership of replacement and precession, and an inescapable discussion in this regard. Writing about the hyperreal – an order of a simulation generated out of such abundance of information and copies that it no longer bears any connection or differential to its referent ("origin or reality") – Baudrillard begins his classic text, *Simulacra and Simulation*,[77] by alluding to the Jorge Luis Borges fable, "On Exactitude in Science," itself a tutor text on replacement. Borges's short story tells of an imagined empire which demanded a map of such exactitude in cartographic science that the map ended up being drawn to the same scale as the empire itself. In essence, the map becomes the ultimate replacement object. As the empire declined, the map – emblematic of the empire's arrogance – correspondingly deteriorated and frayed, even as "some shreds are still discernible in the deserts." Relating the story to modern simulated realities, Baudrillard inverts the allegory to illustrate the power of the hyperreal, whereby it is

76 Andrew Murphie, "Putting the Virtual Back into VR," in *A Shock to Thought: Expression After Deleuze and Guattari*, ed. Brian Massumi (London; New York: Routledge, 2002): 188-214, 189.
77 Jean Baudrillard, *Simulucra and Simulation*, trans. Sheila Faria Glaser (Ann Arbor, MI: The University of Michigan Press, 1995).

"the map that precedes the territory"; the map "engenders the territory." Concomitantly, "it is the territory whose shreds slowly rot across the extent of the map"; it is the real "whose vestiges persist here and there in the deserts."[78]

There are thus three stages to Baudrillard's argument. The first is the *disappearance of difference* – that "sovereign difference" which "constitutes the poetry of the map and the charm of the territory."[79] Like screen boundaries, the "sovereign difference" goes through complex processes in being erased. These processes, according to Baudrillard, are fundamentally those of copying and bombardment of too much information, where models are generated in such quantity and with such reproducibility that they end up being the hyperreal, or "models of a real without origin or reality."[80] The second stage, connected to the first, is that of *precession*, where simulation precedes its original: so absent, adrift or obsolete is the difference between simulacra and the real, that the simulacra becomes realer in effect and reality than the object. This is precession not just in the sense of chronology, but also of generating a different space, where simulation occupies its own dimension of the real, namely, the hyperreal. The third stage is that of *decay*, where, as a consequence of precession, the real deteriorates – it is rotten "like a carcass," and frayed into shreds as "vestiges."[81] By this third stage, replacement is complete, where the hyperreal takes the place of the real – not as a literal replacement to be exactly what the object was, but to take its place as a different object and a different articulation of what matters to people and the reality on which they act. In this sense of replacement thus also lie fundamental contestations of power, for at the heart of Baudrillard's thesis is that a certain model of power has been inverted between object and model, where the latter trumps the former.

The theory of the hyperreal is thus, in essence, a zero-sum relation of replacement. Where an object is replaced (the disappearance of difference), it is gone (the decay of the real) and a different object takes its place (the precession of simulacra). Yet, there the argument ends. What happens after replacement? What can one do in the hyperreal? What creative or generative space can one make of a perfect copy? Baudrillard, where he addresses this (if obliquely), points only to a general void: "It is an

---

78 All quotations in this paragraph are from Baudrillard, *Simulacra and Simulation*, 1.
79 Baudrillard, *Simulacra and Simulation*, 2.
80 Baudrillard, *Simulacra and Simulation*, 1.
81 Ibid.

inevitable consequence of virtuality: there can be no strategy *of* the virtual, since the only strategies now are themselves virtual ones." (emphasis in original)[82] Virtuality takes its purpose from not having any: "The power of the virtual is merely virtual."[83] It spins into a vortex of emptiness: "It gives you everything, but at the same time it subtly deprives you of every thing"; its end point is "panic": "The subject is realized to perfection, but when realized to perfection, the subject automatically becomes object, and panic sets in."[84] Or else virtuality is consumed by a self-devouring greed, per Baudrillard's charge of the telespectator's starvation "on the other side of the screen," as discussed in the introduction,[85] satiated only by a wholesale swallowing of ourselves whereby we "mov[e] around in the world as in a synthesized image."[86]

These are not constructive positions, leading, as they do, to emptiness, dismay and hollowness. Near the end of the introduction of this book, I stated that the erosion of screen boundaries which eliminates differentiations between the virtual and the actual heralds a satiation of simulacra. In VR's case, this totalization arrives through its mode of replacement, where, akin to the taking over by the hyperreal, one reality disappears to – *precedes* – another's decay. If we are to seek an account of VR's relations of reality that may give rise to a more generative space, or that may furnish answers to deeper existential concerns as to the truths of our experiences out of the chasms between representation and reality (the gap of *La Condition Humaine* comes to mind here again), *we need a different framework*.

As such, I argue that this alternative space in terms of the post-screen may instead be articulated not in terms of replacement, but *re-placement*. In this framework lies the imbricated space of sameness *and* difference: where the object is replaced, a different object takes its place; where it is re-placed, it remains *the same yet different*: the same object yet placed in a different space or a different dimension. Beyond simulation, there

---

82 Jean Baudrillard, *Screened Out*, trans. Chris Turner (London; New York: Verso, 2002), 60-61.

83 Baudrillard, *Screened Out*, 60. Baudrillard's next sentence following the quotation is also worth noting: "This is why it can intensify in such a mind-boggling way and, moving ever further from the so-called 'real' world, itself lose hold of any reality principle." (60) The void and emptiness following the domination of the hyperreal is clear.

84 Both quotations from Baudrillard, *Screened Out*, 180.

85 See the introduction, especially text accompanying footnote 93.

86 Jean Baudrillard, "The Virtual Illusion: Or the Automatic Writing of the World," *Theory, Culture & Society* 12 (1995): 97-107, 100, 97. This is also a Platonic fear, where there is no longer an independent immanent truth, or no stable and organized state of being that is independent of thought and representation.

are already several existing analogies for such re-placement: in temporal terms, for instance, a re-placement would be an "anachronism" (same object existing in a different time); in spatial terms, "anamorphosis" (same object viewed in a different angle or distortion); in linguistic terms, "catachresis" (same word employed in a different category or context); in metaphysical terms, "haunting" (same person appearing in a different existential dimension). In relation to VR, the technology re-places the same actuality in a different experience. The real *re-emerges* – the same yet different – in the same slippage as the screen that does not "disappear" but, rather, dis-appears by appearing and re-appearing in paradoxes and flashpoints. The engagement of the boundaries between actual and virtual thus becomes a negotiation rather than negation or elimination or consignment to oblivion.

Hence, VR, as thought through the post-screen rather than its total-izing screen or surround, presents the relations of the actual and the virtual in this different discursive space: not one which merely comments ironically and helplessly on how the real has been replaced by fakes, clones, technical simulacra or even affect – which in themselves have frankly become the clichés of our time[87] – but one that produces a more productive space, out of which emerges contingency, renewal, even resistance. Replacement and re-placement are not mutually exclusive; rather, they are correlative expressions of actual-virtual relations across the diminished boundaries of the post-screen. The copy of the virtual is thus not only produced to deliver a perfect and seamless replacement of the original; it is also a copy for its own sake as a copy, which *precedes* the real for a different kind of energy between actual and virtual realities. The shift in this space for the re-placed copy, then, is also a redemption of sorts for the failure of human creativity in Magritte's *La Condition Humaine*, asserted in equal parts by the jarring and unbridgeable cleft between canvas and landscape. There might be some redress yet from the simulacra.

As with *La Condition Humaine à la* Magritte's painting, the space of re-placement can only be shown through the gaps of VR's screen boundaries.

---

87 One example of such a cliché, if randomly chosen out of many, might suffice to illustrate. I refer, as usual, out of popular culture: in episode 2, "Transgressive Border Crossing," of the 4th season of the television show, "Orphan Black," *Netflix*, 42:52, 2017, a 5-season television series which concerns itself entirely with multiple male and female clones, a new secretive clone is introduced who appears behind a mask...of a sheep, whose clear reference is to Dolly, the world's first cloned mammal. The use of a sheep's face as such a joke demonstrates the sheer banality of replacement, now common enough to be invoked as a visual gag.

In the next three sub-sections, I will illustrate re-placement through three instantiations: (i) what I call VR's *danger paradox*, or the contradictions of danger and safety between the actual and the virtual; (ii) VR as *immersion* in terms of travel, escape and fulfilment; and (iii) VR as *inversion* in terms of witness, empathy and the subjective. Each presents a different demonstration of re-placement of the real in VR as the post-screen. Collectively, though, the argument aims to construct a more creative and productive engagement between the actual and the virtual real where, out of that space, we may re-examine our existential connections and desires for media to better understand who we are.

The Danger Paradox

When the Oculus Rift VR headset launched in 2016, it was anointed "technology's first true breakthrough in bringing virtual reality to a mainstream and commercial audience,"[88] appearing to revive an industry dormant for decades. Its website, sleek and brimming with the technological promises of its wares, hosts a downloadable health and safety PDF leaflet which contains all the usual words of caution and cop outs of legal liability.

One paragraph in particular caught my eye:

Use Only In A Safe Environment: The headset produces an immersive virtual reality experience that distracts you from and completely blocks your view of your actual surroundings. Always be aware of your surroundings when using the headset and remain seated at all times. Take special care to ensure that you are not near other people, objects, stairs, balconies, windows, furniture, or other items that you can bump into or knock down when using—or immediately after using—the headset. Do not handle sharp or otherwise dangerous objects while using the headset. Never wear the headset in situations that require attention, such as walking, bicycling, or driving.[89]

---

88   Josie Ensor, "Oculus Rift's Palmer Luckey: 'I brought virtual reality back from the dead',"
The Telegraph, January 2, 2015, https://www.telegraph.co.uk/technology/11309013/Oculus-Rifts-Palmer-Luckey-I-brought-virtual-reality-back-from-the-dead.html.
89   As quoted from the Oculus Rift Health and Safety warning leaflet, downloadable from https://static.oculus.com/documents/health-and-safety-warnings.pdf.

In short, the whole paragraph reads as a bizarre cautionary that warns of what idiomatically might be called "the bleeding obvious." In effect, it says: "Do not move about or handle dangerous objects with a box over your head where you are unable to see in front of you." Why would anyone need to be told that?

Departing from the conventional answer of the obvious legal liability dodge, I offer an alternative reading – one of the post-screen – where the seemingly facetious warning in the Oculus Rift leaflet is not a straightforward cautionary. Rather, it exposes a paradox, if a familiar one, namely, the contradiction of the VR user being safe from virtual dangers which appear clear and present yet, *across the screen's boundaries*, is really in another kind of danger. It is the same paradox that colours the comedy of the *Uncle Josh* film as discussed in chapter 2, which sets up Uncle Josh's thinking that he is in danger (such as against the oncoming train) when he is actually safe, and vice versa, against the audience's savvier understanding of screen boundaries' interplay of realities.

I call this *the danger paradox* of the virtual – the confounding contradiction of safety and danger across the screen's boundaries. The boundaries thus become a revelatory flashpoint of the relations between the actual and the virtual real, whose true nature emerges *only* in their friction across that space. On either side of the boundary, things are stable, just as Uncle Josh would have stayed out of harm's way had he not disturbed the screen boundaries. Similarly, the cautionary leaflet implies the relative safety of the VR experience *as long as* one does not cross its screen boundaries by remaining seated, being aware of one's surroundings and so on.

However, violate the boundary – tear down the screen as Uncle Josh did, or act against the leaflet instructions – and the relations of the virtual and the actual become exposed, unmasking the true nature of danger. The oddly obvious cautionary leaflet thus conceals an entirely different warning: *do not breach the screen boundary*. In this sense, the danger paradox for VR also brings to light the post-screen's *dis-appearance* of screen boundaries. Even as the whole VR apparatus is geared towards the user's "forgetting" of screen boundaries, the leaflet's message is a reminder that they not only still exist, but moreover will re-appear at these revelatory junctures of friction between the actual and the virtual.

It is in the slippage of this dis-appearance of screen boundaries that the actual real is re-placed. Neither replaced nor discarded, it instead re-emerges out of the virtual across the relational contingency of dis-appearing screen boundaries, brought into relief at flashpoints like the danger paradox: objects that are forgotten until tripped on; unheeded until seen or touched; immaterial until knocked over. They are not the same actual real: to perceive those, the user simply takes off their headset. Nor are they replaced, as they are clearly still there in their original forms. Rather, these objects are the same yet different: the same objects as viewed or touched without the headset, yet different in how they re-emerge out of friction with the virtual.

This re-placement is not limited to actual objects of hidden dangers; the virtual may also re-place actual physical sensations, of which VR users potentially suffer a wide range from their virtual experiences.[90] One common example is cybersickness, which manifests as symptoms similar to classical motion sickness, including nausea, discomfort, headaches and vomiting.[91] Image ruptures, discontinuous graphic space, system crashes and jagged graphics are also common disruptions, as is discomfort from the headset, such as its weight or the annoyance of ill-fitting straps. Even with the lightness of the Google Cardboard viewer,[92] the cardboard exerts pressure against the user's skin, or else its Velcro strap catches on one's hair.

These manifestations are often brought up as evidence for the familiar criticism of how VR still falls short as technology that "feel like life," pointing to sterile complaints of how "you aren't really experiencing total immersion,"[93] which renders the discourse into a space of paralysis of how, in short, VR fails to *replace*. However, if read differently in terms

90   All immersive media are affected by snags. As Janet Murray writes, "immersion is a delicate state that is easily disrupted": "Virtual/Reality," 18. For panoramas, the stillness and silence of the paintings, as "a strange and somewhat unpleasant effect," hampered their immersion, leading Charles Robert Leslie to lament for "the hum of the population, and the din of carriages": *Athenaeum*, February 17, 1949, 173, as quoted in Griffiths, "The Largest Picture," 204-5. For IMAX, distortion, blurring and image "illegibility" is common. Recounting a critical review of an IMAX documentary which complained of "warp perception" and a world "scrolling wildly toward you," Alison Griffiths writes: "Far from heightening reality, movement on this scale and from this perspective renders the image unreadable": Griffiths, "The Largest Picture," 205.

91   See generally Joseph J. LaViola, "A discussion of cybersickness in virtual environments," ACM SIGCHI Bulletin, 32(1) (Jan 2000): 47-56.

92   The Google Cardboard is literally a folded cardboard box-like structure where, along one length, a smartphone delivering the VR visuals is slotted against a pair of stereoscopic lenses. The other length of the box is then pressed against a user's face, usually with the user's hands or via rudimentary straps. See Google Cardboard's website at https://arvr.google.com/cardboard/ for more information; see also footnote 58.

93   Charles Arthur, "The Return of Virtual Reality: 'This Is as Big an Opportunity as the Internet,'" *The Guardian* online, May 28, 2015, https://www.theguardian.com/technology/2015/may/28/jonathan-waldern-return-virtual-reality-as-big-an-opportunity-as-internet.

of post-screen *re-placement*, these intrusions signal another order of reality. Cybersickness, for example, is not simply an actual sensation of unwellness; it is a sensation of unwellness which emerges or makes sense *only* out of the relations between the actual and the virtual, i.e. relations which thus re-place those sensations differently, even as they are the same. The emergence of re-placed reality is not about causality; the issue here is not about what causes cybersickness, but the *totality of relations* which constitutes the existence of cybersickness. Only with that approach can we interrogate the placement of the real and arrive at its re-placement. A phenomenon thus not only exists as itself or its multiple models or copies; it can also exist *differently*. Hence, the re-placement of the actual real also intervenes between the original and its copy: not quite one nor the other, but a third entity which creates a new space between the two.

In this sense, Baudrillard perhaps aggrandizes the dominance of the hyperreal, whose virtuality takes over the actual that then falls into decay. Across the boundaries of the post-screen between the virtual and the real exists a certain context, almost as an aura in the appearance of a kind of force, to the real by which it re-emerges as re-placed, or re-membered as with an inflexion. VR is one exemplar, but, as with the long traditions of media "new"-ness and old-ness, there are also many other examples. One of these, for instance, might be the *trompe l'oeil* – an optical illusion that depicts three-dimensional objects with no obvious differentiating boundaries from its surroundings, and which aims to replace reality with a realistic illusion. A common illustration of a *trompe l'oeil* is an illusion of a realistic-looking hole painted on a solid pavement. The actuality of the pavement becomes re-placed via the virtuality of the illusory hole – the same pavement, yet different: pedestrians, who would normally not even have glanced at that part of the pavement, stop short, look again, and perhaps even walk around to "avoid" the "hole." Like VR, the *trompe l'oeil* relies on the relational qualities between the representational realism of its virtual reality (the illusion of the hole) and the materiality of its actual reality (the solid pavement) for its effect. Only against the actual solidity of the pavement does the virtual illusion of the hole exist; only in the *friction* of that relation between actual and virtual does the hole make sense. And only against the virtuality of the hole does the actuality of the pavement re-appear, even as it disappears, as *emerging anew*.

VR as Immersion: Travel, Escape, Fulfilment

> *No live organism can continue for long to exist sanely under conditions of absolute reality; even larks and katydids are supposed, by some, to dream.*
> ~ Shirley Jackson[94]

As Jackson alludes above, all species dream as a form of escape from their absolute realities.[95] The use of media to enact that liberation – in terms of *replacing* one's actual reality with an alternative one, even with just a dream – plays out, too, as a common thirst, of which VR is just the latest instalment. The popularity of immersive visual attractions in the eighteenth and nineteenth centuries were driven as much by the public's desire for entertainment and novelty (not to forget eroticism) as for sights from faraway countries or historical events. In particular, the Victorian era was marked with an "increasing taste for travel among the public": Bernard Comment writes of how "the yearning for other countries, other places, for changes of scene, developed considerably during the nineteenth century";[96] Evelyn Fruitema and Paul Zoetmulder highlight the nineteenth-century citizen's "visual thirst," whereby "his eager eye remained undernourished."[97]

To such demand, the itinerant peep shows in the nineteenth century thus displayed "sensational topics such as the wonders of China, famous palaces, battlegrounds, or the devastation caused by the Lisbon earthquake"; the Cosmorama in Paris owned paintings "representing the most remarkable sites and monuments of the different parts of the globe."[98] As Rose writes, "[peep shows'] success was, no doubt, owing to the fact that the public had very little opportunity of seeing pictures of any kind, especially in outlying districts, and even in London."[99] The panorama, too, became a veritable means for substituted travel: it suited "Parisians who like to travel without

---

94   Shirley Jackson, *The Haunting of Hill House* (London: Penguin, 1959), 1.

95   It is more than an allusion – research also shows how numerous animal species, including mammals and birds, undergo Rapid Eye Movement (REM) sleep, the stage of the sleep cycle in which the sleeper dreams vividly: see Matthew Walker, *Why We Sleep: The New Science of Sleep and Dreams* (London: Penguin, 2017): 56-77.

96   Comment, *The Painted Panorama*, 132.

97   Evelyn Fruitema and Paul A. Zoetmulder, *The Panorama Pheomenon: Mesdag Panorama 1881-1981*, 5th ed. (The Hague: Foundation, 1981), 30.

98   As cited in Huhtamo, "Toward a History of Peep Practice," 41.

99   A. Rose, *The Boy Showman and Entertainer* (London: George Routledge & Sons, no date of publication), 36.

having to leave home";[100] it enabled "thousands of people to discover, without having to travel, the most celebrated cities, the major seaports and the most interesting countries not only in Europe but also in other areas of the world."[101] Similarly, early cinema promised virtual travel to their viewers as "armchair tourists" to faraway or imagined places,[102] whose camera enables the human body to be mobilized into any place at any time.[103] In Walter Benjamin's oft-quoted words:

> Our bars and city streets, our offices and furnished rooms, our railroad stations and our factories seemed to close relentlessly around us. Then came the film and exploded this prison-world with the dynamite of the split second, so that now we can set off calmly on journeys of adventure among its far-flung debris.[104]

The images of eighteenth, nineteenth and early twentieth century immersive media forms thus slaked a visual thirst for faraway places and historical events, to be a "guide, reminder, or substitute"[105] for travel to them. They *replaced* the viewer's immediate spatiotemporal locality with *another place and time*, dreamt out of colonial conquests, improved transport and increasing reportage of exotic locations.

   In the twentieth century, mediated replacements for travel and escape – indeed, for sightseeing[106] – pivoted to another kind of desire, namely, transcendence of human capabilities, facilitated by the development of mass communication which was seen to extend the senses of the human

---

100 Attributed to Miel's *Salon* of 1817, as quoted in Comment, *The Painted Panorama*, 131.

101 Attributed to Valenciennes in 1800, as quoted in Comment, *The Painted Panorama*, 130.

102 See Alison Griffiths, *Shivers Down Your Spine: Cinema, Museums, and the Immersive View* (New York: Columbia University Press, 2013).

103 Jacques Aumont writes of the same desire in "the unleashed camera" – "a camera enabled not only to see all, but also to see from anywhere": Jacques Aumont, "The Variable Eye, or the Mobilization of the Gaze," in *The Image in Dispute: Art and Cinema in the Age of Photography*, ed. Dudley Andrew, trans. Charles O'Brien and Sally Shafto (Austin: University of Texas Press, 1997): 231-252, 247. Also see discussion in chapter 1, especially text accompanying footnote 84.

104 Walter Benjamin, "The Work of Art in the Age of Its Technological Reproducibility: Second Version," in *Walter Benjamin, The Work of Art in the Age of Its Technological Reproducibility, and Other Writings on Media*, eds. Michael W. Jennings, Brigid Doherty and Thomas Y. Levin, trans. Edmund Jephcott, Rodney Livingstone, Howard Eiland et al. (Cambridge, MA: Harvard University Press), 37.

105 The Metropolitan Museum of Art, *John Vanderlyn's Panoramic View of the Palace*, 11.

106 Also see Giuliana Bruno, *Atlas of Emotion: Journeys in Art, Architecture, and Film* (London; New York: Verso Press, 2002; 2018) on site-seeing via cinema, architecture, psychogeography and emotion.

body.[107] The camera's eye in television and cinema, while still showing visual excitement of unfamiliar sights, presented new rhythms, movements and realities – the kino-eye for the human eye.[108] Yet this is a turn grounded in twentieth-century desire, rather than connected to the specificity of any medium. Even in encountering a nineteenth-century panorama, the modern spectator might relate to expressions of seeing new reality and worlds. Writing of his visit as a child in the 1940s to the panorama of Scheveningen as painted by Hendrik Willem Mesdag in 1881 and still exhibited today in its original location at The Hague, the film director Paul Verhoeven emphasized how the panorama viewing took him to "another world": "It was that experience in the Mesdag Panorama which first made me aware, as a child, that reality can seem more than it is. The Panorama was for me what a Mass is for another, the perception of another world, albeit in a secular sense."[109]

The desire to replace the scene around ourselves for vicarious travel thus shifts to an existential grasping of realities beyond the brute existence of actuality, a theme also made prominent in cyberpunk science fiction from the 1960s and 70s. The opening premise of William Gibson's *Neuromancer*, as a key representative of the genre, is the protagonist's inability to escape "the prison of his own flesh" into a virtual replacement located in the "bodiless exultation of cyberspace,"[110] a replacement which represented to him a life-affirming liberation. More recently, representations of virtual replacement facilitated across screen-based devices also feature large in high-profile blockbuster films such as *Avatar,*[111] *Surrogates*[112] and *Ready Player One* (originally a 2011 novel by Ernest Cline).[113] Significantly, all these stories feature dramatic arcs with central tensions between the failings (and frailties) of the protagonists' actual reality against their omnipotence in the virtual, in which the character,

---

107 See Marshall McLuhan, *Understanding Media: The Extensions of Man*, 3rd ed. (Berkeley, CA: Gingko Press, 2013), especially his arguments on how elements such as clothing and housing are extensions of skin and bodily heat-control mechanisms.

108 For more on the kino-eye, see Dziga Vertov, *Kino-Eye: The Writings of Dziga Vertov*, ed. Annette Michelson; trans. Kevin O'Brien (Berkeley; Los Angeles, CA: University of California Press, 1984): 5-160.

109 "Suspended in time: interview by Yvonne van Eekelen," in *The Magical Panorama: The Mesdag Panorama, An Experience in Space and Time*, trans. Arnold and Erica Pomeran (The Hague: Waanders Publishers, Zwolle / B.V. Panorama Mesdag, 1996), 175.

110 William Gibson, *Neuromancer* (London: Gollancz, 1984), 6.

111 *Avatar*, directed by James Cameron (2009; Los Angeles, CA: 20th Century Fox Home Entertainment, 2010), DVD.

112 *Surrogates*, directed by Jonathan Mostow (2009; Burbank, CA: Walt Disney Home Entertainment, 2010), DVD.

113 *Ready Player One*, directed by Steven Spielberg (2018; Burbank, CA: Warner Bros, 2018), DVD.

in most cases, goes on to fulfil their true destiny. Clearly, more is at stake in virtual replacement besides thrills and entertainment. Like the canvas placed before the window in Magritte's *La Condition Humaine* evidencing the artist's futile attempt to replace its section of landscape,[114] the immersion of VR, as replacement for actual reality, answers deep-seated existential fears of what we are and dreams of who we wish to be.

As with its media precedents, VR fulfils desires for travel and exploration, ostensibly replacing the user's actual reality with the virtual sensorial experiences of remote, exciting and adventurous destinations. References abound in relation to these senses of realized impossibilities in virtual replacement: the *Samsung Gear* VR website, for instance, advertises to its users to "go on adventures you've only dreamt of";[115] the *New York Times* app "[takes] you from the depths of the ocean to the surface of Pluto, via the spire of 1 World Trade Center"; *Orbulus* is "to give us a new perspective on real-world places we can't visit in person"; *RYOT VR* presents "virtual-reality documentaries in far-flung places."[116] Google's *Expeditions* promise to "buddy groups together" and "you'll be able to go [sic] the African plains, inside the ISS, to Jupiter and more."[117] From dinosaur parks to ocean depths to outer space to roller-coaster descents to building rooftops, there are few spaces, imagined or otherwise, for which there is no purported replacement VR experience. In a twist to the theme of global travel, VR also enables the "experience" of worlds that are inaccessible due not to geography, but the physics of human perceptibility.[118] An example is *InCell VR*, a straightforward action/racing game where the viewer is set in the micro world of human cells, viruses and bacteria, in which they have to stop an advancing virus. The app is banal, with its singularly unoriginal gameplay of dodging obstacles and collecting points along the way, but to a certain extent it does enable a physical experience into a humanly imperceptible world, with the user's actuality replaced, as it were, by the molecular.

114 See discussion of this painting's significance in the Introduction.

115 As quoted from the Samsung Gear website at https://www.samsung.com/global/galaxy/gear-vr/.

116 All quotations in this sentence are taken from Stuart Dredge, "10 of the best virtual reality apps for your smartphone," *The Guardian* online, June 13, 2016, https://www.theguardian.com/technology/2016/jun/13/best-virtual-reality-apps-smartphone-iphone-android-vr.

117 Sumra, Husain, "Best Google Cardboard apps: 20 top games and apps for your mobile VR headset," *Wareable*, October 23, 2018, https://www.wareable.com/vr/the-best-google-cardboard-apps.

118 For the idea of the filmic equivalent, see William Brown, "Man Without a Movie Camera – Movies Without Men: Towards a Posthumanist Cinema," in *Film Theory and Contemporary Hollywood Movies*, ed. Warren Buckland (New York; London: Routledge, 2009): 66-85, who argues in his book chapter for the cinema of the humanly impossible.

Yet, as noted, virtual replacement, in all its imperfect substitution, only goes so far. A far more interesting and creative space is the *re-placement* of immersion in the post-screen experience of VR, which I illustrate here via three examples: the VR roller-coaster ride; location-based VR; and a VR re-creation of Eugène Delacroix's *Liberty Leading the People*.

(i)   The VR roller-coaster ride

The first is what has been called the "VR roller-coaster ride." First implemented for the Alpenexpress VR-Ride in Germany in 2015,[119] though now in operation with several rides in theme parks all over the world,[120] customers ride the roller-coaster with a VR headset whose images are synchronized to their actual movements on the coaster. For instance, a virtual image of an airplane cockpit would turn and travel in the same tilts and directions of the actual roller coaster car. Such intertwining of actual movement and virtual image is not new for theme park rides – for instance, the "movie ride," first started in the 1980s, similarly synchronizes actual movement to the image, where "the rider is placed on a hydraulically controlled movable platform or seat that tilts, twists, pitches, and shakes in synchronization with large moving images and environmental sound."[121]

However, the actual movements of the "movie ride" are relatively restricted in a confined space, with the viewer's sensation of movements largely produced *in relation to* the movement of the images, leveraging the mobility/ immobility paradox well-known from cinema.[122] In comparison, the actual movements in the "VR roller-coaster ride" are not only far more extensive, but the rider is also *actually* in movement. Yet in the totalizing virtuality of VR, that actuality of movement also potentially becomes *re-placed*. The movements experienced by the rider are the same with or without the

119 Erik Yate, "Europa Park in Germany Launches World's First VR Coaster," *Behind the Thrills*, September 4, 2015, https://behindthethrills.com/2015/09/europa-park-in-germany-launches-worlds-first-vr-coaster/.

120 With many more to come, as indicated by developments such as the US$1.5-billion Oriental Science Fiction Valley Park, China's first virtual reality theme park currently being built on a 330-acre space in the south-western Guizhou province: see Joseph Campbell, "Virtual reality boom brings giant robots, cyberpunk castles to China," *Reuters*, November 24, 2017, https://www.reuters.com/article/us-china-tech-theme-park/virtual-reality-boom-brings-giant-robots-cyberpunk-castles-to-china-idUSKBN1DO03B.

121 Murray, *Hamlet on the Holodeck*, 49.

122 See in particular Anne Friedberg's elucidation of "virtual mobility" and of cinema's "mobilized gaze," in *The Virtual Window*, especially its Introduction: 1-24; and more generally, Anne Friedberg, *Window Shopping: Cinema and the Postmodern* (Berkeley: University of California Press, 1993).

headset, for the ride travels through the same track. However, the viscerality of the ride, so integral to its thrill, takes on a different content in relation to the "multitudes" of virtuality which now surround the rider. For instance, at one point of a VR ride I took in 2017 at Everland, South Korea's largest theme park, the VR scene mirrored the actual, complete with the ride track stretching before me with its surrounding scenery. Suddenly, a large hand with outstretched fingers appeared in front of me, as if reaching to scoop up the car; I instinctively looked up and was startled to see a giant child, kneeling next to the track and peering down at me. The car rumbled away, and again instinctively I turned my head to look back at her.

All this took place within about five seconds. While I was experiencing the same actual movements in the roller-coaster car with or without the VR headset, those movements became a different expression of reality as situated within and in relation to the virtual. As a result, I did things on the ride I normally would not do without the VR context (such as looking up or behind, which are actions not usually taken on a park ride). The ride's movements in those seconds also took on a different meaning – the car moving away from the frightening giant child became an escape flight, rather than simply rolling onwards on its track for visceral thrill. In the post-screen totalization of the virtual in VR, across its screen boundaries the actuality of the roller-coaster ride becomes *re-placed* in relation to the virtual: the actual locomotion of the ride along its tracks remains the same – in its actual sense, the ride remains unchanged – yet it is different in terms of the reactions it solicited and its changed relational meaning as derived from dimension, scale and perspective. Re-placement in the post-screen through VR thus becomes a contextual and generative space: the roller-coaster's *content* of actuality changes in this space – it evokes new reactions; its actual movements renew and gain new meaning.

(ii)   Location-based Virtual Reality (LVR)

The second example is a variation of the first – namely, the perambulatory version of the "VR roller-coaster ride," or what has been called "location-based virtual reality" (hereafter "LVR"),[123] or, with suitable histrionics, "hyper reality."[124] In LVR, the user, instead of sitting in a roller-coaster car, physically

---

123 Anshel Sag, "Location-Based VR: The Next Phase of Immersive Entertainment," *Forbes*, January 4, 2019, https://www.forbes.com/sites/moorinsights/2019/01/04/location-based-vr-the-next-phase-of-immersive-entertainment/.

124 Edwina Pitman, "The Immersive Hyper Reality World of Star Wars," *BBC News World Service*, last on July 17, 2018, https://www.bbc.co.uk/programmes/w3csww70.

walks around an actual space with a VR headset over their eyes which supplies their virtual reality. The scenario is painted thus: "Imagine walking around a haunted house [as in, haunted in VR], or running around a warehouse with walls that are made of foam, but look like bricks in VR."[125] High-profile LVR attractions to date include The Void's (founded in 2015 and backed by Disney and James Murdoch) much-touted display, *Star Wars: Secrets of the Empire*. First opened to the public in 2017 in the United States but which has since toured other countries, including the UK, the VR experience is set in the fictional world of the *Star Wars* movie franchise. Many other LVR experiences have since been created, with a prediction (dated late 2019) of "location-based entertainment" constituting "around 11%" of the VR industry.[126]

The main feature of LVR is that the actual space is set up *as a space* to correspond with the virtual space. This means that, with an untethered VR backpack on their shoulders and a VR headset over their eyes, the user physically moves through their virtual reality as matched in movement through their actual space. Like the VR roller-coaster ride, the user's actual movement is synchronized with its counterpart in virtual reality. As one reviewer describes his experience in the "Star Wars" LVR: "When I walk forward in *Secrets of the Empire*, I actually walk forward." Unlike the VR roller-coaster ride, whose synchronization is limited to the movement of its locomotion, LVR integrates the virtual with the actual via other sensations, such as the haptic: "Touch the walls of the spacecraft and there are corresponding physical walls, perfectly placed to match the digital view." Or "stepping on…hot lava [from an explosion] I could feel my foot sink into a soft, plasticised substance."[127]

125 Sag, "Location-Based VR," np.

126 Sol Rogers, "The VR Companies Shaking Up Location-Based Entertainment," *Forbes*, November 8, 2019, https://www.forbes.com/sites/solrogers/2019/11/08/the-vr-companies-shaking-up-location-based-entertainment/#43bd5c3014c5. As with many other industries, though, the Covid-19 pandemic has put abrupt and brutal pause to the growth of such VR centres: Janko Roettgers, "The VR Gaming Centers of the Future May Not Survive the Crisis of Today," *Protocol*, June 8, 2020, https://www.protocol.com/location-based-vr-covid-19.

127 All quotations in this paragraph are from Jeremy White, "I was a Stormtrooper for 15 minutes and it was awesome," *Wired.com*, December 16, 2017, https://www.wired.co.uk/article/star-wars-vr-london-secrets-of-empire-void-experience. What is interesting is that the user's actual movements themselves also contain illusions: "we play tricks on your brain that you actually believe – such as something called redirectable walking. You are convinced you are walking down a straight hallway when you are actually walking in a curve": Peter Graham, "The VOID's Star Wars: Secrets of the Empire Coming to Second London Location," *VR Focus*, April 12, 2018, https://www.vrfocus.com/2018/02/the-voids-star-wars-secrets-of-the-empire-coming-to-second-london-location/.

On one hand, the correspondences between actual and virtual space and objects in LVR gesture towards the replacement of the actual, not so much in terms of the objects' realism, but, rather, their *thing-ness* – the VR user would avoid walking into a virtual wall as assiduously as they would an actual one because they know that is a barrier. In that respect, the wall, as a thing, exists as realistically in the virtual as it does in the actual. On the other hand, as with the VR roller-coaster ride, the actual movements of the user are also re-placed – they are no different to any of those movements as normally taken, *and yet they are*. Stepping onto a soft surface entails the same actual movements of stepping onto any surface, yet different in the post-screen virtuality of molten lava – the movement gains a different meaning in relation to the virtual; it emerges out of a different frame of reference.

(iii)   VR re-creation of *Liberty Leading the People*

My third example of re-placed immersion is VR artist Anna Zhilyaeva's re-creation in VR of Eugène Delacroix's painting, *Liberty Leading the People*, in August 2018 at the Louvre Museum in Paris.[128] Specifically, Zhilyaeva re-created the painting, originally a surface in two dimensions, in *three-dimensional virtual space*. The project consisted of Zhilyaeva's "moving" her virtual body "through" the virtualized depths of the painting in VR, and using her controller sticks to "hold" the "palette" in one hand and "apply" the "paint" with the other.[129] As the virtual space gains colour, it gradually "becomes" recognizable as Delacroix's famous painting. In her actual space, Zhilyaeva, with a cabled VR headset strapped to her face, moved through a small cordoned square mapped by the VR system's sensors. A screen installed near her showed her audience what she was doing in VR.

The re-created painting in its virtuality thus re-emerges, though certainly not as a replacement of its original – the computerized, broadly pixellized smears of the Google Tilt brush software are in no way a comparison to

---

128 Emory Craig, "Virtual Reality Artist Live Performance at the Louvre Museum in Paris," *Digital Bodies*, August 27, 2018, https://www.digitalbodies.net/vr-experience/virtual-reality-artist-live-performance-at-the-louvre-museum-paris/.

129 The software which makes this possible is Google Tilt Brush, a painting VR app which enables one to draw in 3D virtual space. Notably, in Zhilyaeva's publicity YouTube video, she gestures to the audience to join her as she then steps "into" the painting's frame; a special effect then dissolves the lower half of her body so that she appears to be entirely "within" the painting. The porosity of post-screen boundaries is again shown here, where, in the post-screen context of VR, even the painting's frame, conventionally rigid and definitional in marking off the painting against its surroundings, becomes fluid and malleable.

Delacroix's fine oils and brush strokes. But that was never the intention of the exercise. The re-creation is showcased not as a visual show of the painting: given its fame, the public already knows what it looks like. Rather, the re-created painting is a re-placement to present another order of the painting's reality, namely, its dimension of depth. The painting remains the same – it *looks* recognizably the same – yet different: as with the re-placement of anamorphosis, the illusory depth of the two-dimensional painting, created out of perspectival cues and the viewer's binocularity, becomes altered and stretched into a three-dimensional space navigable by Zhilyaeva as a virtual body and transformed through her movement and actions in it.[130] Its dimension of depth is raised to another level of perceptibility while retaining its iconic appearance; re-placed as the same, yet different.

The diminished boundaries of the post-screen through VR thus engender a different framework of immersion – not replacement and disappearance, but re-placement and dis-appearance. The two sets of binaries are not in opposition with each other, but shift the virtual and the actual into different contexts, becoming a source of insight into the re-rooting or re-channelling of the VR user's actual reality. The dream of the Holodeck's "total immersion" thus modulates through the post-screen's alternative lens – a different order of immersion in the indiscernibility of the screen via fluid shifts in dimensions from the physical to the virtual, and vice versa. In this sense, the re-ordering of the virtual against the actual in such contextual terms also adjusts virtuality as escape and the overcoming of human limitations: not only might we need to re-think how we are escaping, but perhaps also from what are we doing so.

VR as Inversion: Witness, Empathy, Subjectivity

> *Searching for the medium of technology that will confirm your experience such that*
> *your basic humanity can be recognized.*
> ~ William Jelani Cobb[131]

VR's totalization of the user's senses leads the user not only to be immersed in a different environment, but also to occupy *an alternative subjectivity.* Here, the "forgotten" screen boundaries of the post-screen through VR are for

---

130 Of course, the irony is that that depth itself in VR, as with a 2D painting, remains an illusion – it is a visual perception that arises from stereoscopic effect!
131 Line as spoken in the documentary *13th*, directed by Ava DuVernay (2016), Netflix stream, at 1'30"-1':31".

the user to "see through another's eyes, embodying their experiences, thus 'empathising' with them."[132] This sense of VR is not about being in another world, but being in *another's* world; or from *immersion* to *inversion*, where the user, instead of being "sunk" externally into a different environment, is turned internally to acquire a different subjectivity. In this way, video maker and "VR visionary" Chris Milk, via his VR work and in particular his popular TED talk on the subject, thus also draws on VR to "[connect] humans to other humans in a profound way,"[133] and calls it "the ultimate empathy machine,"[134] an undeniably catchy label which has stuck for better or worse.

In some ways, the replacement/re-placement duality of VR is more straightforward here, if only because replacement (of the user's subjectivity for an Other's) forms so much of the explicit rhetoric for VR as a vehicle for empathy. Various scholars, for instance, note how VR enables the user to "experience the life of someone else by 'walking a mile' in his or her shoes,"[135] or gain "access to the body and mind of another person."[136] In his project, *Carne y Arena*, on migrant crossings across the US border from Mexico, which includes a location-based VR piece, Oscar-winning filmmaker Alejandro González Iñárritu alludes precisely to this replacement of experience in an interview about the project: "It's a re-enactment of [the immigrants'] lives. It's a slice of their nightmare."[137]

Examples of such "replacement" VR projects abound; a few highlights will suffice here as illustration. For instance, in *The Party*,[138] a 2017 *Guardian*

132 Grant Bollmer, "Empathy Machines," *Media International Australia* (2017): 1-14, 1. Bollmer heavily criticizes this position of empathy through VR's embodiment, arguing instead for an alternative "radical compassion" which more explicitly acknowledges the Other, embracing openness to understanding and refusing assimilation into one's own self.

133 Milk, "How virtual reality," at 9′20″.

134 Milk, "How virtual reality," at 3′10″.

135 Quotation attributed to Jeremy Bailenson at the Virtual Human Interaction Lab, as cited in Bollmer, "Empathy Marchines," 4.

136 As quoted from the webpage of "The Machine To Be Another" project, https://docubase. mit.edu/project/the-machine-to-be-another/.

137 Kriston Capps, "The Experience is Virtual. The Terror is Real," *Bloomberg*, June 7, 2018, https://www.bloomberg.com/news/articles/2018-06-07/alejandro-i-rritu-s-vr-film-carne-y-arena-is-pretty-real.

138 For an introduction to this project, see Anrick Bregman, Shehani Fernando and Lucy Hawking, "The Party: a virtual experience of autism – 360 video," *The Guardian* online, October 7, 2017, https://www.theguardian.com/technology/2017/oct/07/the-party-a-virtual-experience-of-autism-360-video.

VR project[139] by Lucy Hawking and Sumita Majumdar, the VR user experiences a dramatized birthday party through the ostensible perspective of a fifteen-year-old girl with autism. The user receives sensorial cues, such as a calibrated auditory overload, as the direct "experience" of the teenager's difficulties in the triggering situation, supplemented by a voiceover of her thoughts on her anxieties and articulations of stress. Another *Guardian* VR project, *6x9: a virtual experience of solitary confinement* (2016), sets the user in the virtual environment of a solitary confinement prison cell.[140] In methods similar to *The Party*, the user "experiences" solitary confinement, if for ten minutes. This includes not only seeing totalizing visualizations of the prison cell's austere and constrained environment as a "direct" experience of the restrictiveness of the practice, but also hearing the thoughts and opinions of psychologists and previous inmates on the resulting psychological damage. At one point, the user experiences a sensation of levitation as demonstration of the hallucinations that often occur from solitary imprisonment.

Nor is this self-proclaimed strategy confined to journalistic political pieces. Art projects similarly seek to evoke this specific mode of "replacement" of perspective. *The Machine To Be Another* (2014) by the international art collective, BeAnotherLab, leverages VR for its user to, as self-proclaimed, literally "see through the eyes of another."[141] The project involves two individuals – an actor with a first-person camera and microphone fixed on them; and a user who dons a VR headset connected to the actor's camera and microphone. With the actor mirroring the user's movements in real-time, the user receives corresponding images from the actor's point of view via their camera, as well as words through their headphones as spoken by the actor in order to "[generate] the perception of someone speaking inside [the

139 As part of ramping up their digital journalism, *The Guardian* in 2016 started producing VR pieces which "follow a first-person narrative on topics the publisher covers in other formats": see Lucinda Southern, "The Guardian Remains Committed to VR, Despite Limited Commercial Opportunities," *Digiday*, October 19, 2017, https://digiday.com/media/guardian-remains-committed-vr-despite-limited-commercial-opportunities/, np. The result is a series of projects which place the user in the unique position or environment of people whose stories the newspaper wished to highlight. For more information, see also *The Guardian* VR webpage at https://www.theguardian.com/technology/ng-interactive/2016/nov/10/virtual-reality-by-the-guardian.

140 As taken from the project page, "Welcome to your cell," https://www.theguardian.com/world/ng-interactive/2016/apr/27/6x9-a-virtual-experience-of-solitary-confinement. See also Caroline Davies, "Welcome to your virtual cell: could you survive solitary confinement," *The Guardian* online, April 27, 2016, https://www.theguardian.com/world/2016/apr/27/6x9-could-you-survive-solitary-confinement-vr.

141 As quoted from the project's listing page on the MIT docubase website: see https://docubase.mit.edu/project/the-machine-to-be-another/.

user's] mind."[142] Hence, the VR user "occupies" the subjectivity of the actor, poised here as an Other, to ostensibly replace their own. To emphasize the difference in subjectivities, the project deliberately casts its pairs along differences in gender, race and disability, such as a male-female pair, or an able-bodied person with someone in a wheelchair.

A variation of this "replacement" paradigm in VR is for the user to get some sense of what the Other is going through by becoming a virtual body which "witnesses" another person's plight (rather than directly "experiencing" it). In a way, this similarly "replaces" the user's subjectivity. Again, there are many examples; a few mentions here will suffice: Nonny de la Peña's *Project Syria*, for instance, created in 2015 and displayed at the World Economic Forum, places the VR viewer in a scene as a "witness" to a child playing on the streets in Syria when a missile strike hits.[143] Chris Milk's own work, *Clouds over Sidra*, a film he made with various collaborators and, like *Project Syria*, was exhibited at the 2015 World Economic Forum in Davos, is very similar. Sidra, a 12-year-old Syrian girl living in a refugee camp in Jordan, narrates her life story as the VR user finds themselves surrounded by scenes recorded from her camp. *Replacement* here of the actual by the virtual is thus not for an immersion in another environment, but an *inversion* into another's perspective. As Milk, a prime advocate for this "replacement" paradigm in service of VR as his "empathy machine," puts it: "When you're sitting there in [Sidra's] room, watching her, you're not watching it through a television screen, you're not watching it through a window, you're sitting there *with her*." (emphasis added)[144]

Yet, as with the criticism of immersive VR which dislodges replacement, critique of VR as such replacement of subjectivity – and there is much of it – similarly upends the paradigm. In a sense, much more lies at stake here in the blitheness of rhetoric such as that of Milk's on replacement in VR. That nonchalance becomes almost a moral arrogance – the assertion that one may conjoin with another's suffering simply through audiovisual media which represents their reality and the erasure of screen boundaries to carry through

---

142 Quotation as taken from their project video, https://vimeo.com/71686981.

143 She also created a similar project involving such witnessing in 2017, *Out of Exile: Daniel's Story*, where the user is an "invisible witness to a family fight in a Georgia living room where a young gay man is insulted and assaulted": quoted from Murray, "Virtual/reality," 24.

144 Milk, "How virtual reality," at 4'48". Janet Murray pours particular scorn, though, on how such "witnessing" could possibly create this effect at Davos, considering the event's politics and (extreme wealth) of the attendees: "the claim that billionaire bankers would experience a fundamental expansion in human sympathy by viewing a 360° film through a headset is pure wishful thinking": Murray, "Virtual/reality," 13.

that media, or that technology can bridge gaps of understanding where human sympathy has somehow failed. Janet Murray's response to Milk's proclaimed effect of *Clouds over Sidra* at Davos is suitably scathing in that regard: "the claim that billionaire bankers would experience a fundamental expansion in human sympathy by viewing a 360° film through a headset is pure wishful thinking."[145] Even as he warms to VR technology's affordances of immersion and transcendence, Bimbisar Irom, writing on such VR films as "humanitarian communication," similarly warns of their countering ideological placings, noting how "the representational strategies of VR are subject to the constraints of ideology and power hierarchies that permeate other representational tools."[146] The sense of the post-screen, in terms of its erased boundaries, thus perhaps manifests here in its most insidious articulation. Namely, the elimination of boundaries between the virtual and the actual translates into a deliberate and casual, even cynical, elision of any moral awareness of their differences, and the wilful discounting of one's own privilege in the few minutes of "replaced" experience.

In this sense, the critical edge of "re-placement" asserts a correspondingly more ethical perspective on the post-screen of VR. Certainly, there is critique of VR which alludes to more nuanced readings of replaced subjectivity in VR. Grant Bollmer, for instance, argues that generating empathy via VR is a limited exercise because VR simply records and re-creates another's experience. It does not – as Bollmer argues should be the mode of empathy – enable the Other "to become conjoined with the experience of the VR user," or "become objects to be used and absorbed." There is no "negative annihilation of the Other." In other words, VR does not *replace* the user with the Other via its erasure of screen boundaries. Instead, Bollmer points to the opposite argument: he calls for *acknowledgement* of VR's screen boundaries; indeed, they are "the surfaces upon which others become visible to our own experience and knowledge"; they "are not barriers to do away with and overcome. Rather, they are the very foundation of any possible relation."[147] This recognition of screen boundaries – or "acknowledgement of distance" – likewise shifts *replacement* of subjectivity to *re-placement*, where the user's subjectivity emerges elsewhere. That elsewhere, according to Bollmer, is that gap of the screen boundary which opens up to

145 Murray, "Virtual/reality," 13.
146 Bimbisar Irom, "Virtual Reality and the Syrian Refugee Camps: Humanitarian Communication and the Politics of Empathy," *International Journal of Communication* 12 (2018): 4269-4291, 4287.
147 All quotations from Bollmer in this paragraph are from "Empathy Machines," 7-12.

understanding, to "acknowledging the limits and the infinite inability to grasp another's experience completely."[148] This, then, is what Bollmer calls "radical compassion" (as opposed to empathy), which "embraces an openness to understanding and refuses assimilation into one's own self... [to be] open to [the experience of another], even if it can never fully grasp it."[149] Read in terms of the post-screen, the VR user's subjectivity is thus also re-placed. It has not been supplanted to become a different subjectivity. Rather, re-emerging in the space of the acknowledged screen boundary, with the refusal of assimilation and recognition of distance, the VR user's subjectivity remains the same, yet also different.

A final project for comment, then, on *Notes on Blindness: Into Darkness* (2016) as a work to illustrate such re-placement in the post-screen of VR, produced by award-winning media studio Archer's Mark and which accompanies a short as well as feature film of the same title. The highly acclaimed project relates the oncoming blindness of John Hull, a professor of religious education at the University of Birmingham, specifically via extensive audio diaries kept by Hull as he started losing his sight.[150] The VR app sets out six chapters, each containing a distinct scene, such as at a concert, in a park, in a shed listening to the rain and so on, and Hull's reflections on being blind in each scene. Objects and people are represented as vague shapes outlined in blue dots; the background is always pitch black. Audial inputs in the form of Hull's voice and the sounds he hears in the VR environment, such as those of birds, footsteps and choral singing, become amplified.

As with the other projects, numerous commentators review *Notes on Blindness* VR as the uncritical replacement of the subjective, such as "to go in-depth on what being blind is like";[151] or "what life is like when you lose your sight completely." [152] Again, the casual reference to the paradigm of replacement of experience is presumptuous to the point of audacity – can one truly know what life is like when one's sight is lost?

Rather, a more ethical reading of *Notes on Blindness* might take shape as the re-placement of Hull's "acoustic world" consciously designed with the amplified sounds of his world – rain, footsteps, the singing of birds – and his thoughts as recorded on his tapes. The totalizing darkness of the VR

148 Ibid.

149 Bollmer, "Empathy Machines," 10.

150 Quoted from the project website, http://www.notesonblindness.co.uk/.

151 Andrea Carvajal, "Notes on Blindness Review," *VR Voice*, August 31, 2019, https://vrvoice.co/notes-on-blindness-review/.

152 Olivia Marks, "What Life is Like When You Lose Your Sight Completely," *Vice.com*, June 30, 2016, https://www.vice.com/en_uk/article/vdqg88/notes-on-blindness-interview.

experience thus re-emerges not as a different subjectivity and certainly not as an empathetic experience of blindness. Instead, it is simply darkness – the same darkness which ensues when we shut or cover our eyes; yet different, as coloured by the project's audiovisual input in the VR experience which gives that darkness an entirely different placement, significance and meaning.

As with immersion, the inversion of VR through its totalization points towards fulfilling a more profound need, albeit not so much about bridging the gap in generating empathy for another. Rather, we may think about VR's experience of inversion as the latest instalment along the trajectory of what media, from autobiography to photography to cinema, has always strived to achieve, namely, an act of documenting to validate human experiences. Or to take Cobb's words per this section's opening quotation: to seek recognition in one's basic humanity. What media delivers is thus a different kind of connective knowledge – not the more bombastic sense of "I understand your suffering" but, more simply, "I understand you suffer."

Defeated by the Ghosts

The post-screen through VR thus discloses different readings of VR as a totalizing environment, opening up spaces of thinking about the simulacra beyond the colonization of the senses and the supplanting of experience. Those are enterprises doomed to fail, no matter how idealistic their goals. Shifting the focus of immersive media environments from their totalization to an understanding of their relational qualities across their boundaries – and thereby the space of the post-screen – effectively breaks up the contestation between the real and its simulacra. That contestation is necessarily one-dimensional because its purpose generally reduces to no more than simply the sake of contesting the real and being that Other environment or perspective as realistically as possible. That near-coercive cycle of the endless copy inevitably collapses into itself: at best, an exhortation of a certain loss – and, as an exhortation, well worth noting – but, in the end, not an outcome which presents any substantive answers. To repeat the driving question of this chapter: what can one do in the space of the perfect copy? Re-placement as an operational framework for the post-screen thus functions as an alternative answer to the question by presenting a more generative space, drawn across re-considering relational qualities of objects between their actual and virtual realities. It channels thinking not about the endless cycle of what disappears and what takes its place, but instead in terms of what appears and dis-appears, and the routes to their revelation.

A final note: the re-emergence or the re-placement of the real is not just about re-discovering or seeing the real anew. As with what media is ultimately about, there is an existential pain at stake. Re-placement is also about re-thinking what we seek in media, or our pursuit through media in achieving arguably its most basic (yet most complex) objective: the expression and reception of meaning which enables not only existence, but the placement of elements in the vast network of elements which make up our world, including nature, animals, technology, tools and, of course, humans. It conveys their distances, their relations, their boundaries of differentiations. Media thus returns us to our obsession with the gap of the human condition *à la* Magritte's *La Condition Humaine* painting, itself a powerful visual statement on the impossibility of achieving that seamless connection between elements, if only, as in its case, between landscape and painting.

Yet that impossibility extends as well to connecting with one another, be they non-human or human. So much of that effort gets swallowed, literally, by the distances betwixt. Kafka writes precisely of that insurmountable distance in his famous letters exchanged with Milena Jesenská, a treasured correspondent and much else. In his despair of communications with her, Kafka creates monsters in the form of "the ghosts" on the prowl which consume (his) letters: "written kisses never arrive at their destination; the ghosts drink them up along the way." The distance to reach another person in order "to attain a natural intercourse, a tranquility of soul" might be conquered through physical means via the train, the car, the aeroplane… "but nothing helps anymore: these are evidently inventions devised at the moment of crashing." And so we continue to write and speak across distance with media – "after the postal system… the telegraph, the telephone, the wireless" – but the ghosts devour them at our expense. Here, again, per the earlier discussion in the introduction, is media as located in the inimitable ache of thirst and hunger. Media places humans into relation with others, but in the particular placing of humans with other humans, it carries something else as a fundamental sustenance and comfort. It is a quality so important that we instinctively fight for it, as do others: "it is this ample nourishment which enables [the ghosts] to multiply so enormously." But, according to Kafka, we are defeated by the ghosts. The more media we have, the more we lose: "[the ghosts] will not starve, but we will perish."[153]

153 All quotations from Kafka in this paragraph are from Franz Kafka, *Letters to Milena*, trans. Philip Boehm (New York: Schocken Books, [1952] 1990), 223.

But such despair comes from the same failure of replacement, the same doom in the attempt to breach the gap between landscape and painting in *La Condition Humaine*. Where re-placement in VR is about the re-emergence of objects in a different dimension, so does it work generally across all media: the voice, the touch, the words, the image, they are not disappearances into the void of the distance, but dis-appearances in another guise. Any other conclusion subjects us to the same disappointment from the shattered dream of the Holodeck – "a far-fetched fiction that can never be achieved"[154] – and to Kafka's despair: "One can think about someone far away and one can hold on to someone nearby; *everything else is beyond human power.*" (emphasis added)[155]

Re-placement draws a different space: it reaches for the same yet different, and VR is merely a thinking device for this idea. Robert, from the start of this chapter, falls *in* love by seeing "multitudes of Amys" all around him, but falls in love by re-placing Amy from friend to lover. John Hull, facing his own oncoming darkness and the unimaginable despair of never seeing in the same way again, in his own context understands the re-placed spaces of and for his blindness: "After all, being human is not seeing, it's loving."[156] In the end, only love is generative, and only love is transformative. In other words, being human is bigger than the sum of all the elements which constitute us – to replace or to copy those elements would always have been futile, and not really even the point.

154 Murray, "Virtual/reality," 12.
155 Kafka, *Letters to Milena*, 223.
156 Line spoken by John Hull at the end of the first episode, "Part I: Memory," of *Notes on Blindness* (2016).

4 Holograms/Holographic Projections: Ghosts Amongst the Living; Ghosts of the Living

Abstract

This chapter explicates holographic projections as the second instantiation of post-screen media. Often mistaken as holograms, these projections of images ranging from Tupac to Julian Assange to holographic protests re-draw the boundaries between life and death, and enable a re-imagination of ghosts, deadness, aliveness and afterlife. The chapter argues for four different moments in a history of ghosts in the media: resurrection; necrophilia; necromancy; and interactivity. The last facilitates spectral life in the post-screen through considering holographic projections of both dead and living figures. In relation to the dead, the post-screen becomes a space in limbo between deadness and aliveness; in relation to the living, the realness of the holographic body stretches in a tetravalence across dual axes of actual/virtual and here/elsewhere, and enlivened in what I call vivification. In these 3D displays on the post-screen of resurrected and vivified bodies, different kinds of life, afterlife and after-death emerge.

Keywords: hologram; holographic projection; death; life; afterlife; ghosts

How We See Ghosts, or, In Love with the Post-Screen

To put it plainly, the man wants to go to bed with a woman who's dead; he is indulging in a form of necrophilia.
~ Alfred Hitchcock[1]

1   As remarked by the director, Alfred Hitchcock, on the truth of the double identity of Judy/Madeleine in his film, *Vertigo*. Taken from an interview with Francois Truffaut in *Hitchcock by Francois Truffaut* (New York: Simon & Schuster, 1985), 244.

Ng, J., *The Post-Screen Through Virtual Reality, Holograms and Light Projections. Where Screen Boundaries Lie*. Amsterdam: Amsterdam University Press 2021
DOI: 10.5117/9789463723541_CH04

"The man who wants to go to bed with a woman who's dead" is Scottie (played by James Stewart) in Alfred Hitchcock's *Vertigo*,[2] a film met with mixed reviews on its release, but today consistently ranked amongst the "greatest films of all time."[3] In the first half of *Vertigo*, Scottie falls in love with a woman, Madeleine (played by Kim Novak), who then dies, or so he is led to think. In the second half, Scottie meets another woman, Judy (also played by Kim Novak), who looks just like Madeleine. Inevitably, Scottie attempts to recreate Madeleine out of Judy, compelling Judy to change her hair, clothes and shoes to resemble his dead lover. The film does not explicate Scottie's motivations: it seems in part a romantic pining, and in part a kind of pathological desire in his uncompromising demands of Judy to alter her appearance. In one of his famous 1962 interviews with François Truffaut, Hitchcock attributes Scottie's fascination for the return of the dead Madeleine to the rather salacious appetite of necrophilia, stoked by the eroticism of the dead and the desire to "go to bed" with them.

What follows, though not discussed, is how shades of that necrophiliac desire land as well on the cinema spectator, themselves watching, even seeking out, onscreen reanimation or, more likely today, computer-generated recreation of dead actors. The screen thus hosts this undercurrent of forbidden yearning – it facilitates the return of the dead, and becomes a permeable threshold between them and the living. After all, it is only a short step across the screen's boundaries to move from one realm to the other. Sometimes, that step is taken literally. In *Minority Report*,[4] another film shot through with desire for the dead, if this time in the form of a more palatable familial yearning, protagonist John Anderton (played by Tom Cruise) longs for his young son, Sean, who had gone missing years ago and is presumed dead or murdered.[5] In the film's science-fiction setting, Anderton projects images of Sean on a "wall-screen," recorded in the past at a beach during which father and son exchanged a casual conversation. Presented initially as a conventional two-dimensional image against the wall, at one point of the projection the image of Sean "peels off" from the wall-screen and "steps" towards John, becoming a fuzzy three-dimensional figure of

2   *Vertigo*, directed by Alfred Hitchcock (1958; Los Angeles, CA: Universal, 2005), DVD.

3   See *Sight & Sound* magazine critics' polls of their Top Fifty Greatest Films of All Time: in 2002, *Vertigo* was placed second; in 2012, it was ranked first. See *Sight & Sound*, London: British Film Institute.

4   *Minority Report*, directed by Steven Spielberg (2002; Los Angeles, CA: 20th Century Fox, 2003), DVD.

5   However, this was never directly confirmed in the film and remains a controversial point: see fan discussion, for instance, at https://www.funtrivia.com/askft/Question20403.html.

light. The "step," fluid and with childlike innocence, recalls (as discussed in chapter 2) Sadako's painful and horrifying crawl across the television screen boundaries in *Ringu*, seeking revenge and death. Sean's movement across the wall-screen's putative boundaries along its two-dimensional surface likewise signals a migration from one realm to another, albeit in a different and more wholesome way. The absent son is now again present *in the father's space*, if only as sparkles of fuzzy light. Anderton, watching the hologram of his son, speaks to the image in the same words as had been recorded from their past conversation. An anachronistic conversation ensues as the Sean-image "replies" in real-time, standing before Anderton as a three-dimensional figure. He is almost as good as returned from the dead.[6] For those few moments, his father smiles and is comforted.

The post-screen thus also lies in this space of eroded screen boundaries whose implications, per this chapter's discussion, are to question and re-draw the boundaries of life and death. Through holographic projections, the post-screen leads to the uneasy and affective co-location of alive beings and bodies from an elsewhere, such as the dead and the living as with Sean and Anderton. The post-screen through holographic projections thus re-visits how we see ghosts, certainly of the dead, though also of the living. The additional twist in *Vertigo* is that Scottie was in love with neither a dead woman nor a living woman. He was in love with an amalgamation of the dead *and* the living, an entity in a coalesced space of being somehow dead yet also alive. In the first half of the film, Scottie was in love with Madeleine, who had already been killed by her husband, yet "alive" via Judy's impersonation (and *yet* again in-between dead and alive in Judy/Madeleine being "possessed" by Carlotta Valdes, Madeleine's dead great-grandmother). In the second half, he falls in love with Judy, very much alive (until the end), yet also "dead" to Scottie first by his being unaware of her existence, and later by his thorough erasure of Judy's identity until she re-appeared to him as Madeleine. His object of affection was a woman who somehow always had one foot in the realm of the living and the other with the dead – in short, who occupied a space of eroded boundaries. Or, we can say, he was in love with the post-screen.

***

---

6   The narrative theme of sons returning from the dead comes to mind, such as *The Monkey's Paw* (New York: HarperPerennial Classics, 2014), the classic short story which pivots on a mother's wish on the magical monkey's paw for the return of her dead son.

Ghosts in the Media: Re-inventing the Afterlife

> ...[O]ne might say that images of the deceased surround us everyday.
> ~ Alexandra Sherlock[7]

The image has always been haunted. It is filled with ghosts and the ghostly, as evident from Maxim Gorky's famous 1896 review of cinema as "the kingdom of shadows" – "it is not motion but its soundless spectre"[8] – to André Bazin's analogy of photography as "the molding of death masks."[9] Gilberto Perez calls cinema itself "the material ghost," where "[t]he images on the screen carry in them something of the world itself, something material, and yet something transposed, transformed into another world."[10] In his 1980 book, *Camera Lucida*, written as much a theoretical inquiry into photography as an emotional tribute to his deceased mother, Roland Barthes describes the photograph as "the living image of the dead."[11]

Key to this haunting is the photochemical image's contradictory expressions of time, specifically *the past time* of the object in the *presentness* of viewing the image. As Barthes puts it, the photograph "establishes not a consciousness of the *being-there* of the thing (which any copy could provoke) but an awareness of its *having-been-there*." (emphasis in original)[12] One of the many fascinations of the photograph to Barthes is its temporality which returns the realities of the past into the present: "for in every photograph there is the always stupefying evidence of this is how it was, giving us, by a precious miracle, a reality from which we are sheltered."[13]

In this temporal convergence of pastness and presentness in the photochemical image also lies their most profound separation – that of between life and death. In this respect, the peculiar temporality of the recorded image which returns past subjects into the present thus also returns the

7   Alexandra Sherlock, "Larger Than Life: Digital Resurrection and the Re-Enchantment of Society," *The Information Society* 29 (2013): 164-176, 165.

8   Both quoted phrases from "I.M. Pacatus" (pseudonym for Maxim Gorky), *Nizhegorodski listok*, 4 July 1896, trans. Leda Swan, in Jay Leyda, *Kino: A History of the Russian and Soviet Film* (London: George Allen & Unwin, 1960): 407-409, 409.

9   André Bazin, "The Ontology of the Photographic Image," trans. Hugh Grey, in André Bazin, *What is Cinema?* (Berkeley and Los Angeles, CA: University of California Press, 1967): 9-16, 12.

10   Gilberto Perez, *The Material Ghost: Films and Their Medium* (Baltimore and London: The Johns Hopkins University Press, 1998), 28.

11   Roland Barthes, *Camera Lucida* (New York: Hill & Wang, 1980), 78-79.

12   Roland Barthes, "Rhetoric of the Image," in Roland Barthes, *Image-Music-Text*, trans. Stephen Heath (London: Fontana Press, 1977), 44.

13   Ibid.

dead to the realm of the living, i.e., as ghosts and apparitions, or as the *resurrected*. Hence, Laura Mulvey takes on, as mentioned, Bazin's invocation of the photograph as a death mask to write specifically of death *and* the ghostly in the photograph: "the deathbed photograph came to replace the death mask. Both record the reality of the dead body and, in preserving it, assume a ghostly quality."[14] Drawing on Barthes's contemplations of the photograph's temporal paradox, Nayef Al-Joulan makes a similar connection: "the opposition between the *here-now* and *there-then* [in the photograph] brings together life and death, or even the dead as living, a paradox related to an essential linkage between time and space." (emphasis in original)[15] The photograph amalgamates its pastness and presentness as *memento mori*: pastness in its reflection and remembrance of death; presentness where for a moment "our death can sit beside us, rather than be a linear construct that we move toward."[16]

With its qualities of animation and kinesis, the *moving* image augments this resurrection. Capitalizing on spooky effects out of combining light and darkness, magic lantern performances, for instance, became one of the earliest entertainments to satisfy Anglo-American fascination with the paranormal in the eighteenth and nineteenth centuries by featuring projections of moving images as visual illusions of "ghosts."[17] One of the most famous shows in the era was the Phantasmagoria by the Brabantine conjuror, Paul Philidor. Performing in the 1780s in Vienna, he "summoned up" revolutionary figures, including dead ones, "by back-projecting images using a lantern mounted on a trolley. Thus the image could at first be made to appear small, and then to grow as if advancing on the crowd."[18] In 1796,

14 Laura Mulvey, *Death 24x a second: Stillness and the Moving Image* (Edinburgh: Reaktion, 2006), 58.

15 Nayef Al-Joulan, *'Essenced to Language': The Margins of Isaac Rosenberg* (New York: Verlag Peter Lang, 2007), 252.

16 Paula Mahoney, "Using Photography as an Analogy in the Experience of Death and Mourning," PhD thesis submitted at Monash University, January 23, 2017, 53.

17 The tradition of magic lantern use and the invocation of the supernatural reaches back even further than the nineteenth century, with scholars such as Athanasius Kircher, a German Jesuit in the mid-seventeenth century, who experimented with projection and magic lanterns to create images of devils and demons so as to have them "mingle" amidst the audience. See Maria Warner, *Phantasmagoria: Spirit Visions, Metaphors, and Media into the Twenty-first Century* (Oxford: Oxford University Press, 2006), and Roberta Hofer, "Metalepsis in Live Performance: Holographic Projections of the Cartoon Band 'Gorillaz' as a Means of Metalepsis," in *Metalepsis in Popular Culture*, eds. Karin Kukkonen and Sonja Klimek (Berlin: De Gruyter, 2011): 232-251.

18 Mervyn Heard, "Now You See It, Now You Don't: The Magician and the Magic Lantern," in *Realms of Light: Uses and Perceptions of the Magic Lantern from the 17th to the 21st Century*, eds. Richard Crangle, Mervyn Heard, Ine van Dooren (Exeter: The Magic Lantern Society, 2005), 16.

Falconi staged a show of "the ghost of [French Revolution figure] Charlotte Corday in the act of stabbing [Jacobin leader Jean-Paul] Marat," appearing as "a luminous body, enveloped in darkness, as large as life, and every feature distinguishable, for the space of 3 or 4 minutes."[19] In 1816, he put on a show in New York "in which a luminous image of [Genevan philosopher Jean-Jacques] Rousseau 'rose majestically and mysteriously from a tomb.'"[20] Another prominent figure on the scene, Etienne-Gaspard "Robertson" Robert, reinvented the show in 1798 by "turning it into an atmospheric, multisensory installation" with "multimedia entertainment," including lantern effects, live action, ventriloquism, sound effects, masks, music, puppetry and shadowgraph sequences.[21] To Robertson, the magic lantern was a "fantascope" – a view into supernatural fantasy – which he innovated to create more vivid effects of the ghostly, such as using oil rather than a candle to increase the brightness of the images.

Twentieth-century media forms prompt further rhetoric of media's affinity with ghosts and the dead. Television, for instance, as Jeffrey Sconce remarks, "brought with it a new form of 'visual' program flow, making it more than an extraordinary medium linking the invisible voices of the living and the dead, the earthling and the alien."[22] Like the magic lantern shows that constitute its predecessor, the moving images of cinema similarly re-animate dead figures as a form of resurrection: "just as the cinema animates its still frames, so it brings back to life, in perfect fossil form, anyone it has ever recorded from great star to fleeting extra."[23] Recalling Bazin's famous description of cinema as "change mummified,"[24] the metaphor specifically points to the moving image as a material preservation of the dead in preparation for their reappearance. As with photography, the invocation of the dead in cinema becomes a calling card to the extent of fetish, such as Paul Willemen identifying "the cinephiliac moment" with "its overtones of necrophilia, of

19   Heard, "Now You See It," 18.

20   Ibid.

21   Andrea Stulman Dennett, *Weird and Wonderful: The Dime Museum in America* (New York: NYU Press, 1997), 118.

22   Jeffrey Sconce, *Haunted Media: Electronic Presence from Telegraphy to Television* (Durham, NC: Duke University Press, 2000), 16. See also generally John Durham Peters, *Speaking into the Air: A History of the Idea of Communication* (Chicago, IL: The University of Chicago Press, 1999), particularly 137-176: "By preserving people's apparitions in sight and sound, media of recording helped repopulate the spirit world," 139.

23   Mulvey, *Death 24x a Second*, 18.

24   Bazin, "The Ontology of the Photographic Image," 15. See also Philip Rosen's direct take on Bazinian ontology of cinema in *Change Mummified: Cinema, Historicity, Theory* (Minneapolis, MI: University of Minnesota Press, 2001).

relating to something that is dead, past, but alive in memory."[25] From this chapter's introductory section, *Vertigo* again comes to mind in re-visiting the filmgoer's own necrophilia alongside the film's (presumed) reading of what cinema ultimately is: the return of the dead.

However, as the twentieth century drew to a close, a different discourse of cinema and ghosts began, where film scholars and critics arriving at cinema's centenary contemplated its ageing and mortality. With cinema already always intersecting between life and death in its re-animation of the dead, these contemplations encompassed cinema's own passing to new technologies – namely, from film to digital[26] – and in the wake of competing institutional forms from television (first terrestrial, then cable); home video; DVDs; internet streaming; to mobile applications today. As Mulvey writes, quoting Chris Petit from his video *Negative Space*: "the cinema is becoming increasingly about what is past. It becomes a mausoleum as much as a palace of dreams."[27] The dirge flowed, and continues to flow, out of popular and critical literature: "the cinema is dead, long live the cinema."[28]

Indeed, digital technologies give rise to new forms of returning the dead. Cinema reanimated dead figures as resurrected; however, those figures were recorded *when they were alive*. Conversely, computer-generated imaging (CGI) and digital editing *re-assemble* images of dead figures with live actors so that, when presented, those images of the dead appear to be recorded *when they were dead*. Alexandra Sherlock describes how popular British comedian Bob Monkhouse, who died in 2003, "re-appeared" in a 2007 television advertisement to raise awareness of prostate cancer, the illness

---

25   Paul Willemen, "Through the Glass Darkly: Cinephilia Reconsidered," in Paul Willemen, *Looks and Frictions: Essays in Cultural Studies and Film Theory* (London: BFI, 1994): 223-257, 227.

26   To many critics and scholars, the advent of the digital sounded a clear death knell for cinema. Mulvey's observation, for instance, is characteristic of this sentiment: "However significant the development of video had been for film, the fact that all forms of information and communication can now be translated into binary coding with a single system signals more precisely the end of an era": Mulvey, *Death 24x a Second*, 18.

27   As quoted in Mulvey, *Death 24x a Second*, 17.

28   See, as a sampling, John Belton, "If film is dead, what is cinema?" *Screen* 55(4) (Winter 2014): 460-470; also Lies Van de Vijver, "The cinema is dead, long live the cinema!: Understanding the social experience of cinema-going today," *Participations: Journal of Audience & Reception Studies* 14:1 (May 2017): 129-144; and Joseph Owen, "Cinema is dead, long live cinema!': an interview with Peter Greenaway at Cairo Film Festival 2018," *The Upcoming*, November 30, 2018, https://www.theupcoming.co.uk/2018/11/30/cinema-is-dead-long-live-cinema-an-interview-with-peter-greenaway-at-cairo-film-festival-2018/. See also Manohla Dargis and A.O. Scott, "Film Is Dead? Long Live Movies: How Digital Is Changing the Nature of Movies," *The New York Times*, September 6, 2012, https://www.nytimes.com/2012/09/09/movies/how-digital-is-changing-the-nature-of-movies.html.

which killed him. In the advertisement, Monkhouse is shown onscreen wearing his trademark out-of-date suit and strolling among gravestones as he quips: "Let's face it, as a comedian I died many deaths. Prostate cancer, I don't recommend. I'd have paid good money to stay out of here. What's it worth to you?"[29] Monkhouse was not recorded making those statements *before* his death as some sort of prescient warning,[30] and it would have been impossible for him to have made those recordings *after* his death. Rather, the images and statements were created after his death by "piecing together various components (archival footage of Monkhouse, a body double, and a voice impersonator) in an almost Frankenstein-like fashion."[31] Sherlock calls this "a sort of modern necromancy," "as though [Monkhouse] was speaking to us from beyond the grave, advising us of the danger of prostate cancer."[32] With the advent of digital technologies, the moving image's relationship to the dead thus shifts from resurrection and necrophilia with the dead via its image to a kind of augmented necromancy of the dead speaking to *and* appearing before the living.[33] The latter, as I will argue, and its inherent shift are keys to heralding the more radical tectonics between the dead and the living in relation to the post-screen through holograms and holographic projections.

29   Dan Bell, "Bob Monkhouse back from grave to promote prostate cancer fight," *The Guardian* on-line, June 11, 2007, https://www.theguardian.com/media/2007/jun/11/advertising.medicineandhealth.
30   *Cf* Yul Brynner's equally famous anti-smoking advertisement, in which he made the statement: "Now that I'm gone, I tell you: Don't smoke, whatever you do, just don't smoke." It appears that he had made this statement posthumously, but he was actually alive when he made it on "Good Morning America" on ABC News. However, the impact of his statement certainly lay in the posthumous nature of his statement. As a spokesman was quoted saying: "There's nothing more forceful than when someone dead looks into the camera and says: 'Don't smoke. I did.'" All quotations in this footnote from Barron H. Lerner, "In Unforgettable Final Act, a King Got Revenge on His Killers," *The New York Times*, January 25, 2005, https://www.nytimes.com/2005/01/25/health/in-unforgettable-final-act-a-king-got-revenge-on-his-killers.html.
31   Sherlock, "Larger Than Life," 165. Sherlock's metaphor of "piecing together" via Frankenstein also recalls a similar idea from another -stein, namely Sergei Eisenstein's ideas of montage, where a similar sense of afterlife – in the form of alternative, more potent meanings – emerges out of the piecing together – indeed, collision and conflict – of images: see Eisenstein, "Montage of Attractions" and "The Montage of Film Attractions" in Sergei Eisenstein, *S.M. Eisenstein: Selected Works: Vol. 1: Writings, 1922-34*, ed. and trans. Richard Taylor (London: BFI, 1988), 33-58.
32   Sherlock, "Larger Than Life," 164.
33   While my concern here is with visual media, note that *audio*-only communications with the dead have their own long history of media, notably Victoria-era media such as "the 'spiritual telegraph,'" as well as more modern technologies of wireless technology and radio: see generally Sconce, *Haunted Media*, 21-91. On spiritualism and telegraphy in particular, also see Richard J. Noakes, "Telegraphy Is an Occult Art: Cromwell Fleetwood Varley and the Diffusion of Electricity to the Other World," *The British Journal for the History of Science* 32(4) (Dec., 1999): 421-459.

But a further twist to Monkhouse's necromancy has yet to be noted. Monkhouse was not only appearing and communicating to his audience "from beyond the grave"; he was also doing so "from beyond the grave" *with an open acknowledgement of his death...*when *dead.* In other words, this posthumous appearance is not made in the conventional sense of an afterlife, which typically refers to life after death in the "next" world, be that heaven, hell or otherwise – hence the significance of "speaking to us from beyond the grave" as a communication from that "next" world. The twist here is that Monkhouse's posthumousness is not merely of life after death, but of life after death in a mediated and *amplified* form. This form is key to the import (and impact) of Monkhouse's cancer warning which was delivered in an "after*life*" that is arguably more powerful, more vital and more dominant in death than in life – in short: *being more alive* when *dead.*[34] In such a reading, death becomes, in Jacques Derrida's terms, "a sort of non-event, an event of nothing or a quasi-event which both calls for and annuls a narrative account."[35] In his essay, "Before the Law," in which Derrida reads Kafka's fable of the same title, Derrida takes on Freud's account of the murdered father (by the son) to essentially return law to fiction. The father's death, out of which law is instated ("the two fundamental prohibitions of totemism, murder and incest"), becomes a non-event, or an "event without event," precisely because *the dead father holds even more power* [*than when he was alive*]"; he is "more dead alive than *post mortem.*" (emphasis added)[36] The significance of Monkhouse's posthumous warning is thus drawn not only from the necromancy of the CGI wizardry[37] in "re-assembling" him after

---

34 See also Ackbar Abbas's account of this posthumousness in his lecture, "Posthumous Socialism," October 23, 2017, Global University for Sustainability, https://our-global-u.org/oguorg/en/professor-ackbar-abbas-posthumous-socialism/. We may also recall echoes of such posthumousness from Walter Benjamin's essay, "The Task of the Translator," where he describes a translation as issuing "not so much from its life as from its afterlife," "mark[ing] their [i.e., the works of literature] stage of continued life." (*Illuminations: Essays and Reflections*, trans. Harry Zohn, ed. Hannah Arendt (New York: Schocken Books, 1968), 254) In this context, Benjamin, too, describes vis-à-vis translation a posthumousness (of the original work) which takes on a certain more-alive-when-dead vitality, where "in [these translations] the life of the originals attains its latest, continually renewed, and most complete unfolding." (255)

35 Jacques Derrida, "Before the Law," in Jacques Derrida, *Acts of Literature*, ed. Derek Attridge (New York; London: Routledge, 1992), 198.

36 Ibid. Hence, "the origin of the law" is also a non-event, or "nothing new happens": "Thus morality arises from a useless crime which in fact kills nobody, which comes too soon or too late and does not put an end to any power. ... In fact, it inaugurates nothing since repentance and morality had to be possible *before* the crime." (emphasis in original), ibid.

37 Note how wizardry may also connote (black) magic associated with necromancy and the supernatural: see, for instance, Zakiya Hanafi, *The Monster in the Machine: Magic, Medicine, and*

his death, but from this amplified power of being "more dead alive than *post mortem.*" In this sense, CGI heralds an important turning point regarding the spectral vitality of images (and one to which we will again turn in relation to the post-screen through holographic projection) – a vitality not just of the afterlife from re-animations and re-constitutions of the dead, but also the specific more-alive-*when*-dead after*life* of digital imagery.

Bob Monkhouse was a sign of what is to come. Numerous other advertisements have since followed the format, with the dead returning to "endorse" modern products or be placed in modern settings anachronistic to their lifetimes. These were, of course, commercial backings to which the dead actors had never actually agreed or been contracted, thereby also putting a late twentieth-century neoliberal spin to necromancy, subjecting even communications from the dead to market capitalism.[38] In 2014, a 19-year-old Audrey Hepburn was re-created with body doubles and CGI effects to appear in a chauffeured open-top car along the Amalfi coast, brandishing a Galaxy chocolate bar.[39] Whisky company Johnny Walker did the same with Bruce Lee, who had died in 1973, re-creating him for a 2013 advertisement of Lee in contemporary Hong Kong via performance capture of a body double, digital facial mapping and CGI technologies.[40]

the Marvelous in the Time of the Scientific Revolution (Durham, NC: Duke University Press, 2000), for a treatise specifically on the relationship between magic, nature and monsters. Carrying from that, connotations of magic also invoke cinema's own myth as a "medium for magic in modern times": see Rachel O. Moore, *Savage Theory: Cinema as Modern Magic* (Durham, NC: Duke University Press, 1999).

38 The issue of consent, or ethics of posthumous usage rights, is indeed a pressing one: is there a legal and/or moral obligation to seek consent from the deceased's estate in using their recorded images and voices to "reconstitute" the deceased, particularly for profitmaking enterprises? See, as one article out of numerous, Laura Barton, "Back From the Black: Should Amy Winehouse and Other Stars be Turned into Holograms?," *The Guardian* online, October 19, 2018, https://www.theguardian.com/music/2018/oct/19/amy-winehouse-stars-turned-into-hologram-virtual-reality, which sums up the issue thus: "The decision to turn [Amy Winehouse] into a virtual reality experience has divided fans. Is she being vividly celebrated, or ghoulishly pushed back on stage without her consent?", np. See also Kirsten Rabe Smolensky, "Rights of the Dead," *Hofstra Law Review* 37(3) (2009): 763-803. Prior to his death in 2014, the actor Robin Williams apparently established a trust to which, on his death, he passed on his name, signature, photograph and, most importantly, likeness. This makes it unlawful, without the trust's consent, to recreate the last with CGI for any posthumous appearances: Benjamin Lee, "Robin Williams restricted use of his image for 25 years after his death," *The Guardian* online, March 31, 2015, https://www.theguardian.com/film/2015/mar/31/robin-williams-restricted-use-of-his-image-for-25-years-after-his-death.

39 Mike McGee, "How we resurrected Audrey HepburnTM for the Galaxy chocolate ad," *The Guardian* online, Oct 8, 2014, https://www.theguardian.com/media-network/media-network-blog/2014/oct/08/how-we-made-audrey-hepburn-galaxy-ad.

40 This performance capture apparently included 250 different expressions taken from the grid of the body double actor, Danny Chan Kwok Kwan: see *South China Morning Post*, "Video:

In films today, CGI re-creations of dead actors are likewise commonplace. Sometimes, it is a way to complete scenes on actors' deaths mid-way through shooting. Examples include the recreation of the actor Brandon Lee after his tragic death while filming Alex Proyas's *The Crow*,[41] with images lifted digitally from previously recorded scenes and superimposed onto subsequent ones. The actor Oliver Reed, who also died in the middle of the film shoot, had digital renderings of his face added to body doubles in *Gladiator* to complete later scenes;[42] the same kinds of CGI work were applied to Paul Walker for *Fast and Furious 7*.[43] On other occasions, digital recreation, especially of famous dead actors, is used to publicize films in highlighting their technological wizardry. Examples include digitally combining archival footage of a long-dead Laurence Olivier with that of a body double for *Sky Captain and the World of Tomorrow*,[44] as with Marlon Brando for *Superman Returns*.[45] The most high-profile case to date is probably the "necromantic cinematic feat" of re-creating Peter Cushing to "reprise" the character he had played in the original 1977 film for the *Star Wars* sequel, *Rogue One: A Star Wars Story*.[46] The result was Cushing, "looking" relatively alive and well, seen onscreen "acting" "near seamlessly" with contemporary actors in the sequel, "even though he has been dead for more than 20 years."[47]

On television, the years 2006-2007 featured a rash of what has been termed "posthumous duets," where dead musicians are shown "performing" in real-time with living musicians.[48] The duet typically takes place first with

Bruce Lee stars in a Johnny Walker advertisement," YouTube video, 5'11", July 10, 2013, https://www.youtube.com/watch?v=tz25tcFSusc. Again on the issue of consent from dead actors, see also the phenomenon of "Bruceploitation" from Brian Hu, "'Bruce Lee' after Bruce Lee: A Life in Conjectures," *Journal of Chinese Cinemas* 2(2) (2008): 123-135.

41   *The Crow*, directed by Alex Proyas (1994; London: Entertainment in Video, 2003), DVD.

42   *Gladiator*, directed by Ridley Scott (2000; Los Angeles, CA: Universal Pictures, 2004), DVD.

43   *Fast and Furious 7*, directed by James Wan (2015; Los Angeles, CA: Universal Pictures, 2015), DVD.

44   *Sky Captain and the World of Tomorrow*, directed by Kerry Conran (2004; Los Angeles, CA: Paramount (Universal Pictures), 2015), Blu-Ray.

45   *Superman Returns*, directed by Bryan Singer (2006; Burbank, CA: Warner Home Video, 2006), DVD.

46   *Rogue One: A Star Wars Story*, directed by Gareth Edwards (2016; Burbank, CA: Walt Disney Studios Home Entertainment, 2017), DVD.

47   Joseph Walsh, "Rogue One: the CGI resurrection of Peter Cushing is thrilling – but is it right?," *The Guardian* online, Dec 16, 2016, https://www.theguardian.com/film/filmblog/2016/dec/16/rogue-one-star-wars-cgi-resurrection-peter-cushing.

48   See Shelley D. Brunt, "Performing Beyond the Grave: The Posthumous Duet," in *Death and the Rock Star*, eds. Catherine Strong and Barbara Lebrun (New York; London: Routledge, 2016): 165-176. Indeed, music has had a longer history (*cf* cinema) of combining dead and living singers,

the living performer and a tribute artist as a body double. The recording is then digitally edited and manipulated so that what viewers ultimately see on the broadcast programme is the dead singer appearing to be "live" onstage with the living performer.[49] Writing on such posthumous collaborations, Stanyek and Piekut note these "new arrangements of interpenetration between worlds of living dead" as the "intermundane," where "the living do not one-sidedly handle the dead, but participate in an inter-handling, a mutually effective co-laboring."[50] Again, the afterlife of CGI ghosts persists here as more-alive-when-dead: theirs is not a passive haunting, but a sense of being in some kind of active partnership.

As the twenty-first century dawns on the rise of social and interactive media, the paradigm for the dead onscreen shifts once more – from their resurrection in the production of nineteenth century images to the necro-philia of twentieth-century cinema to the necromancy of CGI composites at its turn and, now, to the ephemerality of performance and interactivity in the twenty-first. Not so do ghosts arise out of preservation via the image or, in Bazinian terms, their mummification.[51] Rather, in the digital age of being "always on," the constant (inter-)activity of social media – posting, updating, uploading, re-tweeting, chatting, commenting, tagging, liking and so on – becomes the marker of life and afterlife.[52] Hence, ghosts return

or at least their recordings thereof: Stanyek and Piekut suggest that "the first posthumous collaboration occurred when dead tenor Enrico Caruso's voice from a 1907 recording was dubbed with live musicians for a 1932 release": see Jason Stanyek and Benjamin Piekut, "Deadness: Technologies of the Intermundane," *TDR: The Drama Review* 54 (1) (2010): 14-38, as quoted in Brunt, "Performing Beyond the Grave," 167.

49   In 2006, the BBC television programme, *Duet Impossible*, shown over the year's Christmas season, presented numerous posthumous duet pairings, including The Sugababes with Dusty Springfield (1939-99) for "Dancing in the Street"; Katie Melua with Eva Cassidy (1963-96) for "Somewhere over the Rainbow," and so on: Brunt, "Performing Beyond the Grave," 171. In April 2007, an episode of *American Idol* broadcast a similar posthumous duet between Celine Dion and Elvis Presley on the thirtieth anniversary of the latter's death, combining an on-stage body double of a tribute artist for wide shots with rotoscoping from previous recordings to replace the double's face with images of the real Elvis: ABC News, "Elvis on 'Idol:' How It Was Done," *ABC News*, January 8, 2009, https://abcnews.go.com/GMA/story?id=3087711.

50   Stanyek and Piekut, "Deadness," 14.

51   See text accompanying footnote 24 of this chapter.

52   Sometimes, this "marker of life" becomes literal, such as the case of a young mother whose family alerted police of her being in danger only after not receiving the mother's daily Facebook updates of her baby: see Jessica Green, "Mother,19, lay dead in her flat for two days before her body was found next to her dehydrated five-month-old son after fears grew because she hadn't posted her daily Facebook baby pictures," March 3, 2019, *MailOnline*, https://www.dailymail.co.uk/news/article-6766231/Mother-lay-dead-flat-two-days-body-five-month-old-baby.html. Conversely, see Slavoj Žižek's ideas of "inter-passivity" (as the "uncanny double" of interactivity):

to the living via their digital traces of emails, text messages, selfies, videos and other social media postings. Moreover – and here heralding again the ghosts of the post-screen – they return onscreen to *interact* with the living, rather than just appearing before them. Speculative apps, such as DeadSocial, promise to enable clients to "while still alive, write and schedule messages that will be pushed to your social media accounts after [their] death[s]."[53] One particularly poignant account describes the dead returned to interact with the living through video gameplay. In 2014, an appeal, posed through a YouTube video by PBS Game/Show, asked the question "can video games be a spiritual experience?" The appeal attracted a response from a user who related how he and his father used to play Xbox games together before the latter died when the user was six years old. Years later, the user, now an adult, reloaded a racing game called RalliSport Challenge which he used to play with his father. He discovered that the game had saved his father's winning lap as a ghost car so that even after the ensuing years, as he puts it, "his [father's] ghost still rolls around the track today";[54] his father is "back" to play against him. The user ends his story by describing how he "played and played, and played" until he was nearly able to beat the ghost car, at which point he "stopped right in front of the finish, just to ensure [he] wouldn't delete it."[55] The interactive ghost of the father thus lives on, and never loses the race. That ghost also indicates another kind of break between audiovisual recording and interactive media: the dead returning not as an image of likeness, but as an *event* which affects the present.

Increasingly, the mobile phone – with its small, portable and connected screen – replaces the big, immobile and isolated[56] screen of cinema as the threshold to ghosts, as the bereft use "the emotional and affective power

---

"The obverse of interacting with the object…is the situation in which the object itself takes from me, deprives me of, my own passivity, so that it is the object itself which enjoys the show instead of me, relieving me of the duty to enjoy myself": see Žižek, *How to Read Lacan* (New York: WW Norton & Company, 2007), 28.

53 Sidney Fussell, "This site lets you control your social media profiles after you've died," *Business Insider*, May 10, 2016, https://www.businessinsider.com/what-is-dead-social-2016-5?r=US&IR=T.

54 As quoted and reported in Mary-Ann Russon, "Ghost in the Machine: Teenager Races Against Deceased Father's Ghost Car in Old Xbox," *International Business Times*, July 24, 2014, https://www.ib-times.co.uk/ghost-machine-teenager-races-against-deceased-fathers-ghost-car-old-xbox-1458145.

55 Ibid.

56 Although see Franco Casetti, "Return to the Motherland: the film theatre in the postmedia age," *Screen* 52:1 (2011): 1-12 for analysis on, among others, the connectedness of large and small screens in modern film viewing practices.

of the mobile"[57] to memorialize, grieve and remember the dead through their social media postings, chat and other interactive forms. In the 2013 television episode, "Be Right Back," of the Charlie Brooker-created *Black Mirror* science-fiction series, a deceased character, Ash, returns to "life" first as an onscreen presence via text messages and phone conversations with his widow.[58] An "upgrade" takes the form of Ash returning in a synthetic body with software which specifically "learns" his speech and behaviour patterns from his social media postings so as to speak and act like him.[59] Where, in the age of digital reproduction, CGI and digital effects were used to manipulate and re-assemble audiovisual recordings of the dead, the science-fiction setting of "Be Right Back" imagines re-production anew out of social and interactive media, where a loved one returns from the dead at the front door, in synthetic flesh and with algorithmically produced behaviour.

A history of ghosts from screens can thus be drawn across four fundamental moments – resurrection; necrophilia; necromancy; and interactivity. In each instance, *screens actively facilitate ghosts*: they not only define the boundary between life and death inviting return and re-appearances, but colour their interrelations and the meanings of each and of the afterlife. As the rest of this chapter will show, the eroded boundaries of the post-screen through holograms and holographic projections herald different kinds of spectral life, being and bodies in different interactions with the living. The dead continue to return, and edge ever closer to us.

The Post-Screen Through Holograms/Holographic Projections

On September 7, 1996, American rapper and actor Tupac Shakur (1971-1996) was shot in a drive-by shooting and died six days later in hospital. Young, controversial and "wildly popular," his death was mourned by millions of rap fans.[60] On April 15, 2012, Tupac "returned" from the dead, "appearing" on the

---

57 Kathleen Cumiskey and Larissa Hjorth, *Haunting Hands: Mobile Media Practices and Loss* (New York: Oxford University Press, 2017), 121.

58 *Black Mirror*, "Be Right Back," season 2, ep. 1, *Channel 4*, first aired February 11, 2013, written by Charlie Brooker, directed by Owen Harris.

59 At the episode's ending, Ash is eventually consigned to the attic with the family's other buried secrets and forgotten antiques. There is a death, yet, to the afterlife from a social media resurrection: to be ignored, dismissed, and cast out from all interaction with the living.

60 See Pamela Constable, "Rapper Dies of Wounds From Shooting," *Washington Post*, September 14, 1996, https://www.washingtonpost.com/news/arts-and-entertainment/wp/2016/09/13/tupac-shakur-died-20-years-ago-today-read-the-front-page-story-from-1996/?noredirect=on&utm_term=.0c70c3b8c5a8. See also generally Michael Eric Dyson, *Holler If*

stage at the Coachella Valley Music and Arts Festival by performing his song "Hail Mary" before entering into a duet with rapper Snoop Dogg to perform "2 of Amerikaz Most Wanted." Around a year and a half later on September 6, 2013 and across the globe from Coachella, mega Taiwanese star Jay Chou likewise sang a duet at his live concert in Taipei with a "resurrected" Teresa Teng (1953-1995), another Taiwanese singer immensely popular across East and Southeast Asia from the 1970s to 1990s and who had died of respiratory illness in 1995 to similar outpourings of grief from fans around the world.[61] Both concert appearances immediately became sensations across social and reported media.

While many media accounts referred to these concert illusions as holograms, as many correctly clarified that they were not true holograms. Rather, they were optical illusions projected onstage through CGI, strategic lighting and clever use of reflections and surfaces, creating *a holographic effect* whereby the dead singer appeared naturally (i.e. viewed without any 3D glasses or optical visual aids) in colour, sound, three-dimensions, animation and thus with an overall relative amount of realism. The base technique for these holographic illusions first appeared in the nineteenth century, debuting in the Christmas season of 1862-3 at the Royal Polytechnic in London's Regent Street precisely to project ghosts. Dubbed "The New Phantasmagoria" (as a sequel to the lantern-based Phantasmagoria show in the 1790s by the Brabantine conjuror, Paul Philidor, mentioned earlier), the show became an instant hit: "If ever there was a 'boom' that was one. Within a few months ghosts sprang up at theatres and halls in nearly all parts of London."[62] This technique, then, is the aptly named Pepper's Ghost, an illusion jointly patented in 1863 by two English engineers, John Henry Pepper and Henry Dircks (although the financial rights and, as is evident today, the fame of the name accrued to Pepper).

Here is how to create a Pepper's Ghost projection: a transparent reflective pane, such as one made out of glass or clear plastic, is tilted towards the

You Hear Me: Searching for Tupac Shakur (New York: Basic Civitas, 2001). As Regina Arnold writes, citing Dyson, "Shakur is widely considered the world's most influential rapper, and one of the most important figures in the genre's brief history": see "There's a Spectre Haunting Hip-hop: Tupac Shakur, Holograms in Concert and the Future of Live Performance," in *Death and the Rock Star*, eds. Catherine Strong and Barbara Lebrun (New York; Oxon: Routledge, 2015): 177-189, 179.

61   CGTN, "23 years on: Death can't do Teresa Teng and her Chinese fans part," *China Daily*, January 29, 2018, http://www.chinadaily.com.cn/a/201801/29/WS5a6edc6aa3106e7dcc1376ad_1. html.

62   R. McDonald Rendle, *Swings and Roundabouts: A Yorkel in London* (London: Chapman and Hall, 1919), 184-15.

audience at the front of a stage. With appropriate darkness onstage, it thus appears invisible to the audience. The subject of the illusion, by way of an actor or a projected image, is positioned at an angle below the pane, concealed from the audience and brightly illuminated. The result is that the tilted pane reflects the light rays bouncing off the brightly lit subject to the audience. Due to the transparent pane's reflective qualities combined with the darkness onstage, the audience sees the subject as a mirror reflection appearing *behind* the pane. With no knowledge of its actual existence below the stage as hidden from their view, the audience thus sees the subject as having magically appeared out of thin air, achieving its ghost-like effect.

The import of the post-screen through the Pepper's Ghost illusion is that the projected image appears *with no visible boundaries around it*, so that the virtual image merges seamlessly with its surrounding actual reality. Set as a performance onstage, the ghost and the living share both the same stage-space and the same presentness. *They share the post-screen.* And where screen boundaries contain the haunted image, the post-screen through real-time performative holographic projections thus re-draws the boundaries between life and afterlife. Specifically, it drives two kinds of spectral life: those *amongst* the living, and those *of* the living.

Holographic Projections (1): Ghosts Amongst the Living – Limbo Between Deadness and Aliveness

Without screen demarcations, real-time holographic projections of the dead in the post-screen appear as *ghosts amongst the living.*[63] They become the mediated equivalent of a supernatural return, typically ritualized in myriad cultures from the *Día de Muertos* (Day of the Dead) in Mexican culture to the Gaelic festival of Samhain to Obon, a traditional Buddhist festival celebrated as the Japanese Day of the Dead. In this sense, Singaporean media scholar Liew Kai Khiun, too, describes the holographic projections of Teng as *ying hun*, or "the Chinese concept of 'shadow and soul,'" which "provide[s]

63 This effect acquires particular leverage in the UK Channel 4-commissioned programme series, *Ghost*, which, according to Channel 4's press release, uses "cutting edge holographic technology" to record "six terminally ill people" so that they may re-appear to their loved ones "in vivid, three-dimensional, holographic form, allowing them to appear as if from beyond the grave": see "Channel 4 announces Specialist Factual Commission, Ghost," *Channel4.com*, January 30, 2019, https://www.channel4.com/press/news/channel-4-announces-specialist-factual-commission-ghost-wt. As of writing, it is unclear, though, if this programme series was or will ever be released.

the sensation of immanence that is not alien to Sinitic religious cultures accustomed to marking the presence of departed relatives and venerated ancestors returning as 'ghosts' during festive periods."[64] As does Ralph et al., who interpret Tupac's "hologram" as part of the Afro-Atlantic ritual tradition of *palo monte*, which facilitates the return of the dead in vicious subservience pacts, a reading ultimately and rightly concerned with the posthumous exploitation of Tupac.[65]

But the "border-less" post-screen space of holographic projections more than rubs out the distinction between the world of the living and that of the dead for their intermingling; it also facilitates the real-time interactivity which increasingly colours twenty-first century ghosts. In the cases of Teng and Tupac, their ghostliness is imbued with an alive-ness arising not from the frenzy of social media activity, but from the *liveness* of the setting of performance, where performance, as is life itself, exists in the ephemerality of the present.[66] Performance also takes on the vitality and presence of its onstage bodies. As Henri Gouhier writes in *The Essence of Theater*:

> The stage welcomes every illusion except that of presence; the actor is there in disguise, with the soul and voice of another, but he is nevertheless

---

64   Liew Kai Khiun, "Shadow and Soul: Stereoscopic Phantasmagoria and Holographic Immortalization in Transnational Chinese Pop," in *Asian Perspectives on Digital Culture: Emerging Phenomena, Enduring Concepts*, eds. Sun Sun Lim and Cheryll Soriano (New York; London: Routledge, 2016): 152-168, 165-166. Actually, rather than being marks of Sinitic culture, these Ghost Festivals as observed to mark "wandering ghosts" have their origins in Taoist and Buddhist religions, such as the Mahayana scripture. Variations of these festivals in celebrating and appeasing the return of ancestral spirits are also observed in Asian countries as diverse as Japan (*Chugen*), Indonesia (*Sembahyang Rebutan*), Vietnam (*Tết Trung Nguyên*), Cambodia (*Pchum Ben*) and Sri Lanka: see generally Chow Shu Kai, *Investigation of Ghost Month – Zhong Yuan, Ullambana and Hungry Ghost Festivals* (Hong Kong: Chung Hwa Books, 2015); and Rita Langer, *Buddhist Rituals of Death and Rebirth: Contemporary Sri Lankan Practice and Its Origins* (New York; London: Routledge, 2007).

65   This would be the reading of the "enslaving [of] Tupac's digitized remains," where his hologram becomes the embodiment in which the dead rapper is harnessed to perform/make money for his master who has "provided them with a new form of sublunary existence and the capacity to once more agentively interfere in the world of the living": see Michael Ralph, Aisha Beliso-De Jesús and Stephan Palmié, "Saint Tupac," *Transforming Anthropology* 25(2) (2017): 90-102, 97.

66   Peggy Phelan's oft-quoted definition comes to mind: "Performance's only life is in the present. Performance cannot be saved, recorded, documented, or otherwise participate in the circulation of representations of representations: once it does so, it becomes something other than performance": see Peggy Phelan, *Unmarked: The Politics of Performance* (London: Routledge, 1993), 146.

there and by the same token space calls out for him and for the solidity of his presence.[67]

Performance thus asserts life through both the presentness and physicality of bodies; it "implicates the real through the presence of living bodies."[68]

In that sense, the holographic ghosts of Teng and Tupac in the post-screen appear to be not only amongst the living, but also to be peculiarly alive as pseudo-bodies performing live with living performers and in front of a living audience.[69] The projection of Tupac, for instance, "faced" the audience; was choreographed to be "rapping" to them; and even "addressed" them with suitably expletive-laden greetings and gestures. With no visible boundaries between them, the virtuality of the image merges with the actuality of the stage event; the deadness of one amalgamates into the aliveness of the other. As with the interactivity of social ghosts, Teng and Tupac also "participated" in real-time "interactivity" with the living performers. In their duet, Snoop Dogg and Tupac appeared reciprocal with "exchanged" glances, "shared" conversation snippets and "synchronized" dance movements. Teng and Chou, while keeping further apart, also frequently "gestured" to each other and "exchanged" gazes during their duet. Stanyek and Piekut, picking up on the familiar critique of posthumous exploitation, write of how this liveness of ghostly performance "is nothing more than a transitive effect of deadness, and deadness is nothing more than the promise

---

67   As cited in Bazin, *What is Cinema Vol. 1*, 95.

68   Phelan, *Unmarked*, 146. Phelan's ideas of the body in performance are actually more complex than mere presence: she subsequently writes about how that body – "metonymic of self, of character, of voice, of 'presence'" – also "disappears and represents something else – dance, movement, sound, character, 'art,'" and so on (150). Having said that, the body in performance is also changing as robot artists and performances become increasingly sophisticated. See, for example, Ai-da, "the world's first ultra-realistic humanoid AR robot artist," who also performs as a performance artist: https://www.ai-darobot.com/jointhemovement.

69   In a separate, though related, argument, we can extend this thinking into how holographic projections in performance can also signal the liveness of *performers who had never been alive*. This could refer to performers who are entirely fictional constructs, such as the animated Japanese pop star, Hatsune Miku, who "sings" via a software voicebank and typically appears as a young girl whose most distinctive features are two "long, turquoise blue" twintails (see, for e.g., Cara Clegg, "Is Hatsune Miku without twintails still Hatsune Miku," *SoraNews24*, May 8 2015, https://soranews24.com/2015/05/08/is-hatsune-miku-without-twintails-still-hatsune-miku/). At concerts, Miku is "brought to life" and onto stage via 3D projections, notwithstanding she does not have a corporeal existence. Another possibility of "performers-who-had-never-been-alive" is virtual bands, such as the British music band Gorillaz, who, while they are voiced by humans, usually take the visual form of animated characters which, in live concerts, appear as holographic projections: see Hofer, "Metalepsis in Live Performance," 232-251.

of recombinatorial and revertible labor."[70] Yet this deadness could only have been recombined and reverted in its labour because it exists in the post-screen with redefined boundaries between deadness and aliveness, and a re-drawing of our imaginations of ghosts through which the dead return, appearing increasingly more alive *when* dead.

Accommodating both the dead and the living with no apparent separation between them, the post-screen onstage becomes a curious space in limbo, not quite of one realm nor the other. The echoes here with similarly neither-dead-nor-alive limbo figures, such as zombies and vampires, are unmistakable. Representations of zombies and vampires consistently present tensions between being human and non-human: on one hand, they display human-like visages and movement; on the other, they originate from the dead and are marked with non-human characteristics such as uncontrollability or thirst for blood. Ultimately, these stories interrogate our identity as humans, and lay bare the muddied lines between the dead and the living, just as post-screen spaces underscore their same re-scrambled boundaries. As with how cinema and photography changed our understanding of time and space, the post-screen through holographic projections and their accompanying problematizing of boundaries similarly engages in these larger issues. It re-interrogates the conceptual placement of the dead and the living, a question which correspondingly gains increasing relevance in tandem with medical advancement such as life-support technologies.[71] The post-screen thus reflects the complexities in the binaries between the dead and the living as with the virtual and the actual, and, ultimately, where and how living and aliveness should be defined.

A postscript, then: relating to the space of limbo between the dead and the alive, we might also recall a short story to illustrate that mysterious in-betweenness – Edgar Allan Poe's "The Facts in the Case of M. Valdemar," first published in December 1845, and also already mentioned in chapter 1.[72] In the story, its unnamed narrator regales the reader with an account of how he had hypnotized his friend, M. Valdemar, at precisely such a disjuncture between life and death, namely, the moment just before he passed. M. Valdemar then reposed in that unholy suspension – "mesmerized in articulo

---

70 Stanyek and Piekut, "Deadness," 32.

71 See Sarah O'Dell, "'The Facts in the Case of M. Valdemar': Undead Bodies and Medical Technology," *Journal of Medical Humanities* 41 (2020): 229-242.

72 Edgar Allan Poe, "The Facts in the Case of M. Valdemar," in *Collected Works of Edgar Allan Poe, Volume III: Tales and Sketches 1843-1849*, ed. Thomas Ollive Mabbott (Cambridge, MA: Belknap Press, 1978): 1233-43.

mortis"[73] – for seven months, appearing as if in sleep, albeit still breathing, responding faintly to questions, and with flesh intact.

The concert "resurrections" in the post-screen similarly appear in such delicate temporary suspension between present and past, life and death. On one hand, they appear alive in the same space as the living – they speak, sing and dance. As with M. Valdemar, their bodies are unimpaired by decay and other processes of death. They interact with the living, if only for three to four songs. On the other hand, Poe's story might offer a more fitting conclusion: seven months after hypnotizing Valdemar at the point of his passing, the narrator finally awakens his friend, only for Valdemar to collapse into a heap of heavily decayed, rotten, putrid flesh. There is, perhaps, yet an unseen and unheard coda to performances by the returned dead.

Holographic Projections (2): Ghosts of the Living – Vivification of the Virtual Real

While Pepper's Ghost illusions of *the dead* give rise to ghosts being amongst the living, its projections of *the living* emanate another kind of spectral life – namely, ghosts *of* the living. In 2006, six years before Tupac's "resurrection" at Coachella and possibly as the earliest display of its kind, a 3D holographic projection of British model Kate Moss (not yet dead) closed out Alexandra McQueen's "Widows of Culloden" fashion show in Paris.[74] Widely publicized and received with acclaim at the time, the holographic display was brought back in 2015 as part of the "Savage Beauty" exhibition of McQueen's work at the Victoria and Albert Museum in London. Per its original Victorian technique, painstaking efforts were made to conceal the screen's presence so that the holographic image of Kate Moss did not appear to be contained within any boundaries. The image of Moss thus emerged *in the post-screen*, with no visible separation from the audience's space. As the projection production designer noted:

> The swirly material of [Moss's] dress meant that we needed to do a lot of masking and distortion to make sure no one saw where it stopped and

---

73 Poe, "The Facts," 1233.

74 Véronique Hyland, "Alexander McQueen's Kate Moss Hologram Is Being Revived," *The Cut*, October 9, 2014, https://www.thecut.com/2014/10/mcqueens-kate-moss-hologram-is-being-revived.html.

the illusion wasn't shattered...One of the big challenges was making sure you didn't see the edge of the frame of the Kate Moss footage.[75]

In this instance, the subject is not dead. The concealment of screen boundaries per the post-screen thus re-draws relations not between the dead and the living, but across another kind of estrangement – physical distance. Unlike an image on a two-dimensional screen contained within visible boundaries, the holographic subject appears to be *virtually here* amongst the present audience even as they are *actually elsewhere* at the time. The realness of the holographic body in the post-screen can thus be construed in a tetravalence across dual axes – actual/virtual; here/elsewhere. In the unbounded space of the post-screen, the holographic body is melded across this tetravalence, amalgamating shreds of the corporeality of its actual body from its *elsewhere* across distance off-stage with the immateriality of its virtual body *here* onstage. Navigating across this tetravalence, the holographic body of the living thus takes on not the life of resurrection, but of another kind of spectral vitality. As with twenty-first century ghosts, this vitality manifests via real-time interactivity with the audience and qualities of the virtual body in volumetric, sensual and realistic terms. I call this enlivenment of the virtual real in the post-screen its *vivification*; its quality of vitality its *vividness*; and the resulting projection as a spectre of the living, the *vivified*.

As with holographic illusions of the dead, such vivification of the living in holographic projection derives primarily from the interactivity of live performance and stage settings. A prime example is American singer Mariah Carey's Christmas concert in November 2011 for a Saatchi & Saatchi campaign created for Deutsche Telekom, in which she appeared as a 3D holographic projection to "perform" simultaneously in five European cities.[76] As with Tupac, the "alive-ness" of Carey as a holographic performer was clinched

75   Press release by production designer Joseph Bennett, "Creative Review: How We Made Alexander McQueen's Kate Moss Hologram," no date, http://josephbennett.co.uk/press/creative-review-how-we-made-alexander-mcqueen-s-kate-moss-hologram.

76   These cities were Zagreb, Podgorica, Krakow, Skopje and Frankfurt. The concert was part of T-Mobile's advertising campaign to "boost awareness across Eastern Europe" during the Christmas run-up, which is key to sales: see press release by Musion Holograms Pvt. Ltd, no date, http://musion.co.in/indexfcbe.html?portfolio=mariah-carey-hologram. Carey did not appear as a true hologram, but as a 3D holographic image; yet the media, as with Tupac, conflates the two: see, for eg, The Drum, "Mariah Carey hologram performs simultaneously in five cities for Deutsche Telekom campaign," *The Drum*, November 18, 2011, https://www.thedrum.com/news/2011/11/18/mariah-carey-hologram-performs-simultaneously-five-cities-deutsche-telekom-campaign.

in her "interactivity" with the audience. At one point in the "performance," the holographic image of Carey "addressed" the audience:

> This is incredible. At the exact same time, I am simultaneously live in Poland, Germany, Macedonia, Croatia and Montenegro. It feels like the whole entire universe! It's team effort. We're all connected. Isn't this great? Say hiiii!"[77]

At the last, the projected Carey "inverted" the microphone and "pointed" it to the audience, a familiar gesture by live performers as a signal for the audience to respond. In this case, notwithstanding she was not actually on the stage, nor was the microphone actually pointed at them, nor was it even actually there, the audience responded with an actual and enthusiastic collective "hi" back to the image in the realness of Carey's presence, *as if she was actually there.* In a sense, responding to images onscreen is a common reaction, such as shouting at sports referees on television even though they are obviously elsewhere. In this case, though, the "here-ness" of the living Carey emerges from not only the context of performance, but also the post-screen. With no obvious differentiating boundaries around the image against its surroundings, the figure appears to be alive and *with* the audience in space and time. Their interactivity in this space thus draws on a certain kind of spectral life not out of the dead but of the living, of whom their holographic projections on the post-screen are neither of the alive nor the dead, but of the *vivid.*

As with Bob Monkhouse for necromancy, Mariah Carey's 2011 concert in Europe was a sign of the times for vivification, where various subsequent examples of vivified interaction are reinforced by the paradox of their virtual holographic bodies *here* onstage against their actual bodies *elsewhere.* In the April 2019 Indonesian national elections, then incumbent president Joko Widodo (who later won the election) had 3D holographic images of himself projected onto invisible mobile screens, addressing crowds across the country's vast archipelago containing 190 million registered voters.[78] Similar to Carey's concerts, the post-screen vividness of Widodo was palpable in the interactions between "him" and his audience. In one rally, his projected

77   See GSMOnline.PL, "T-Mobile – Mariah Carey holographic concert Poland Cracow making of Christmas TV ad," YouTube video, 4:55, November 17, 2011, https://www.youtube.com/watch?v=d13eeV5j_uo at 2'59".

78   Agence France-Presse, "President's hologram hits Indonesia's election campaign trail," *The Guardian* online, March 28, 2019, https://www.theguardian.com/world/2019/mar/28/president-hologram-indonesia-election-campaign-joko-widodo.

image "asked": "In a few days, it will be time to choose. Do you want to choose an optimist, or choose a pessimist?" As with Carey's "say hi" request, his audience similarly responded enthusiastically: "Optimist!"[79] Widodo was only the latest in a string of politicians who used holographic imagery for greater public outreach. The current Indian Prime Minister Narendra Modi similarly used projection technology in many of his election rallies between 2012 and 2014, where his likeness was projected "live, in 3-D" to appear and sometimes give speeches. Reports describe his "lifelike performance" to have been "greeted with a mix of awe and disbelief," with voters often staying behind after rallies "to check behind the dais to see if he was really there."[80]

There are also other kinds of "interactions" with holographic projections. In 2014, Julian Assange, at the time resident at the Ecuadorian embassy in London under political asylum,[81] "appeared by hologram" for an interview with filmmaker Eugene Jarecki at The Nantucket Project, an annual conference in the eponymous town in Massachusetts in the United States "about what matters most."[82] "Beamed in" from the embassy, the organisers

79   Saifulbahri Ismail, "Indonesia's presidential candidates turn to holograms to reach out to supporters," *ChannelNewsAsia* video, 3:31, April 11, 2019, https://www.channelnewsasia.com/news/asia/indonesia-s-presidential-candidates-turn-to-holograms-to-reach-11430794. There are many other instances of holographic projections by politicians, though whose responses by the audience is not as obvious: in 2014, the Turkish Prime Minister Recep Tayyip Erdogan projected a "towering, photon-based figure" of himself to address a political party meeting (see Matt Ford, "Giant Hologram of Turkish Prime Minister Delivers Speech," *The Atlantic*, January 27, 2014, https://www.theatlantic.com/international/archive/2014/01/giant-hologram-of-turkish-prime-minister-delivers-speech/283374/). In 2017, French politician Jean-Luc Melenchon projected himself making a speech at a rally in Paris while launching his presidential campaign at another rally in Lyon at the same time (see Maya Nikolaeva and Catherine Lagrange, "France's far-left Melenchon uses hologram to spread election message," *Reuters*, February 5, 2017, https://www.reuters.com/article/us-france-election-idUSKBN15K0Q7).

80   "Magic Modi," as he became known, first used holographic projection technology in the 2012 Gujarat state assembly election and again in the 2014 general election in India to become Prime Minister, by the end of which the estimate was that he had appeared as a "hologram" at 1,450 rallies, reaching "more than 14 million extra voters": see Dean Nelson, "'Magic' Modi uses hologram to address dozens of rallies at once," *The Telegraph*, May 2, 2014, https://www.telegraph.co.uk/news/worldnews/asia/india/10803961/Magic-Modi-uses-hologram-to-address-dozens-of-rallies-at-once.html.

81   Julian Assange faced extradition to two countries – Sweden over sexual assault allegations, and the United States over allegations of conspiracy to download classified US databases for Wikileaks, an information organization founded by Assange. On April 12, 2019, Assange was arrested at the embassy after Ecuador revoked his political asylum. As of writing, he is still in UK prison undergoing extradition hearings: see Patrizia Rizzo, "Who is Julian Assange and Where is He Now?", *The Sun*, August 17, 2020, https://www.thesun.co.uk/news/12427118/julian-assange-wikileaks-prison-ecuador-embassy-britain-trump/.

82   As taken from their project webpage, https://nantucketproject.com/.

projected a 3D holographic image of Assange onto a stage to appear as if sitting opposite Jarecki, with Assange answering questions from both Jarecki and the audience in real-time. Notably, Jarecki and Assange ended the interview "with a hologram high five," where the two speakers, one physically and the other virtually onstage, manoeuvred their hands to meet in mid-air as an interactive "high five" gesture.[83] Assange's virtual body thus takes on a certain vitality in the unique interactivity of its here-ness – *it becomes vivid*: one that appears and interacts *virtually here* onstage with an actual presenter and audience, but *actually there* elsewhere. The vivification of Assange becomes more plangent in the context of his political asylum: in the post-screen of the Nantucket stage, he is not only a virtual body, but a vivid one which has overcome the physical distances and challenges of his politicized exile, and amplifies his political defiance.

On one level, all these examples appear to be simply glorified substitutes of the figures in question – notably celebrities and politicians hungry for publicity – to expand outreach and grab headlines about their technological mastery. In that sense, the phenomenon echoes Tom Gunning's conception of early cinema's reception as a "cinema of attractions," where early cinema served to grab attention and be celebrated for "its ability to *show* something." (emphasis added)[84] To that extent, these holographic projections could also be seen to be their own kind of "spectacle of attractions," harnessing the projections' visibility, exhibitionism and technological wonder.

Yet, on another level, their tetravalence across here/now and actual/virtual also extends the vivification of their holographic virtuality in line with the sensual physicality of their actual body in the image. Film studies has already contributed much to thinking about the haptic and the body in the moving image, such as Laura U. Marks's theory of "haptic visuality";[85] the argument of "the tactile eye" from Jennifer Barker on cinema that "gets beneath the skin, and reverberates in the body";[86] and particularly earlier scholarship such as Vivian Sobchack's groundbreaking work on drawing between cinema and phenomenology to think about the moving image

83   Sarah Begley, "Julian Assange Speaks in Nantucket – as a Hologram," *Time.com*, September 29, 2014, http://time.com/3442834/julian-assange-hologram/.

84   Tom Gunning, "The Cinema of Attractions: Early Film, Its Spectator and the Avant-Garde," in *Early Cinema: Space-Frame-Narrative*, eds. Thomas Elsaesser and Adam Barker (London: BFI Publishing, 1990): 56-62, 57.

85   Laura U. Marks, *The Skin of the Film: Intercultural Cinema, Embodiment, and the Senses* (Durham, NC: Duke University Press, 2000).

86   Jennifer M. Barker, *The Tactile Eye: Touch and the Cinematic Experience* (Berkeley, CA: University of California Press, 2009).

being itself "a sensing, sensual, sense-making subject."[87] Elsewhere, I have argued that the spectatorial gaze in films exhibiting handheld digital camera aesthetics expresses for the camera its own body, knowledge and agency.[88] In the vein of thinking through ever more complex conflations of visual knowledge and the body, these arguments propose that, in specific kinds of cinema, the image extends beyond a visual representation of experience; rather, it is accorded a physical and multi-sensorial embodiment.

Other studies have expanded these ideas of grounding senses of the body in media to the *digital*, such as in the niche but growing practices of user-generated computer-generated imagery (CGI) pornography. Pornography as a general genre is inherently tied to the visual, physical and visceral experience of the human body, with its sensory excesses, affect, and capture on film as "living expression[s]."[89] In CGI porn, as Rebecca Saunders argues, the digital images become bodies which transfer a different kind of knowledge, veracity and arousal, where it is no longer about "evidentiary proof of a veracious sexual event."[90] Instead, other desires come into play, such as "the erotic pull of the blatantly artificial," and arousal out of users' alternative desires "from the capacity to relate to and expand the diegetic worlds of fond, familiar characters in previously unsanctioned, pornographic ways."[91] Saunders specifically ties these modes of arousal to the digital medium itself, where CGI porn is no longer about "representations of real sexual interaction" but, rather, about "libidinally engaging, in part, with the capabilities and limits of the medium," or "a new erotic imperative to engage with the digital animated medium itself."[92] The digitality of the medium thus contains or evokes a body that is sensual and arousing in different ways – transmedial, transgressive, adapted, artificial, manipulatable, ripe with the potential of what may be created with technology. In that sense, we might also think of the evolution of the body as influenced by technology,

87 Vivian Sobchack, *The Address of the Eye: A Phenomenology of Film Experience* (Princeton, NJ: Princeton University Press, 1991).

88 Jenna Ng, "The Handheld Digital Camera Aesthetics of *The Blair Witch Project* and *Cloverfield* (via *Strange Days*)," *16:9* 7(32) (June 2009), http://www.16-9.dk/2009-06/pdf/16-9_juni2009_side11_inenglish.pdf.

89 Jean Mitry, *Semiotics and the Analysis of Film* (London: Athlone, 2000), 32.

90 Rebecca Saunders, "Computer-generated pornography and convergence: Animation and algorithms as new digital desire," *Convergence: The International Journal of Research into New Media Technologies* 25(2) (2019): 241-259, 248.

91 Indeed, "the notion of mastery, with regard to the ability to animate a character to behave in ways beyond their 'will' is frequently asserted as part of the thrill of [user-generated computer generated pornography]": see Saunders, "Computer-generated pornography," 249.

92 Saunders, "Computer-generated pornography," 250.

such as in transhumanism, where technological processes are capable of modifying its biological determinations.[93] Here, then, is the converse, where technology itself *contains* a body which operates in sensuality imbricated between the actual and virtual, entangled between the carnality of sex and the digitality of the medium.

These increasingly complex enfoldments of the haptic and the sensual in the digital virtual body also appear in popular culture. An example is "Black Museum," another episode out of the TV series of *Black Mirror*.[94] The episode relates, among other stories, how an executed prisoner, Clayton Leigh (played by Babs Olusanmokun), became "resurrected" as a hologram in a museum, having signed over his "post-death consciousness"[95] to the proprietor in exchange for money provided to his family. However, on his holographic "resurrection," Clayton's virtual body took on the immateriality of a holographic projection which, unexpected to him, could still feel actual pain as if from his physical body. The episode's horror thus lies in how that sensorial "loophole" became a sadistic offering by the museum, where its visitors, by pulling a lever, caused Clayton to experience the agonies of the electric chair all over again. The sensorium of the virtual holographic body is thus not only manifest, but its realness is also effectively exploited for the episode's horror.

These instances of the haptic and the sensuous in the digital image – the imbrications between the actuality of the physical body and the virtuality of the digital – thus shore up the vivification of holographic projections. Clayton's actuality of pain in his virtual body is, of course, speculative; the argument here is not to suggest such a literal meaning out of the body's actuality.[96] Rather, the idea is about continuing the discourse of the body vis-à-vis media and mediation not as a biologically predetermined entity

---

93 See in general *The Transhumanist Reader: Classical and Contemporary Essays on the Science, Technology, and Philosophy of the Human Future*, eds. Max More and Natashi Vita-More (Oxford: Wiley-Blackwell, 2013).

94 *Black Mirror*, "Black Museum," season 4, ep. 6, *Netflix*, first released December 29, 2017, written by Charlie Brooker, directed by Colm McCarthy.

95 Ira Madisson III, "'Black Mirror's' Season 4 Finale 'Black Museum' Offers a Horrifying Critique of American Racism," *Daily Beast*, January 5, 2018, https://www.thedailybeast.com/black-mirrors-season-4-finale-black-museum-is-a-horrifying-critique-of-american-racism.

96 The use of the neurological – in experiencing pain – in pushing the actuality of the body in Clayton's hologram lends to the quality of vividness a sense of regression, one which falls back to defining the human body by its neural pathways and sensorial content, rather than a radical transformation of the body to think of it on new planes of expression, or even to overcome its biological, neurological and psychological limits. The vivid body here, while still mapped onto the axes of reality and space, continues on an oft-trodden discourse of the body's construction.

but as a discursive construct, one that is certainly informed by regulatory norms and power relations,[97] but also by practices of re-construction and re-constitution by contemporary media. In that sense, the post-screen continues critical speculation of the nature of the virtual body and the extents to which media technologies may form, deform and de-form it; these continuing evolutions of bodies across institutions, information and individual identity, in turn, inform their biopolitics, as Foucault and Deleuze have already shown.[98] With expanding use of facial recognition technologies and biometric security today, it is the post-screen which registers these increasingly complex gradations of the virtualized body, and their ever closer imbrications between the virtual and the actual. The signification of the screen and its boundaries – the subversion of which is not only in terms of how they are actively concealed, but also the relations and engagement across them – are hence part of these shifting constructions.

Moreover, the tetravalence across which these holographic vivid bodies lay is itself a fluid configuration whose values *shift* in their particular contexts. One result is the amplification of both the spectral life of the holographic body's vividness "here" *as well as* the physical body's actuality "elsewhere." An example is the "hologram protest," or "ghost rally." As with Assange's case, the political context of protests intensifies the vividness of the holographic body. A "hologram protest," as the name suggests, is a holographic projection of protesters onto hidden screens in the street, almost always in defiance of authorities' bans on actual protests or gatherings. "The world's first hologram protest," as anointed by the media,[99] took place on April 12, 2015 in Madrid to protest the imminent imposition of a controversial Citizens Safety Law. Dubbed "the Gag law" by its critics, the regulation would render it illegal, among other restrictions, for citizens "to gather in front of government buildings without permission from

---

97   Michel Foucault, *Discipline and Punish: The Birth of the Prison* (New York: Pantheon Book, 1977).

98   See Foucault, *Discipline and Punish*; and Gilles Deleuze, "Postscript on the Societies of Control," *October* 59 (Winter 1992): 3-7.

99   See, for instance, Simon Tomlinson, "The world's first HOLOGRAM protest: Thousands join virtual march against law banning demonstrations outside government buildings in Spain," *Daily Mail*, April 14, 2015, https://www.dailymail.co.uk/news/article-3038317/The-world-s-HOLOGRAM-protest-Thousands-join-virtual-march-Spain-against-law-banning-demonstrations-outside-government-buildings.html; Harry Readhead, "The world's first hologram protest was held in Spain this weekend," *Metro*, April 13, 2015, https://metro.co.uk/2015/04/13/the-worlds-first-hologram-protest-was-held-in-spain-this-weekend-5148663/; and Jethro Mullen, "Virtual protest: Demonstrators challenge new law with holograms," *CNN*, April 12, 2015, https://edition.cnn.com/2015/04/12/europe/spain-hologram-protest/index.html.

authorities,"[100] with the definition of "government buildings" ranging from universities to hospitals. After months of actual protests, on the evening of April 12, 2015, "between 9pm and 10.30pm," the self-named No Somos Delito ("We Are Not a Crime") group, comprising of various organizations and social movements,[101] projected "more than 2,000 holograms" onto a near-invisible screen in front of the Spanish parliament building.[102] Created by civil rights organizations, artists and others around the world, the effect was a parade of dimly illuminated and ghost-like human figures filing before the government building in the style of a conventional protest march. The next year in February 2016, ten months after the protest in Spain, Amnesty International Korea in Seoul organized a similar protest which took place in front of the Gwanghwamun, or the main gate of the Gyeongbok Palace (Blue House), after their request to hold a live rally there was rejected by police.[103] The 30-minute protest in Seoul projected protesters

> ... chant[ing] slogans like 'Guarantee peaceful assembly' and 'We are not illegal,' [sic] also holding a banner reading 'Assembly is a human right.' Some walked in silence, wore masks and held flowers to their chests. The holographic work ended with five participants urging the government to respect their constitutional rights.[104]

Like all the holographic projections discussed so far, the post-screen's imperceptibility of boundaries is key to the re-drawing of liveness in the holographic bodies' "here-ness." This is despite the size of the screens, such as in the case of the South Korean protest, where it was as large as ten meters

100 Zachary Davies Boren, "Spain's hologram protest: Thousands join virtual march in Madrid against new gag law," *The Independent*, April 12, 2015, https://www.independent.co.uk/news/world/europe/spains-hologram-protest-thousands-join-virtual-march-in-madrid-against-new-gag-law-10170650.html.

101 Lily Hay Newman and Juliana Jiménez Jaramillo, "Spain is Banning Protestors in Front of Parliament, So Activists Sent Holograms Instead," *Slate*, April 15, 2015, https://slate.com/technology/2015/04/no-somos-delito-organizes-hologram-protest-spanish-gag-laws.html.

102 Carlos Córdoba Vallet, "Activists mount hologram protest against Spain's 'gag' law," *El País*, April 14, 2015, https://elpais.com/elpais/2015/04/14/inenglish/1429013219_604510.html.

103 Also see S.J. Kim, "Hologram protest," YouTube Video, 3:22, February 24, 2016, https://www.youtube.com/watch?v=Ob7QBwCoFQ4.

104 Kim Se-jeong, "'Ghost protest' faces police scrutiny," *The Korean Times*, February 24, 2016, http://www.koreatimes.co.kr/www/news/nation/2016/02/116_198941.html.

long by three meters wide.[105] The images of these "hologram protesters" were made to appear as if they were "on the scene" or "amongst the public" – on the streets, in the square, before the palace. As Esteban Crespo, director of the Madrid protest, comments: "You have [to] make all sorts of calculations, keeping in mind the size of images and the street. The goal was to make it look like these holograms were actually on the street itself."[106] One journalist described the Madrid protest thus: "Hordes of demonstrators seemed to appear on the street – the scrim was basically invisible – in wispy, flickering white forms."[107]

As with the 3D projections of Mariah Carey and the other celebrities and politicians, a part of these holographic projections is undoubtedly for spectacle, in this case to draw attention to their cause. Having said that, these "protest holograms" also signal a spectral life of vividness unique to how they appear as images, and in this case, enwrapped as well in political messaging. The vividness of the holographic protesters, cut between their tetravalence of here/elsewhere/actual/virtual, is *thickened* in its political context – they are virtually *here* only as a counter-purpose, because they are forced to be actually elsewhere. That actuality of the-body-elsewhere thus gets drawn with and into the holographic bodies' vividness. On one hand, taking on Foucault's ideas of discipline and punishment,[108] that body is a clear apparatus of discipline through which governance and political control operate with laws that block gatherings for protest and stop public assembly of bodies. On the other hand, those actual-bodies-elsewhere are also no longer simply bodies under discipline. Rather, in tandem with the virtual protest, they become *post-screen entities* which counter their inherent anatamo-politics of discipline and punishment, even as – or only because – they collide in the tensions of the actual/virtual/here/elsewhere tetravalence in which they are entangled. Here, the tetravalence across the post-screen shifts in favour of the actual-elsewhere axis, whereby, to become a counter anatamo-political artefact, the actual-body-elsewhere enhances the vividness of the holographic-projection-here.

In this vein of amplified play between virtual here-ness and actual elsewhere-ness, the 3D projections of Holocaust survivors at the Illinois Holocaust Museum and developed by the USC Shoah Foundation offers

---

105 AP, "South Korea Hologram Protest," *The Kyunghyang Shinmun*, February 25, 2016, http://english.khan.co.kr/khan_art_view.html?code=710100&artid=20160225195252A&medid=AP.

106 As quoted in Jonathan Blitzer, "Protest by Hologram," *The New Yorker*, April 20, 2015, https://www.newyorker.com/news/news-desk/protest-by-hologram.

107 Blitzer, "Protest by Hologram," np.

108 See generally Foucault, *Discipline and Punish*, 1977.

a different illustration of the holographic body's tetravalent vividness. In October 2017, the Museum debuted 3D projections of thirteen recorded Holocaust survivors, mostly from English-speaking countries,[109] to offer visitors "a real-time conversation with the likeness of a survivor."[110] The Museum sat the survivors through a week of extensive interviews about their Holocaust experiences, recording their answers on high-definition video in a "half-dome studio"[111] in front of fifty-plus cameras.[112] Their recorded images and responses became part of a unique exhibition, where the visual recordings of the survivors would appear as "holograms"/holographic projections on the stage of the Museum's auditorium before an actual audience of visitors. The audience could then ask the "holograms" questions about their lives as Holocaust survivors, akin to a real-time Q&A session. Voice-recognition technology and machine learning piece together and produce responses out of the survivors' recordings, so that it appears the "holograms" are "replying" to live questions in real-time.[113]

Again, as with Tupac, Carey and other holographic projections onstage, the key effect of aliveness here arises out of the post-screen subversion of screen boundaries which normally demarcate the image against its surroundings. The subject onstage thus appears to be in the same spatio-temporal space as with the audience. As one visitor commented: "It really does look like she is sitting on the stage in front of you." However, the tetravalence of the holographic body's vividness takes on a different twist here, namely, *an ambiguity as to whether the subject is dead or alive.* Holographic projections of the dead present them as ghosts returned to the living; of the living, they appear as vivid bodies across here/virtual/elsewhere/actual. Of these Holocaust survivors, though, that status is unclear. To a question posed by a visitor, "how old are you now," the answer from the recordings is returned with deliberate circumvention: "I was born December 1st, 1935, and so please figure out." At some point, that calculation will turn up a number that is beyond reasonable human mortality. In the meantime, though, the ambiguity is arresting, where the holographic image *here* is compounded by the actuality of a body *elsewhere* that could be in the

---

109 These include the United States, Canada and Britain.

110 Nova Safo, "US museum debuts first 3-D holograms of Holocaust survivors," *The Times of Israel,* October 28, 2017, https://www.timesofisrael.com/us-museum-debuts-first-3-d-holograms-of-holocaust-survivors/.

111 Ibid.

112 See more information at the Illinois Holocaust Museum and Education Center webpage at https://www.ilholocaustmuseum.org/abe-ida-cooper-survivor-stories-experience/.

113 Safo, "US museum," np.

world of the living or in the afterlife. Confounded by this ambiguity, the holographic projections of these survivors in the post-screen take on *a third kind of coexistence* that vacillates between the vivified and the dead: a state of the not-dead, and, in a way, perhaps never will be dead. The piece of history they carry *as bodies* is too important, and as *holographic bodies* thus emanate an eternal resonance in the peculiar space of the post-screen. To that extent, the museum's motivation for this project – "to preserve accounts of a fast-disappearing generation"[114] – also rings clarion. In this mediated form, the museum applies the brakes to that generation's and those survivors' disappearance. It suspends their bodies, as with M. Valdemar, mentioned earlier via Poe's short story, in perfect repose between the three-way tension between the dead, the living and the vivified, from which they do not awaken.

In these wider senses, the post-screen through holographic projections thus offers the body yet further articulations of the posthuman and the post-organic. These conceptualizations also extend the current discourses of the body being freed from biological constraints, such as with enhanced senses and capabilities.[115] They continue decentring humanist agendas away from, among others, gender, race, species and sexuality, and towards new imaginations of the imbrications between bodies and technologies as articulated between the actual and virtual, and between different spatio-temporal configurations. In turn, these reflections map onto the most profound of axles around which all life revolves – the chasm between the living and the dead, and the abject passing from one to the other. In his essay, "Death Every Afternoon," film critic André Bazin protests against the reproducibility of cinema for enabling the repetition of death which should otherwise be "the unique moment par excellence": "thanks to film, nowadays we can desecrate and show at will the only one of our possessions that is temporally inalienable: dead without a requiem, the eternal dead-again of the cinema!"[116] Holographic projections as post-screen media achieve more than simply repeating the moment of death. In these 3D displays and their strategies of constructing resurrected and vivified bodies, different kinds of life, afterlife and after-death emerge.

---

114 All quotations in this paragraph are taken from Safo, "US museum," np.

115 See, for instance, Judith Halberstam and Ira Livingston, *Posthuman Bodies* (Bloomington, IN: Indiana University Press, 1995); or work by performance artist Stelarc, which uses various kinds of body mods to stretch human capabilities – see http://stelarc.org/projects.php.

116 André Bazin, "Death Every Afternoon," in *Rites of Realism: Essays on Corporeal Cinema*, ed. Ivone Margulies, trans. Mark A. Cohen (Durham, NC: Duke University Press, 2002): 27-31, 31.

A Funny Thing Happened on the Way to Substitution

A final thought, then, as a coda on this chapter's central mission in connecting the screen and the post-screen with death, living and resurrection. One key rationalization of the post-screen through this book is thinking about the screen in relation to the impossible: the impossibilities of travelling far afield and of being someone else, per the last chapter on Virtual Reality; here, the impossibilities of traversing to the realm of the dead, of seeing again the dead as alive, or of being in different places at the same time. The screen is thus an interface not only to information, but also to the metaphysical and the imagined; the post-screen only re-imagines those relations.

In that respect, media technologies are, and have always been, our substitutes for impossibilities, if only as imperfect ones. Jonathan Safran Foer makes this precise point:

> Most of our communication technologies began as substitutes for an impossible activity. We couldn't always see one another face to face, so the telephone made it possible to keep in touch at a distance. One is not always home, so the answering machine made a message possible without the person being near their phone. ... These inventions were not created to be improvements on face-to-face communication, but a declension of acceptable, if diminished, substitutes for it.

The movement from screen media such as photography and cinema to post-screen media such as holographic displays can also thus be seen as a wider trajectory towards increasingly less-diminished and more-acceptable substitutes for impossibilities, particularly of the returned dead.[117] However, there is a twist in Foer's account of technology: *"But then a funny thing happened: we began to prefer the diminished substitutes."* (emphasis added) He continues:

> It's easier to make a phone call than to make the effort to see someone in person. Leaving a message on someone's machine is easier than having a phone conversation – you can say what you need to say without a response;

---

117   As a journalist puts it (with emphasis mine): "The point of improving the technology isn't to make anyone think the holograms are real, *but to make them feel a more irresistible and automatic emotional connection*": see Kaitlyn Tiffany, "No industry is weirder than the dead celebrity hologram industry," *Vox*, October 23, 2018, https://www.vox.com/the-goods/2018/10/23/18010274/amy-winehouse-hologram-tour-controversy-technology.

it's easier to check in without becoming entangled. ... Each step 'forward' has made it easier – just a little – to avoid the emotional work of being present, to convey information rather than humanity.

According to Foer, the consequence to the "ease" of substitution is this: "The problem with accepting – with preferring – diminished substitutes is that, *over time, we too become diminished substitutes*. People who become used to saying little become used to feeling little." (emphasis added)[118]

Foer's cautionary is clear: there is a price to substitution with technology, whereby impossibilities get fudged into somewhat acceptable possibilities. But that price is paid out of some tenet of the human condition. In that respect, the task in this chapter of thinking about death and the post-screen is also to think about and account for the price of realizing impossibilities, namely, the value of pain, loss and grief, and ultimately the question of profoundly understanding the self out of the impossibilities of understanding. These are the impossibilities which do not end in being overcome, and never get finalized. But they are the impossibilities which allow us insight and intuition. The post-screen is thus more than just rubbing out the demarcations of actual and virtual worlds, or even to call out the falsities of grasping for certain conveniences. It is also a comment on the new ghosts which confront us, and the statements they make about the living, and about living.

---

118   All quotations from Foer in this paragraph are from Jonathan Safran Foer, "Technology is diminishing us," *The Guardian* online, December 3, 2016, https://www.theguardian.com/books/2016/dec/03/jonathan-safran-foer-technology-diminishing-us. On this specific point here, though, Foer's correlation between brevity and feeling might be too hastily drawn. It is not a dialectic between the amount of chat and the amount of feeling; those values might shift – whereby people who become used to saying a lot become used to feeling little – and still be a valid conclusion.

4A (Remix) True Holograms: A Different Kind of Screen; A Different Kind of Ghost

Abstract

This chapter discusses true holograms as a "remix" discussion from Chapter 4 of ghosts out of the post-screen. Commonly confused with holographic projections, true holograms are two-dimensional images naturally viewed (i.e. without optical aids) as 3-dimensional objects. Leveraging theoretical sources such as Deleuze's notion of "the brain is the screen" and Vilém Flusser's ideas of point culture and linearity, the chapter argues for the post-screen through the true hologram whose ghosts are not of the spectral return of the dead, but digital apparitions via which the human mind ideates and projects realities. These digital ghosts thus return with a necromancy of their own on the terms of zero-dimensionality and post-rationality, confronting us with new problems of reality and questions about ourselves.

Keywords: hologram; screen; brain; screen; ghost; Flusser; crisis of linearity

Screens and Ghosts, or, the Window and the Guy in the Basement

In *Parasite*,[1] the widely acclaimed and first ever non-English language film to win the Academy Award for best picture, there are two houses. In one lives a poor family; in the other, a rich family.

The house with the poor family is semi-basement, with one window which opens onto the outside world at street level. The architectural window becomes a screen through which the family, enclosed in the darkness of their largely subterrestrial home, sees sights of the outside street. But, as with

---

1   *Parasite*, directed by Bong Joon-Ho (2020; London: Curzon Artificial Eye, 2020), DVD.

Ng, J., *The Post-Screen Through Virtual Reality, Holograms and Light Projections. Where Screen Boundaries Lie.* Amsterdam: Amsterdam University Press 2021
DOI: 10.5117/9789463723541_CH04A

the post-screen, the window's boundaries are open, porous and permeable, so the window also lets in fumigation fumes and, at the film's climax, rain floods which pollute and destroy their home. We can say that this is a different kind of screen.

The house with the rich family has a hidden basement with a guy in it. He had once emerged from the basement to steal food from the kitchen and frightened the rich family's little son, who refers to the episode as having "seen a ghost." This guy turns out to be the original housekeeper's husband who had been secretly living in the basement bunker for over four years hiding from loan sharks – indeed, a man with a ghosted identity.[2] But he is not a conventional ghost traversing supernatural boundaries; this is a sort of post-screen ghost who crosses numerous different kinds of boundaries. The first is the architectural boundary between the house and the basement – an elaborate set-up requiring the removal of a shelf and a wall to be dragged open. The second is the environmental boundary from the light-filled, glass-paned, spacious house to the underground, bunker-like and claustrophobic basement. The third boundary is the class divide from the wealthy elite of the rich family to the economically straitened of the poor working class, first felled in the financial crisis of 1997 and now living in the shadows.[3] The final boundary the ghost crosses is its taking of different bodies – the original "ghost," the housekeeper's husband, is later killed, but the father of the poor family promptly takes his place in the basement, hiding as much from the law for his own crimes as from the economic inequalities in the world "above" which squeeze his existence into its phantom state. We can say that this is a different kind of ghost.

In the last chapter, the post-screen through holographic projections was discussed as a space for ghosts from the dead and ghosts of the living. This chapter, by way of a "remix" discussion, segues chapter 4's thinking about ghosts and the post-screen through another medium of the ghostly, and one that is often confused with holographic projections (though in technical terms they are completely different) – namely, *true holograms*. As this chapter will argue, true holograms are also of the post-screen which produce apparitions. However, as with screens and ghosts in

---

2   The husband having vanished without a trace also becomes an unwitting take on the contemporary use of the English phrase "to ghost" someone, as in to cut off all communication in a relationship with no explanation.

3   See analysis in S. Nathan Park, "'Parasite' Has a Hidden Backstory of Middle-Class Failure and Chicken Joints," *Foreign Policy*, February 21, 2020, https://foreignpolicy.com/2020/02/21/korea-bong-oscars-parasite-hidden-backstory-middle-class-chicken-bong-joon-ho/#:~:text=In%20 Parasite%2C%20the%20king%20castella,but%20their%20implications%20are%20chilling.

Parasite, this chapter's argument will demonstrate the post-screen of the true hologram as one with a different kind of screen, and with a different kind of ghost.

***

True Holograms

Unlike holographic projections which appear as optical illusions on deliberately hidden screens and whose boundaries are made imperceptible as thresholds between life and various meanings of afterlife, the true hologram is not viewed on a screen or within conventional screen boundaries. Rather, the true hologram is a two-dimensional image that the viewer naturally sees – i.e. without any headgear or eyewear – as a three-dimensional virtual object coherent in their own space. By coherence, this means that the object changes vis-à-vis its position relative to the viewer,[4] resulting in the viewer's sense not only of the object's volume but also that it shares their space without distortion. In comparison, an image displayed ordinarily on a screen looks the same from every direction (other than for perspective); it remains within the screen. This iteration of a naturally seen three-dimensional object in a shared space between the hologram and the viewer thus replaces the conventional material existence of the screen for the image and the boundaries around it. With the hologram visible in this way sans screen and boundaries, it becomes *the post-screen*. But, as we shall see, it is haunted by a different kind of ghost.

First, a brief explanation of the hologram's technical qualities to better understand the holographic post-screen. The way a hologram changes to produce its 3D effect ranges from the simplistic to the complex. A simplistic change may be by way of just two movements, such as those of souvenir hologram cards which give two different perspectives of its object, depending on the angle one leans the card. Another example is the hologram on most

---

4   Note that the viewer can only see these changes from a restricted range of positions, as determined by the size of lens used. Jean Baudrillard uses another metaphor – the fractal – to describe this state of incremental change in viewing the hologram: "In the hologram, this perfection of the virtual image, all parts are microscopically identical to the whole, so we are in *a fractal deconstruction of the image*, which is supplanted by its pure luminous definition." (emphasis added): see "The Virtual Illusion: Or the Automatic Writing of the World," *Theory, Culture & Society* 12 (1995): 97-107, 104.

credit cards which similarly changes shape in two ways, depending again on how the card is tilted.[5]

The change may also be more complex, such as multiplex holograms (or "holographic stereograms") which are made by combining cinematographic and holographic techniques to display a short animated image loop. The holographic film is then rolled into a rotating 360° cylindrical form and/or the viewer simply walks around the hologram, resulting in the image appearing to move as an animated three-dimensional object. Famous examples of such holograms include *The Kiss* (Lloyd Cross and Pam Brazier, 1973), where the eyes on the female subject's face seem to follow the viewer as the latter moves. Moreover, the subject's fingers on her mouth (if the viewer moves continuously from one side to the other of the hologram) have also been photographed into a small moving sequence so as to appear to extend and blow a kiss to the viewer.[6] Another example is the hologram of the singer Alice Cooper as produced by Salvador Dalí and Selwyn Lissack, also in 1973. It is displayed as a strip of holographic film "rolled up to form a cylinder, which is lit from beneath by...an ordinary light bulb."[7] The cylinder is then placed on a rotating base; as the cylinder rotates, the image of Cooper, appearing in the middle of the cylinder, likewise changes (to a stationary viewer) to reveal all sides of the singer, producing the hologram's 3D effect. Umberto Eco describes this quality of the hologram's three-dimensional space, combined with movement, as such: "It isn't cinema, but, rather a kind of virtual object in three dimensions that exists even where you don't see it, and if you move you can see it there, too."[8]

---

5 Jens Schroeter calls this "the flip-book" technology, where "observers have to take the book and flip the pages by themselves for the images to move. If the observers do not move their hands, the images do not move either": see Schroeter, "Technologies Beyond the Still and the Moving Image: The Case of the Multiplex Hologram," *History of Photography* 35(1) (February 2011): 23-32.

6 Again, taking from Schroeter's taxonomy, we can employ his label for this kind of hologram as "the so-called lenticular images" (as opposed to the "flip-book" technology): "These are images in which there are arrays of lenses directly on the surface of the image. Under specific conditions, one can record several different images of one phase of movement onto the image plane. When you move the image or when you move in front of the image, a small moving sequence can be seen": see Schroeter, "Technologies," 24.

7 Schroeter, "Technologies," 26.

8 Umberto Eco, *Travels in Hyperreality* (San Diego, CA: Harcourt Brace, 1986), 304. Notably, we need not look only to holograms (and movement) for natural 3D visualization; various stereo image presentation systems are also being developed to create the same effect, one example being Vision Engineering's Deep Reality Viewer, which uses 3D display technology in a 3D imaging microscope that "creates stereo high definition 3D images, without using a monitor or requiring operators to wear headsets or specialist glasses: images 'float' in front of a mirror": see Ian Bolland,

This sense of the hologram as a naturally perceived and coherent three-dimensional object is also how, at least in popular culture, they take on a mythic but factually erroneous quality of being volumetric light projections that *simply appear out of thin air*. The most famous example of this mythic hologram is the Princess Leia projection featured in the 1977 *Star Wars* film,[9] beamed out by the R2D2 droid as a looping vision of light. There are also more recent (and, incidentally, also female) holographic figures which simply appear out of thin air, such as the "purchasable holographic companion," Joi (played by Ana de Armas), in *Blade Runner 2049*.[10] Unlike Leia, Joi as a hologram is equipped with sophisticated hardware that enables her "to extensively sense the environment [she] was in and translate that into data for the artificial intelligence to 'experience' life along with the user." She also possesses software to communicate, interact and record data, "allowing her 'personality' to change over time."[11] However, notwithstanding her technical sophistication, as with Leia, Joi appears as a three-dimensional projection of light again visible simply out of thin air. Even though she clearly "require[s] some kind of projecting device to manifest to the viewer/user"[12] – and whose device subsequently becomes an important plot point[13] – the projection does not seem to require being projected *onto* any sort of screen or surface. Similarly, the science-fiction world of *Minority Report*,[14] a film also mentioned in the last chapter and replete with screens (and ghosts), features a shopping mall walkway which bombards John Anderton (played by Tom Cruise) with holograms of models, scenes and objects, all addressing him with personalized messages to grab his attention. They appear three-dimensional yet, as with Joi and Leia, are seemingly projected out of thin air.

"On closer inspection: A sneak peek at Vision's DRV-Z1," *Med-Tech News*, June 25, 2019, https://www.med-technews.com/features/on-closer-inspection-a-sneak-peek-at-vision-s-drv-z1/.

9   *Star Wars: A New Hope*, directed by George Lucas (1977; Los Angeles, CA: 20th Century Fox, 2004), DVD.

10   *Blade Runner 2049*, directed by Denis Villeneuve (2017; Culver City, CA: Sony Pictures, 2018), DVD.

11   All quotations relating to Joi in this paragraph are from Fandom, "Off-World: The Blade Runner Wiki," no date, https://bladerunner.fandom.com/wiki/Joi.

12   Ibid.

13   In short, Joi ends up being a vital companion who accompanies K in his quest to find the replicant child; in order for her to do so, in their first scene together K gifted Joi an Emanator, a portable projector of sorts, so that she may travel with him. The projector also later plays an important part as K transfers Joi into the Emanator, only to have his enemies crush it, destroying Joi too.

14   *Minority Report*, directed by Steven Spielberg (2002; Los Angeles, CA: 20th Century Fox, 2003), DVD.

However, this sort of "free space" projection of light remains largely science-fiction.[15] The appearance of the hologram has nothing to do with ethereal light projection which magically appears. Rather, it is due to the physics of the two-dimensional surface of the holographic film or plate: "Holograms appear to the eye to be three-dimensional, but 'all of the magic is happening on a 2-D surface.'"[16] A hologram, as with a photograph, is created by recording light, though not light as formed by a lens, but by splitting a light field (typically sourced from laser light) with a beam splitter. The divided light is then directed, usually with mirrors, to various locations. Crucially, the interactions between the two light beams create interference patterns which, together with light reflecting off the object, are recorded by a camera onto the light-sensitive emulsion surface of a holographic plate. The emulsion is then developed as with photographic film (though again with some differences), and darker and lighter areas of the emulsion (corresponding to how much light each has received) become the interference fringes of the plate. Creating a hologram thus produces a unique interplay between the reflection and diffraction of light off the surface of the holographic film on which the object was photographed, resulting in an interference pattern on the plate. When a holographic light source, usually monochrome or pure white light, lands correctly on the interference pattern, it reconstructs the wave form of light as if the light had come from the object itself. Combined with the viewer's own cognitive processes of perceiving shadows, distances and parallax, the result is that the viewer, without optical aids, sees the two-dimensional plate or film as a three-dimensional object. In other

15   Albeit this "free space" projection is getting closer to reality with current groundbreaking work by, among others, researchers at Brigham Young University, led by Daniel Smalley, on what is called "volumetric display" which creates exactly this sort of "three-dimensional hologram appearing in thin air" imaging (by controlling dust-like particles in the air with laser light): see Seth Borenstein, "Better than Holograms: A New 3-D Projection into Thin Air," *Phys.org*, January 24, 2018, https://phys.org/news/2018-01-holograms-d-thin-air.html. As a nod to the popular myth of holograms as perpetuated by the *Star Wars* movie, the scientists involved have nicknamed their research the "Princess Leia Project" – see Todd Hollingshead, "Better than a hologram: BYU study produces 3D images that float in 'thin air'," *BYU News*, January 24, 2018, https://news.byu.edu/news/better-hologram-byu-study-produces-3d-images-float-thin-air.
16   Borenstein, "Better than Holograms," np; the quotation within the quotation is from Daniel Smalley, the lead researcher on the project. This also echoes the holographic principle as discovered by theoretical physicists, where information is found to be stored in the two-dimensional boundary, such as the event horizon of a black hole, which, on radiation, *becomes* three-dimensional: see Leonard Susskind, "The World as a Hologram," *Journal of Mathematical Physics*, 36 (11) (1995): 6377-6396 or, for a more journalistic take, Paul Sutter, "Are We Living in a Hologram," *Space.com*, January 29, 2018, https://www.space.com/39510-are-we-living-in-a-hologram.html.

words, the object appears to the viewer as if it is – to play on Barthes's words – *actually-being-there*, as opposed to (in reference to the past-present temporality of the photograph discussed in the last chapter) the viewer's consciousness of the object's *"having-been-there."*[17]

Compared to Virtual Reality and holographic projections, true holograms thus signal a different relationship between image and screen as symptomatic of the post-screen. In the previous two chapters, the post-screen was discussed in relation to continuous edge-less media environments, be those the totalizing immersive surrounds of VR or the invisible surfaces and borders of screens in holographic projections as hidden through reflection and concealment. Here, the true hologram presents a tripartite converse of the post-screen: not a screen purporting to present objects in a world with no boundaries around the screen, but *an object purporting to present a world with no screen.*

This conclusion needs to be further unpacked. Firstly: *what kind of object?* This would be *the virtual object* which, when the hologram is lit with holographic light, appears in a three-dimensional form that the viewer is able to see naturally without any optical aids or headsets. (In this respect, the hologram differs from 3D cinema or VR, which require visual aids such as 3D glasses or a VR headset). Secondly, *what kind of world?* This would be the viewer's space into which the virtual object extends, thus appearing, as mentioned, to have a quality of "actually-being-there." Sometimes, they even appear to interact with the viewer, as with Cross and Brazier's *The Kiss*, with the kiss blown in their direction. (In this respect, the hologram differs from images seen on a 2D surface, such as screens and photographs, which do not change in relation to the viewer's position, and which invoke the quality of its object "having-been-there"). The third question, then: *why no screen?* This is because the hologram can be viewed naturally as an image presenting a virtual 3D object. Hence, it does not require a screen for its display, at least in the conventional sense. (In this respect, the hologram differs from projected or emitted images which need to appear *on* a surface of some kind).

A Different Kind of Screen: Brains, Nerves, Thought

Yet is there really no screen? At this juncture, it would be apposite to ask once more: what is a screen? Is it merely or always a literal two-dimensional

---

17   Roland Barthes, "Rhetoric of the Image," in Roland Barthes, *Image-Music-Text*, **trans. Stephen Heath** (London: Fontana Press, 1977), 44.

flat surface on which an image appears? If so, then there is no screen for the hologram. But what if we re-think (as, indeed, others have) the screen as something more conceptual and more abstract – or an epistemological flashpoint to create and/or re-create a viewer's visual imaginary? In other words, the screen not as an external material surface for projection or an assemblage of electronic circuitry, but as an internal mental space in which an image exists? This question could also be re-phrased as how the mind, in the first place, relates to images. Where does the image exist – internally in the mind, or externally on an outer surface? To this question, two dialectical positions are generally offered. The first is *internalism*, where all consciousness resides in the human brain, so that the image is seen "with my mind as much as, if not more than with my eyes." (Or "what we know of the world is not garnered directly from external stimuli but construed from internal representations of reality in our minds.")[18] The second position is *externalism*, where consciousness exists more widely in the body and in environmental objects so that the image is apprehended through a "cross-modal" account of our environment across touch, hearing and other senses. (Or, as Pepperell puts it, in terms of how "the development and operation of perception... is entirely dependent on direct engagement with the external world.")[19]

Other positions intervene between this internal/external opposition. Hans Belting, for instance, argues for a revised accounting of iconology which understands images not as a product of the internalist or externalist model of mind, but as an image-medium-body triad.[20] In this triad, images do not exist by themselves, but "happen" in their transmission through an agent (as the medium) and in the body's perception of them; with the rise of neuroscience, the last also now involves the neurological processes of the brain.[21] Robert Pepperell, asking the question particularly germane here, "where's the screen?", accepts both internalism and externalism, and,

18 Robert Pepperell, "Where's the Screen? The Paradoxical Relationship Between Mind and World," in *Screen Consciousness: Cinema, Mind and World*, eds. Robert Pepperell and Michael Punt (Amsterdam; New York: Rodopi, 2006): 181-198, 184.

19 Pepperell, "Where's the Screen?", 185.

20 Hans Belting, *An Anthropology of Images: Picture, Medium, Body*, trans. Thomas Dunlap (Princeton: Princeton University Press, 2011).

21 Although Gregory Flaxman notes that many early film theorists have also commented on what he calls "the psychomechanics of cinema": "One finds variations on this theme in the effusions of early French critics who hailed the rush of cinematic affects (Jean Epstein, Elie Faure, various surrealists), in Soviet theorists of montage who dwelt on the physiology and psychology of cinematic stimuli (Dziga Vertov, Sergei Eisenstein), and in various other theorists who sought to describe the sensation of cinema (Ricciotto Canudo, Émile Vuillermoz, Jean Goudal, Walter

as "a dialethic [sic] state," concludes that "the screen is perceived 'in here' and 'out there' *at the same time*" (emphasis in original):[22] it "exists in some Schrödingerian state of indeterminacy, being both distinct from and unified with the audience, experienced both inside the viewer's head and outside it."[23] These positions also form part of the larger discussion of the imbrications between humans and technology in hybrid media environments of actual and virtual realities.[24] The image – in its conception, formation and consequences thereof – thus becomes variously embroiled in assemblages or ensembles of phenomena, visuality, technology, body, brain processes and, crucially, screens. In such an ensemble or, to take Foucault's word, apparatus (*dispositif*), also lie implications such as the cinema aesthetics of what Patricia Pisters calls "the neuro-image,"[25] as well as various organizations and structures of power, control, behaviour and politics.[26]

Without subscribing to any particular school of thought, what is conceivable of the hologram is that, even without a material screen on which it appears, a screen still exists for it through the myriad interconnections between hologram, retina, perception, brain and the nervous system. Enfolded in this patchwork, the screen transforms: it becomes organic, moulded with the circuitry of the brain rather than of electronics or optical physics. In that respect, film philosophers have also long mooted a merger between brain and screen. Gilles Deleuze, for example, argues for "a cinema of the brain" in his 1985 book, *Cinema 2: The Time-Image*,[27] as he points to how cinema, "because it puts the image in motion, or rather endows the image with self-motion, never stops tracing the circuits of the brain."[28] The "cinema of the brain" thus contains the "cerebral components" of "the point-cut, relinkage and the black or white screen,"[29] "the cerebral mechanism, mental

---

Benjamin)": see Flaxman, "This Is Your Brain On Cinema: Antonin Artaud," in *Film as Philosophy*, ed. Bernd Herzogenrath (Minneapolis: University of Minnesota Press, 2017): 66-89, 76-77.

22   Pepperell, "Where's the Screen?," 192.

23   Pepperell, "Where's the Screen?," 193.

24   See, for instance, Stéphane Vial's *Being and the Screen: How the Digital Changes Perception*, trans. Patsy Baudoin (Cambridge, MA: MIT Press, 2019).

25   Patricia Pisters, *The Neuro-Image: A Deleuzian Film-philosophy of Digital Screen Culture* (Palo Alto: Stanford University Press, 2012).

26   See Pasi Väliaho's *Biopolitical Screens: Image, Power, and the Neoliberal Brain* (Cambridge, MA: MIT Press, 2014) for a detailed exposition of the workings of these images in terms of what he calls the "biopolitical visual economy" via video games, virtual reality and others.

27   Gilles Deleuze, *Cinema 2: The Time-Image*, trans. Hugh Tomlinson and Robert Galeta (Minneapolis, MN: The University of Minnesota Press [1985] 1997), 205.

28   Ibid.

29   Deleuze, *Cinema 2*, 215.

functioning, the process of thought."[30] In this, Deleuze echoes Antonin Artaud's essay from almost sixty years before the publication of *Cinema 2*, titled "Sorcery and Cinema" (circa 1928).[31] In this essay, barely a thousand words long, Artaud describes the unique nature of cinema as having a "whole element of contingency and mystery... that isn't found in the other arts." In particular, Artaud charges that cinema, with its power of visual expression and meaning, "reveals a whole occult life, *one with which it puts us directly in contact.*" It is not a vehicle for narrative: "to use [cinema] to tell stories, a superficial series of deeds, is to deprive it of the finest of its resources, to disavow its most profound purpose." Rather, Artaud looks to cinema as a resource "made, above all else, to express things of the mind, the inner life of consciousness... restores them to us with their matter intact, without intermediate forms, without representations."[32]

Both theorists converge in arguing for a direct connection – without intermediation, without representation – between cinema and thought, where the former traces, reflects and visualizes the latter.[33] Deleuze concocts a memorable phrase to sum up this argument: "*the brain is the screen.*"[34] The material screen on which film images appear thus becomes metaphorical as the brain: as Deleuze writes, the screen "no longer seems to refer to the human posture, like a window or a painting, but rather constitutes a

30   Deleuze, *Cinema 2*, 209.

31   Antonin Artaud, "Sorcery and Cinema," first printed in full in Antonin Artaud, *Oeuvres complètes*, tom 3 (Paris: Gallimard, 1961); referred here in *The Shadow and Its Shadow: Surrealist Writings on the Cinema*, ed. Paul Hammond (Monroe, OR: City Lights Books, 1978), 103-105. See also how Gregory Flaxman makes the same connection between Deleuze and Artaud in thinking through the latter's philosophy of cinema: Flaxman, "This Is Your Brain," 66-89.

32   All quotations in this paragraph by Artaud are from "Sorcery and Cinema," 103.

33   In that respect, both theorists naturally also ascertain this visualization of thought to be what they desired cinema's aesthetic future to be, given its expressive power. It is noted that thinking about the hologram have also taken up those lines of the holographic image reflecting thought or mental processes: Jens Schröter, for example, reflects on the surrealist intentions of Dalí's Alice Cooper hologram to "relate...to a notion of the unconscious, beyond reality in the realm of the surreal," noting it "as an image and metaphor for subconscious or unconscious processes below everyday reality": see "Technologies," 29. In that sense, there are already discernible parallels in the literature between characterizations of cinema and holograms in how their images may relate to the nature, visualization and movement of thought, conscious or otherwise.

34   Deleuze first states this phrase in a "remixed" interview with Pascal Bonitzer et al. as a response to why he chose to study philosophy through thinking about cinema. His answer? Because "thought is molecular": "The circuits and linkages of the brain don't pre-exist the stimuli, corpuscles, and particles [grains] that trace them. Cinema isn't theatre; rather, it makes bodies out of grains......Cinema... never stops tracing the circuits of the brain." Interview originally published in *Cahiers du cinéma* 380 (February 1986); here as taken from Gregory Flaxman, *The Brain is the Screen* (Minneapolis, MN: University of Minnesota Press, 2000), 366.

table of information, an opaque surface on which are inscribed 'data.'" It "functions as instrument panel, printing or computing table"; it becomes "the brain-information," "the brain-city."[35] Even as it manifests as an external surface, its ontological contradiction is that the cinema screen becomes swallowed anyhow into this interior chasm of brain lobes, nerves, tissue, synapses, spinal cord, cells, neural pathways. The screen, the shot and the brain all somehow integrate into each other; *all boundaries are dissolved.*

With the true hologram where the image does not actually appear but is directly perceived by the brain as an object, this "brain-screen" of the post-screen becomes consummate. The subversion of the screen's boundaries here is thus different from the cases of VR or holographic projections – they are not eroded in the totalization of the media environment or concealed through staging and lighting. Instead, in the convergence of philosophies of image perception and imaginations of the "cinema of the brain," they disappear into the folds of the brain's linkages, synaptic gaps and thought processes. They thus manifest a different kind of post-screen, and with that, a different kind of ghost. The question, then, is: what kind of ghost? The next and final section answers this.

A Different Kind of Ghost: "A Memory, A Daydream, A Secret,"[36] or, Digital Apparitions

> *A ghost can be a lot of things. A memory, a daydream, a secret. Grief, anger, guilt. But in my experience most times, they're just what we want to see...... Most times, a ghost is a wish.*
> ~ Dialogue line from *The Haunting of Hill House*[37]

Across the boundaries of the conventional screen, the image shows the dead reanimated, exhibiting movement as life. In the post-screen of concealed screen boundaries in live holographic projections, the images of the dead share an interactive space with the living in performative settings which become a limbo between deadness and alive-ness. On the same tangent, the images of the living appear as vivified entities across their ambiguities of time and space.

---

35 Deleuze, *Cinema 2*, 265 and 267.
36 Dialogue line from *The Haunting of Hill House*, "Steven Sees a Ghost," directed and written by Mike Flanagan, aired October 12, 2018, on Netflix.
37 Ibid.

Conversely, the hologram is apprehended naturally by the viewer as a coherent 3D object without any sort of surface or screen. There thus does not appear to be any commensurability between holograms vis-à-vis the image and its implications of the afterlife. This is because, as with screens, we also need to re-think our concept of ghosts.

The most common idea of ghosts is in terms of the afterlife of the dead, whose relationship with media (and with the post-screen) was leveraged at length in the last chapter. However, we can also think about ghosts in another sense: one deriving not from the dead (nor even from the living) but as an entirely different creed, namely, *the ghosts via which the human mind ideates, describes, apprehends, beholds and projects their realities*. Or what Vilém Flusser calls in digital media theory "digital apparition." In his essay titled with the same phrase, Flusser argues that computation has provided us with a new "exact calculatory thought" with which to behold our world as "worlds that we ourselves have designed, rather than something that has been given to us, like the surrounding world."[38] As a result, these purported, projected and designed worlds in their computational forms take on a contingency of existence and imagination that Flusser connects to "digital apparitions":

> Even later, the far-reaching suspicion emerged that perhaps the entire universe, with all its fields and relations, from Big Bang to heat death, might be a projection which calculatory thought attempts to retrieve 'experimentally.' Ultimately, computers demonstrate that we can not only project and win back this one universe, but that we can do the same with as many as we want. In short: our epistemological problem, and therefore also our existential problem, is *whether everything, including ourselves, may have to be understood as a digital apparition* [emphasis added].[39]

Flusser's argument is that if reality is realized through "computations of points," or as "'digital computations' of swirling point-potentialities," then "everything is digital, i.e., that everything has to be looked at as a more or less dense distribution of point elements, of bits." The hologram, however, is not computational, but is realized through photographic recording, physics,

---

38 Vilém Flusser, "Digital Apparition," in *Electronic Culture: Technology and Visual Representation*, ed. Timothy Druckrey (London: Aperture, 1996): 242-5, 242.
39 Flusser, "Digital Apparition," 243.

optics and the viewer's own mental processes.[40] But this is irrelevant because Flusser's thinking of "digital apparitions" connects to what elsewhere he calls "second degree imagination." In turn, this new "second" order of imagination is articulated through "the new codes" which refer primarily to computer codes, but also to a more essential ontology of points, or point elements, or "punctual elements":

> Here is how second degree imagination works: clear and distinct ele-
> ments, of which rational thought is composed, are being pulled from
> their linear structure in order to be inserted into other structures. They
> form thus mosaics, generally of two dimensions (as in computer screens),
> but may equally acquire additional dimensions (as in the case of moving
> holograms).[41]

Flusser calls this configuration of points and elements "zero-dimensional structures, since they are composed of punctual elements and intervals."[42] In his articulation of this ontology, Flusser harks to his wider argument of the "crisis of linearity,"[43] first pointed to in various writings in 1985-86,[44] and then outlined in a lecture series at São Paolo in 1986, later published into a volume titled *Into Immaterial Culture*. With respect to this crisis, Flusser argues that human expression and existence become increasingly abstract as technological advancements remove accumulating dimensions of concreteness. Hence, humans, who "exist within a situation with four dimensions" via, for instance, tool-making, "produced instruments, and lived within a three-dimensional circumstance, within objective culture": "The flint knife is frozen circumstance: still... The temporal

---

40   Confusingly, Flusser does mention the hologram as part of a list of digital apparitions: "Why is it that we distrust these synthetic images, sounds, and *holograms*? Why do we disparage them as 'apparitions'? Why are they not real for us?" (emphasis added): 242. My conclusion is that Flusser refers here to the hologram not in the technical sense of the true hologram, but, rather, as a general reference to a synthetically realistic image, or, perhaps, even to holographic projections, which do get referred (confusingly and inaccurately) a lot as holograms.

41   Vilém Flusser, *Into Immaterial Culture*, ed. and trans. Rodrigo Maltez Novaes (Milton Keynes: Metaflux Publishing, [1986] 2015), 30.

42   Flusser, *Into Immaterial Culture*, 30-31.

43   Per the original title of his essay, "Krise der Linearität," in *Absolute Vilem Flusser*, eds. Nils Roller and Silvia Wagnermaier (Freiburg: Orange-Press GmBH, 2003). Translated into English as "Crisis of Linearity" by Adelheid Mers, in *Boot Print* 1:1 (2006): 19-21.

44   Flusser already discussed the abstraction of reality in his books, *Into the Universe of Technical Images* (Minneapolis, MN: University of Minnesota Press, 2011), originally published as *Universum der technischen Bilder* (Berlin, Germany: Andreas Müller-Pohle Stargarder, 1985); and *Towards a Philosophy of Photography* (Göttingen, West Germany: Andreas Müller-Pohle Stargarder, 1984).

dimension was abstracted from the knife." Humans then "retreated from the circumstance into his subjectivity" for a "distanced view of the world," where "the world is no longer palpable, the hands no longer reach it. This is a world that is only apparent for the eyes." In this respect via, for instance, painting and cartography, humans then introduced "a bi-dimensional zone of first degree imagination," or "images that represent the concrete world." Text, or lines of writing, then introduces "a unidimensional zone of alphanumerically codified explications," but which "rupture the veil of imagination: the alphabetic and numeric codes, and their use, allowed for the development of discursive reason, which is the ability to analyse, to critique, to enumerate, to align, and to calculate."[45]

The current zone we are in, according to Flusser, is of "the post-rational" – the zone of zero-dimensionality consisting of "computed and digitally codified images,"[46] but in which also includes point-by-point recorded media of "techno-codes," such as photography,[47] cinema and, indeed, holograms. Photography, cinema and even holographic projections, as discussed, produce the ghosts of the dead via reproducing images of their likenesses. Friedrich Kittler, in his book *Gramophone, Film, Typewriter*, published in German coincidentally in the same year (1986) as Flusser's "Into Immaterial Culture" lectures at São Paolo, referred precisely to such ghosts of the dead out of reproducible media. In thinking similar to Flusser's divergent zones of uni- and zero-dimensionality, Kittler specifically contrasts ghosts out of writing – "the regime of the symbolic"[48] – against the reconstructing or reproducing of bodies through media "of physical precision": "once...the dead and ghosts become technically reproducible, readers and writers no longer need the powers of hallucination. Our realm of the dead has withdrawn from the books in which it resided for so long."[49] Kittler quotes from media theorist Rudolf Arnheim: "[media reproductions] are not only supposed to resemble the object but rather guarantee this resemblance by being, as

45   All quotations in this paragraph are from Flusser, *Into Immaterial Culture*, 26-32.

46   Flusser, *Into Immaterial Culture*, 32.

47   See Flusser's explanation of the photograph as a technical image: "A photograph is not the image of the facts at hand, as was the case with the traditional image, but rather the image of a series of concepts, which the photographer has come up with in the scene that signifies the facts at hand": from *Writings*, trans. Erik Eisel (Minneapolis, MN: University of Minnesota Press, 2002), 40.

48   Friedrich A. Kittler, *Gramophone, Film, Typewriter* (Palo Alto: Stanford University Press, 1999), 8.

49   Kittler, *Gramophone*, 10.

it were, a product of the object in question, that is, by being mechanically produced by it."[50] Buttressed by this guaranteed reference "to the bodily real" through mechanical reproduction, media thus "always already provide the appearances of specters."[51] Media does not, as he puts it, "have to make do with the grid of the symbolic."[52]

However, Flusser's thinking of the zero-dimensionality of point elements and of techno-images proposes a more radical thinking of this "post-rationality," namely, the imposition of the "second degree imagination." Out of this alternative imagination from point culture emerges "a new layer of consciousness – with new codes, and therefore, with new categories of thought, evaluation, and action"[53]… and new ghosts. With the "second degree imagination" visualized by new images of the brain and the nervous system, the "digital apparitions" are thus able to appear. The hologram, *as a particular iteration of the post-screen* in merging the screen with the brain and in directly connecting point information and individual consciousness, is thereby the precisely appropriate vessel for these new ghosts, as opposed to the media vehicles of photographs or cinema contained as they are within screen boundaries. These are not ghosts of the dead, but ghosts out of a different order of reality and apprehension of reality, namely, the "emerging alternative worlds" that we have created out of the images of zero-dimension, *including ourselves*: "the 'digital apparition' is the light that illuminates for us the night of the yawning emptiness around and in us. *We ourselves, then, are the spotlights that project the alternative worlds against the nothingness and into the nothingness.*" (emphasis added)[54] Both kinds of ghosts problematize the real, as ghosts do, but there is an ontological shift here of ghosts from the spiritual dimension of the dead to the ghosts from the zero dimensionality of the living. If linearity exorcises or demystifies the ghosts of the dead from the two-dimensional image – with its accordant powers of magic and wizardry – *what happens in this instantiation of the post-screen is their return as digital apparitions with a necromancy of their own on the terms of zero-dimensionality and post-rationality, confronting us with new problems of reality and questions about ourselves. Cf* specifically with Kittler's thinking, the ghosts

50   Rudolf Arnheim, as quoted in Kittler, *Gramaphone*, 11-12.
51   Kittler, *Gramophone*, 12. Kittler himself notes the connection between media and ghosts via linking the invention of the Morse alphabet in 1837 to "the tapping specters of spiritistic séances sending their messages from the realm of the dead": ibid.
52   Kittler, *Gramophone*, 11.
53   Flusser, *Into Immaterial Culture*, 35.
54   Flusser, "Digital Apparition," 245.

of Flusser's "second degree imagination" (and of the true hologram) thus for the first time break free of the problematic of mummification that for so long dominates the ontology of the image. Even interactivity, as via the more recent images out of gaming and social media, can be read as an extension of that preservation paradigm. In a way, the ghost out of interactive media is no more than the mummy in Mahfouz's short story as awoken not quite to speak but, in a similar spirit, to post, comment and affect the present.[55] Instead, these new apparitions are an entirely different beast – they are the afterlife of rationality, certainty and the symbolic as set loose by this point culture of zero-dimensionality. They are the hauntings of, as Flusser puts it, "our epistemological problem, and therefore also our existential problem," which is, as it has always been, about how we should understand ourselves. They are, in short, a different kind of ghost.

A final connection between ghosts and the understanding of ourselves, if from an entirely different source located miles away in intellectual discipline from media theory. In the web television series, *The Haunting of Hill House* (2018), a character, per the opening quotation of this section, explains ghosts as memories, daydreams, secrets, emotions, desires; "most times, [ghosts are] just what we want to see......a ghost is a wish."[56] This 2018 web television series was re-imagined from Shirley Jackson's 1959 novel of the same title, widely regarded as one of the finest horror novels of the twentieth century, and Jackson herself much admired generally for her masterful writing of horror and suspense. The novel's main innovation, picked up as well by the TV adaptation, was to entwine the supernatural elements such as the haunted house with the characters' psyches and inner lives: what they desired; what they lacked; what they feared; the secrets they kept; the illnesses they had. Those were the true ghosts of Hill House – *not from the dead, but the bodies*

---

55   We may recall the diatribe of the literally awoken mummy in one of Arab writer (and Nobel Literature prize winner) Naguib Mahfouz's early short stories, who rises only to rain admonishment on the living Pasha for his arrogance and unkindness – itself a gesture to the temporality of the photograph in Mahfouz's use of Egypt's Pharaonic past to politicize its problematic present: see Naguib Mahfouz, "The Mummy Awakens," trans. Raymond Stock, *The Massachusetts Review* 42(4) (Winter 2001/2002): 507-523. For a purely technical point of view, see as well how researchers have literally recreated the voice of a 3,000-year-old mummy with "a 3-D printer, a loudspeaker and computer software" out of CT scans of the mummy's mouth and throat, combined with an electronic larynx: Nicholas St. Fleur, "The Mummy Speaks! Hear Sounds From the Voice of an Ancient Egyptian Priest," *The New York Times*, January 23, 2020, https://www.nytimes.com/2020/01/23/science/mummy-voice.html.

56   *The Haunting of Hill House*, "Steven Sees a Ghost," directed and written by Mike Flanagan, *Netflix*, 12 October 2018.

and minds of the living. The post-screen, then, via the true hologram thus also becomes a space in which media theory asks its questions afresh in the face of a new media logic of point culture and zero-dimensionality: where is the screen, and where are its ghosts? In this chapter, my answer is this: like the ghosts of Hill House, the screen and its ghosts ultimately rebound and are re-bound to us, the viewer.

5   Light Projections: On the Matter of Light and the Lightness of Matter

Abstract

This chapter explicates light projections as the third instantiation of the post-screen. The chapter's argument premises on light being the matter of light, namely, light as transformational in the convertibility between materiality and immateriality; matter and energy; rigidity and fluidity, whereby the corporeal is not only rendered impalpable but – its body energized, vitalized and volatile – turned inside out; consumed; deposed. The chapter thus argues for the post-screen through contemporary light projections and projection mapping technologies as the transformation of a variety of surfaces – the urban (e.g. building façades); the amorphous (e.g. water droplets and ash); the biological (e.g. bodies and faces) – giving rise not only to dynamic interrelations between materiality and immateriality, but also a radical convertibility between matter, energy, solidity, mass, and body that signals a moment of media history today.

Keywords: light projections; projection mapping; energy; mass; materiality; immateriality

The City Rises

In 1910, the Italian painter and sculptor Umberto Boccioni completed his painting, *The City Rises*.[1] A monumental work spanning almost ten feet across, it is a major statement of Italian Futurism and its main themes: technological progress, speed, energy, movement, violence, destruction and youth.[2] Boccioni's city rises with the relentless drive for modernization that

1   Umberto Boccioni, *The City Rises*, 1910, oil on canvas, 199 cm x 301 cm, Museum of Modern Art (MoMA).
2   On Futurism generally, see Marjorie Perloff, *The Futurist Moment: Avant-Garde, Avant Guerre, and the Language of Rupture* (Chicago: University of Chicago Press, 2003).

Ng, J., *The Post-Screen Through Virtual Reality, Holograms and Light Projections. Where Screen Boundaries Lie*. Amsterdam: Amsterdam University Press 2021
DOI: 10.5117/9789463723541_CH05

coloured the Futurist movement, its vision writ large across the smokestacks and construction scaffolding in the painting's background as, in the foreground, human (read as male) labour harness horses and machines towards their dream of industrial and technological development.

But it is the painting's light that catches my attention. Boccioni uses the technique of Divisionism which separates colour for optical interaction to achieve maximum luminosity.[3] His rapid directional brushstrokes and the brilliant colours cast across the painting emphasize the power, dynamism and destructive promise of the rising city. But light also represents several beliefs held by Boccioni and Filippo Tommaso Marinetti (who led the Futurist movement and authored its first manifesto). One of these was how, following "recent scientific and pseudoscientific theories of matter,"[4] such as those by Gustave Le Bon's *The Evolution of Matter* (1905), a text well known to the Futurists, the structure of matter could "dissociate" into energy. Matter became "particles endowed with immense speed, capable of making the air a conductor of electricity, of passing through obstacles, and of being thrown out of their course by a magnetic field." According to Le Bon, matter could dematerialize into electricity, light, heat or other "unstable forms."[5] The spillage and collision of light in *The City Rises* thus reflect not only the city's dynamism, but also its *transformative* capacities in terms of the dissipation, even vaporization, of matter and materiality – from stability and fixity to energy and volatility. They set a literal scene for the reversibility or convertibility between matter and energy – materiality and immateriality – which, as we shall see, also substantiates the matter of the post-screen through light projections.

These interplays in the city under the transformation of light thus constitute the theoretical underpinnings of this third and final iteration of *the post-screen through light projections*, completing its triangulation with the other two iterations through virtual reality (VR) and holographic projections. The chapter will argue that contemporary illumination and projection mapping technologies diminish the perceptibility of screens' boundaries by *fusing* image and light into the materiality and solidity of the structures on which the light is projected. This fusion thus forms a virtual

3   See Charles Blanc, *The Grammar of Painting and Engraving* (Chicago: S.C. Griggs and Company, 1891). For a critique of Divisionism and its proponents' understanding of colour theory, see generally Alan Lee, "Seurat and Science," *Art History* 10 (June 1987): 203-24.

4   Christine Poggi, *Inventing Futurism: The Art and Politics of Artificial Optimism* (Princeton, NJ: Princeton University Press, 2009), 101.

5   All quotations from Le Bon in this paragraph are from his *Propositions of Force and Matter*, as quoted in Poggi, *Inventing Futurism*, 102.

skin which, energized by light, deposes of the structure's materiality, whereby the structure is turned inside out and itself consumed – *swallowed whole* – by the light. The process of transformational light on the post-screen through projections and projection mapping is thus of vitality and dynamism, but, more than that, also a kind of volatility that comes from devouring and senseless ingurgitation. Hence, this third instantiation of the post-screen emerges as not merely an ambiguous surface on which the light reposes, but an unbounded territory for the image in the city which, like Boccioni's colour in his painting, takes on a destructive life in its morphing, dematerialization, instability and gluttonous desire. With this metamorphic energy, the city rises again, this time drawn not from physics and pseudo-science, but the unique coalescence between screens, images, media and materiality.[6] As solidity diminishes, something else appears: a new space for the scribing of solidarity, but also for the deliquescence of solidity.

***

The Light Rises, or, Light as the Matter of Light

> *There is a certain essentialism inherent to Architecture, a certainty that after the weather has washed the informing decorations off the stones of the Acropolis, erased the strident colours from the striated columns of Durham Cathedral, that the play of light, from the rising of the sun to the setting of the moon, across the remaining 'dumb platonic solids', will animate the structures for us; imbue them with a poetic that chimes with our inescapable one-ness with the universe.*
> ~ Leon van Schaik[7]

Light transforms life. Van Schaik writes, as above, of the "play of light" as a force of life-giving animation, almost mythical in its cosmic connection. But light also transforms in palpable socio-cultural ways. For example, Wolfgang Schivelbusch, in his vaunted account of the industrialization of light, describes how various developments of street lighting changed the life of the European city. These developments range from the use of lanterns

---

6   William J. Mitchell's *City of Bits: Space, Place, and the Infobahn* (Cambridge, MA: MIT Press, 1995) also comes to mind here as a related idea, albeit connected more to digital telecommunication and electronics rather than specifically to media environments.

7   Leon van Schaik, "The Lightness in Architecture," in *Urban Screens Reader*, eds. Scott Mcquire, Meredith Martin and Sabine Niederer (Amsterdam: Institute of Network Cultures, 2009), 75.

displayed outside each Parisian house "to identify itself" in the sixteenth century to lanterns fixed on the streets in the late seventeenth century to the use of reflector lanterns (or *réverbère*, its original French name) in the eighteenth century with its technical advancements of multiple wicks and dual reflectors.[8] Each level of street lighting beckoned an observably different aspect of city life. For instance, street lanterns tightened absolutist police authority as their lighting allowed authorities to patrol and control the streets at night – "the lanterns showed who lit the streets and who ruled them."[9] Street life also changed with the phenomenon of "lantern smashing" which appeared in eighteenth-century Paris, where *réverbères* were frequently destroyed as "a small act of rebellion against the order that it [the street lantern] embodied," and whose destruction became "a collective, plebeian movement" during the nineteenth-century revolutions and rebellions in Paris.[10]

Conversely, commercial and advertising lighting in the cities of eighteenth-century Europe transformed night life for shopping, entertainment and pleasure. As Schivelbusch (again) writes, night life "derives its own, special atmosphere from the light that falls onto the pavements and streets from shops..., cafés and restaurants, light that is intended to attract passers-by and potential customers."[11] The English pleasure gardens in the 1700s such as those at Vauxhall and Ranelagh "only really came alive at night," lit with the attractions of "concerts, illumination and fireworks."[12] The bright lights of the city ushered the late hours of baroque culture into a nocturnal whirl of operas, theatre, late evening meals, soirées and other late night pleasures for the leisured classes. Commercialized and festive illumination enabled shops, warehouses and entertainment businesses to run deep into the night for the masses.

However, it was electrification which capped the transformation of the lit city. Electrical light was first pioneered in 1800 by Humphrey Davy as arc lighting, "produced by a discharge of electric current between two carbon electrodes."[13] In 1879, Thomas Edison developed the carbon filament lamp which, on its showing at the Paris Electricity Exposition in

8   Wolfgang Schivelbusch, *Disenchanted Night: The Industrialization of Light in the Nineteenth Century*, trans. Angela Davies (Berkeley and Los Angeles, CA: The University of California Press, 1995), 82.
9   Schivelbusch, *Disenchanted Night*, 87.
10   Schivelbusch, *Disenchanted Night*, 100.
11   Schivelbusch, *Disenchanted Night*, 142.
12   Schivelbusch, *Disenchanted Night*, 140.
13   Schivelbusch, *Disenchanted Night*, 52.

1881, introduced the electric bulb to the wider public. Replacing the carbon filament with tungsten brought about the greater brightness of modern illumination.[14] As central electricity stations, and then remotely built power stations, supplied electricity not only to a single town but whole regions, the city transformed yet again. As with gaslight but with greater effect and popularity due to its relative safety, lack of odour and centralized supply, electric lighting galvanized shopping areas as "shopkeepers understood lighting as a weapon in the struggle to define the business centre of the city, dramatizing one sector at the expense of others."[15] Among numerous other changes, electricity supplies also re-constituted the city's economic structures: Schivelbusch, for instance, describes how "the concentration and centralisation of energy in high-capacity power stations corresponded to the concentration of economic power in the big banks."[16]

Of greater interest, though, for this chapter is how electrical light transformed the city's *materiality*, particularly, as Boccioni envisioned (via Le Bon) in *The City Rises*, in terms of evanescing the solidity of its buildings and structures as a way of expressing its politics, aliveness and energies. These connections between light, energy and the city occur as common themes in various accounts and descriptions. As Scott McQuire writes, "[t]he experience of the modern city seen under electric lights conferred a novel sense of mutability on the previously immutable and monumental... To some observers, light seemed capable of dissolving their mass entirely."[17] A key transfiguration of the city by electrical light is thus to render, as McQuire puts it, the city's "growing sense of architectural ephemerality."[18] By its electrically lit urban buildings, McQuire argues that the city becomes fluid, immaterial, ethereal, oneiric.[19] He quotes from Ezra Pound (amongst

14 Schivelbusch, *Disenchanted Night*, 58.

15 David Nye, *American Technology Sublime* (Cambridge, MA: MIT Press, 1994), 177.

16 Schivelbusch, *Disenchanted Night*, 74.

17 Scott McQuire, *The Media City: Media, Architecture and Urban Space* (London: Sage, 2008), 122.

18 Ibid.

19 This, in turn, ties into large amounts of literature on the city as more generally occupied by dreams and ghosts – see, for instance, Steve Pile, *Real Cities: Modernity, Space and the Phantasmagorias of City Life* (London: Sage, 2005) – and the persuasive interrogation of cities' realness, or the consideration of their being "a state of mind" in terms of, for example, "a kind of psychophysical mechanisms in and through which private and political interests find not merely a collective but a corporate expression": Robert E. Park, Ernest W. Burgess and Roderick D. McKenzie, *The City* (Chicago: The University of Chicago Press, [1925], 1984), 1-2. The references quoted in the text, however, more specifically refer to connections between the city and its immateriality in direct relation to media and its effects.

others): "It is then [in the evening, when the lights come on] that the great buildings lose reality and take on their magical powers. They are immaterial; that is to say one sees but the lighted windows."[20] In her inimitable 1925 novel, *Metropolis*, Thea von Harbou describes how Georgi first saw the city of Metropolis – "wonder of the world" – almost blinded and drugged by the city's light and energy: "by night shining under millions and millions of light," Metropolis was "the ocean of light which filled the endless trails of streets with a silver, flashing lustre"; there was "the will-o'-the-wisp sparkle of the electric advertisements"; "an ecstasy of brightness."[21] In these imaginations, the materiality of the urban melts away in the energy of almost too-bright light, as ice by fire. Sergei Eisenstein, too, alludes to this liquefied quality of the lit city in his description of the "modern urban scene, especially that of a large city at night"[22] as he points out its "absence of perspective" where

> all sense of perspective and of realistic depth is washed away by a noc-
> turnal sea of electric advertising... these lights tend to abolish all sense
> of real space, finally melting into a single plane of colored light points
> and neon lines moving over a surface of black velvet sky.[23]

Hence, besides spurring the myriad palpable social, cultural, economic and architectural transformations of the city, projections and emanations of light also transform and convulse the city into ontological flux between the materiality of monuments, bridges and other physical structures, and the immateriality of flow, momentum and dynamism of energy. Therein as well lie the nuanced differences between *light as the matter of light* and *light as lighting*, whereby, *cf* the latter, the former constitutes the basis of the argument of the post-screen through light projections. *Light as lighting* is the concern of analyses such as those of Schivelbusch's, namely, the transformation of night into day whereby light reveals the environment

---

20 As quoted in McQuire, *The Media City*, 122. The imagery of ethereality in these lines also recall Ezra Pound's other famous Imagist poem, evoking the same sense of the apparitional and the impalpable: "The apparition of these faces in the crowd;/ Petals on a wet, black bough," from "In a Station of the Metro" in Ezra Pound, *Personae: The Shorter Poems of Ezra Pound* (London: Faber & Faber, 1952), 4.

21 All quotations in this paragraph are from Thea von Harbou, *Metropolis* (Cabin John: Wildside Press, [1963] 2003), 35-6.

22 Sergei Eisenstein, *The Film Sense*, ed. and trans. Jay Leyda (New York: Meridien Books, 1957), 98. More precisely, Eisenstein's comparison of the city at night was made to jazz, where similarly conventional perspective is lost as "in jazz all elements are brought to the foreground," 96.

23 Ibid.

around one which would otherwise be in the dark, and in so doing triggers profound social, cultural and economic transformation.

In comparison, *light as the matter of light* argues for thinking through the whole relationship between matter and energy and, in turn, the play between the material (i.e. of matter) and the immaterial (i.e. of light/energy). Matter is thus not merely revealed by light; it is also both *dissolved* – i.e. converted into energy per Einsteinian physics – and *constructed by* light; or de-materialized and materialized. In turn, this interplay between dissolution and construction becomes crucial to the conceptualization of the post-screen through light projections. In this conceptualization, the fluid and dissolving boundaries of matter against energy form the conceptual framework through which the boundaries of the image may be similarly conceived in alternative relationships against object, or as screen against image.

This interplay of materiality and immateriality per *light as the matter of light* may already be seen in numerous examples across existing architectural and artistic spheres. One example out of many is the use of electrical light to literally substitute the solid volumes of urban structures, where buildings are not dissolved *by* light but "constructed" *of* light. Materiality thus shades into immateriality in paradoxical interplay – visible yet incorporeal; concrete structure against ethereal reflection. Perhaps the most prominent instance of such paradoxical construction is the *Tribute in Light* installation set up in New York City to memorialize the 9/11 attacks, whereby two giant beams made up of eighty-eight 7,000-watt light bulbs and reaching "up to four miles into the sky"[24] are set up at the site of Ground Zero in New York City to commemorate and represent the absent World Trade Centre towers. The specific immateriality of light thus becomes "a building material" in itself.[25] There are many other examples; a further one from a different era might be Robert Krebs's *Sky-Pi*, a ten-day long light project created as part of the Greater Philadelphia Cultural Alliance's 1973 May Festival. Using a composition of lasers and strategically placed reflective surfaces, Krebs's *Sky-Pi* beamed various lengths of laser light connecting the mile-long stretch between the Philadelphia Museum of Art and City Hall, described "to form

---

24 Clayton Guse, "The 9/11 Tribute Lights Have Returned to NYC," September 6, 2018, *Timeout New York*, https://www.timeout.com/newyork/news/the-9-11-tribute-lights-have-returned-to-nyc-090618.

25 Anne Friedberg, *The Virtual Window: From Alberti to Microsoft* (Cambridge, MA: MIT Press, 2006), 151-2. McQuire also points out this "long line of light-based architecture," citing Albert Speer's "Dome of Light" (or "cathedral of light") as created from a series of anti-aircraft searchlights encircling the Nuremberg rally of 1935: *The Media City*, 113.

an intricate lattice that almost abolishes any sense of bodily space."[26] Instead, the lines of lights themselves, *as matter* in their immateriality, become the markers and constructors of physical space. In the domain of art, we might also think of Robert Irwin's 1998 (and re-exhibited in 2015) art work, *Excursus: Homage to the Square*, for which Irwin created sixteen (and later eighteen) chambers,[27] all divided by scrim-like walls made of delicate, translucent fabric. As with the use of light which perplexes the materiality of building structures in *Tribute in Light* or even Albert Speer's "Dome of Light,"[28] light similarly accords *Excursus*'s chamber walls a peculiar texture and luminosity in the work's deliberate play between materiality and immateriality, whereby "[t]he material [of these scrim-walls] is fundamentally luminous in the way it reflects and absorbs the natural and artificial light that constitutes an important part of the installation's architecture."[29] Not only do the walls take on light to layer and confound their own solidity, but the light itself, as it falls on the fabric, also acquires its own materiality: "Filtered through scrims that are essentially veils, light itself appears layered, coated, and textured."[30]

In these myriad manifestations across diverse contexts, *light as the matter of light* is thus seen to be transformative of mass and materiality, convertibly connective to energy in flux against immateriality. To now tie together Boccioni, light-city-life-transformation and, of course, the post-screen, of primary interest in this chapter is the work of light in both the transformative constitutions of screens and screen boundaries, and their ensuing paradoxes not only of materiality and immateriality, mass and energy, but also the themes by now familiar in this book: space and non-space; animation and being inanimate; skin and object; two and three-dimensionality. As light subverts screen boundaries in the city, the post-screen emerges out of not only the urban environment, but also the space of these multiple binaries and the folds between their paradoxes. Like steel on flint, the sparks of life and energy in the interplay of such subversion and paradox become important constitutions of the post-screen. Where Pound and Harbou et al. write of the lit city's fluidity and dream-like state of dazzle and blinding

26 Jennifer M. Rice, "The Evolution of a Laser Artist," *Optics and Photonics News* (1999): 20-23, 23.

27 The installation in 1998 appears to have 18 chambers, while its exhibition in 2015 features 16.

28 See L. Krier, *Albert Speer: Architecture: 1932-1942* (New York: Monacelli Press, 2013).

29 Giuliana Bruno, *Surface: Matters of Aesthetics, Materiality, and Media* (Chicago: The University of Chicago Press, 2014), 73.

30 Bruno, *Surface*, 74.

brilliance, the post-screen likewise breaks down the city's brute solidities through media, light and diminished boundaries. What also bears out in this constitution of the post-screen through light projections is a strange parallel between the development of media and that of science, where Einsteinian physics of matter, light and energy – immortalized via "the most famous equation in the world"[31] of e=mc² from Einstein's 1905 paper and, incidentally, practically contemporaneous with Boccioni's *The City Rises* completed in 1910 – breathes the same air as the post-screen. As does physics render the material into the immaterial – of subatomic particles and quantum mechanics – so does media, *qua* the post-screen, convert the solid into energy, as likewise pictured in *The City Rises*. The same intuition thus runs through physics and art history and, via the post-screen, through media too: the reversibility or *convertibility* between matter and energy which literally scribes new rules of constituting, thinking about and looking at our material realities.

Cities of Screens

> *When the screen becomes our dominant quotidian interface, there is an understandable desire to extend that interface into the public, urban sphere – so that all surfaces, animate and inanimate, likewise become screens.*
> ~ Abigail Susik[32]

The city is replete with screens. One need only stand for a few minutes at the Shibuya Crossing in Tokyo or in the middle of Times Square in New York to appreciate the density of screens in urban surroundings as illuminated with a constant flow of still and moving images, invariably crammed with advertising, and always in fierce competition for attention. Screens appear in myriad other places in the city too – along passageways in subway stations, buses and trains, hotels and shopping centres, as well as in the form of large screens for events in sports stadia, outdoor concerts in parks and open air film screenings, among many others. Their omnipresence, however, seems to backfire as the sheer multitude of street screens conversely causes them to fade instead into visual background noise.

31 David Bodanis, *E=mc²: A Biography of the World's Most Famous Equation* (London; New York: Bloomsbury, 2009).
32 Abigail Susik, "Sky Projectors, Portapaks, and Projection Bombing: The Rise of a Portable Projection Medium," *Journal of Film and Video*, 64(1-2) (Spring/Summer 2012): 79-92, 84.

Christoph Kronhagel describes Times Square as "a cacophony where nothing is attuned to anything else...merely generat[ing] a white blur."[33] As Erkki Huhtamo writes: "Passers-by glance at the screens, but don't get easily 'absorbed' into them. The wall-mounted screens form an ambience rather than a set of targets for sustained attention."[34] Malcolm McCullough takes up the entire theme of ambience in his book, *Ambient Commons*, to argue for a new era of contextual media and information environments which warrant a different kind of attention and tuning in, or at least "to re-examine the urban citizen's distraction."[35] A chief element of this environment of ambience is the electrically lit city as "glowing forms," where "with electrification, walls were not only written on, but lit up as well," and as media façades, "when huge electronic displays become a persistent part of physical architecture."[36]

As these scholars have shown, the wall-mounted or fixed screen forms a part of the long history of public media displays, reaching back from large banners displayed for travelling shows in the nineteenth century to broadsides in sixteenth-century Britain to signboards by ancient Romans "to identify craftsmen's workshops and various services."[37] However, Huhtamo, with characteristic nuance from his media archaeological approach, connects the "proto-screen" not only to these billposters but also the practice of what is called "placard advertising" in England, where companies bought legal rights to use divided "lots" of space for their advertising in a bid to bring some order to "decades of billposting anarchy," where broadsides in England in the first half of the nineteenth century were pasted and layered over all available surfaces with unrestraint.[38] As Huhtamo points out, a screen is "an information interface," and so "should function both as a frame and a gateway through which messages are transmitted and retrieved."[39] A

33 Christoph Kronhagel with Phil Lenger, "Designing for Commerce in Public Spaces," in *Mediatecture: The Design of Medially Augmented Spaces*, ed. Christoph Kronhagel (Vienna: Springer-Verlag, 2010): 172-179, 174.

34 Erkki Huhtamo, "Messages on the Wall: An Archaeology of Public Media Displays," in *Urban Screens Reader*, eds. Scott McQuire, Meredith Martin and Sabine Niederer (Amsterdam: Institute of Network Cultures, 2009): 15-28, 15.

35 Malcolm McCullough, *Ambient Commons: Attention in the Age of Embodied Information* (Cambridge, MA: The MIT Press, 2013), 23.

36 McCullough, *Ambient Commons*, 143-152.

37 Huhtamo, "Messages on the Wall," 16.

38 Huhtamo, "Messages on the Wall," 17-18. See also Erkki Huhtamo, "Pre-envisioning Mediatecture: A Media-archaeological Perspective," in *Mediatecture: The Design of Medially Augmented Spaces*, ed. Christoph Kronhagel (Vienna: Springer-Verlag, 2010): 20-27.

39 Huhtamo, "Messages on the Wall," 17.

broadside or billposter by itself as pasted on a public wall, while a medium of textual and visual message, is indeed not quite a screen in the sense of images and text contained in an enclosed frame. Yet as has also been the thesis of this book, a screen is the conceptual vehicle for a frame which contains information within it, and in particular the boundaries of which acquire significance in terms of what they include, exclude, protect against and gets leaked through. In that respect, the billboard, in terms of where and how its frame divides the advertising against the rest of the wall, is certainly germane to a prototypical urban screen in terms of the ubiquitous rectangles of light that we see across today's cityscapes, flashing advertising, news, art and other information.[40]

Most fixed screens in the city – attached to subway walls, shop windows and so on – have obvious boundaries, marking the borders around the information they contain against their surroundings. Where urban screens edge closer to the post-screen is in their not being self-contained units with clear boundaries, but *re-purposed* from the surfaces, façades and framings of buildings in the urban environment. The post-screen in this case thus emerges where a particular surface in the city doubles up or becomes read as a display of information akin to a screen, and whose boundaries, then, merge between the virtuality of the display and the physicality of its surroundings.

Such re-purposing may take place in one of three ways. The first is how, as McCullough points out, the architectural and aesthetic aspects of buildings become screens by their *architectural framing* of information and content: "Because a [building] façade may bear inscriptions, whether in stone, calligraphy, fresco, flyposting, neon, or LED meshes, its full extent also becomes a frame."[41] As we think about a building façade being a screen in terms of its framing, Anne Friedberg's work in her book, *The Virtual Window*, comes to the fore in her argument relating the perspectival apertures of architectural window frames via openings in walls and rooms to cinema and virtual computational (Microsoft) Windows onscreen.[42] Uta Caspary, citing Martin Pawley, also explicitly parallels building façades (as media architecture) with gothic cathedral windows by way of how they

---

40   See also Nikos Papastergiadis et al., "Mega Screens for Mega Cities," *Theory, Culture & Society*, 30 (7/8) (2013): 325-341, where the authors, in their case study of linking large screens between Melbourne and Incheon to explore "the creation of an experimental transnational public sphere," discuss urban screen use beyond advertising, specifically in three forms, or what they call "alternative models": (i) public space broadcasting; (ii) civic partnership; and (iii) art.
41   McCullough, *Ambient Commons*, 154.
42   Anne Friedberg, *The Virtual Window: From Alberti to Microsoft* (Cambridge, MA: MIT Press, 1999).

both frame and present information: "Both are perceived as originating in a radical change – societal as well as technical and artistic – caused by the advent of new information technologies for the public or a mass audience."[43] This idea is also clear in the self-explanatory title of Denise Scott Brown and Robert Venturi's 2004 book, *Architecture as Signs and Systems*, which argue for architecture "as sign or communication." Whether thinking about the hieroglyphics on the surfaces of ancient Egyptian temples or the sculptures in Greek and Roman temples, buildings framed as both architectural façade and aperture constitute what Brown and Venturi call "billboards for a proto-Information Age,"[44] and in that respect a clear pointer towards being screens for information or communication. Hence, even before being filled with any projected or electronic light for practical display of visual information, a building façade can itself already be framed as a screen on an abstract level by way of its architectural framing and features.

The second and more contemporary way in which buildings may re-purpose into screens is where the façade itself becomes literally *lit as a screen*. Where a single wall-mounted screen in its relatively modest dimensions might form an electronic window on the building surface, with electrification and, more significantly, the development of the light-emitting diode (LED; as well as its associated technologies and materials such as gallium for its semiconductors),[45] many buildings today have large areas of their façades illuminated with LED displays, turning into what various scholars call "media façades."[46] In essence, the building wall, in its entirety, becomes *itself* a screen. In other discourses, the term "mediatecture" has also caught on – a portmanteau word of "media" and "architecture" – for this phenomenon of "something else" that is "no longer just film or pure architecture, design or communication," but encapsulates all of them in a practical and disciplinary amalgamation. As "a mediator between the worlds of built and physical realities on the one hand and of imagined identities and visions on the

43   Uta Caspary, "Digital Media as Ornament in Contemporary Architecture Facades: Its Historical Dimension," in *Urban Screens Reader*, eds. Scott McQuire, Meredith Martin and Sabine Niederer (Amsterdam: Institute of Network Cultures, 2009): 65-74, 67.

44   All quotations in this paragraph are from Robert Venturi and Denise Scott Brown, *Architecture as Signs and Systems: For a Mannerist Time* (Cambridge, MA: Belknap Press, 2004), 24.

45   See Sean Cubitt, "LED Technology and the Shaping of Culture," in *Urban Screens Reader*, eds. Scott McQuire, Meredith Martin and Sabine Niederer (Amsterdam: Institute of Network Cultures, 2009): 97-107.

46   McCullough, *Ambient Commons*, 148. See also the section on "Media Façades" in Kronhagel (ed.), *Mediatecture*, 109-241.

other,"[47] "mediatecture" thus already contains fragments of the imagined and the immaterial in its built materiality. Here is where we get closer to the idea of the post-screen city of energizing and dematerializing light.

Examples of "media façades" abound from around the world as grabbing attention with skyscraper-height swathes of electrical light becomes a kind of marker of world stage prominence. In Jeddah, Saudi Arabia, the King Road Tower building currently holds the record for the world's largest LED screen, featuring 9,850 square metres of LED screen stretched over twenty-one floors on the north and south façades and sixteen floors on the west façade.[48] Another notable example is the Grand Indonesia Tower in Jakarta, a fifty-seven storey skyscraper covered with two LED videoscreens which together comprise of approximately 5,500 square metres of LED coverage along its exterior wall.[49] In 2016, a 3,065 square metre LED screen was installed on the 163-storey Burj Khalifa in Dubai, completed in 2010 as the tallest building in the world (and, at time of writing, still is) to cover its outer façade.[50] On 31 May, 2019, Samsung presented "massive five-screen LED displays" along a face of the twenty-five storey One Times Square building in New York, measuring "more than 1,180 square meters when combined."[51] Urban LED screens also exist out of illuminations of other kinds of surfaces. The "sky screens," for example, in the Chinese cities of Suzhou (located in the Harmony Times Square) and Beijing (The Place mall) are LED "video walls" which run along the underside of overhead walkway covers stretching across shopping malls and plazas, a common architectural feature in East Asia to shelter shoppers against sun and rain.

The expansion of LED screens on urban buildings shows no sign of abating – scrolling through the Pinterest account of "LED Screen" reveals seemingly never-ending arrays upon arrays of media walls and façades from around the world. There is multiple rationale for their popularity – lit buildings attract night-time shoppers, tourists and passers-by; earn revenue through

---

47   Harold Singer, "Origins of Mediatecture in ag4," in *Mediatecture: The Design of Medially Augmented Spaces*, ed. Christoph Kronhagel (Vienna: Springer-Verlag, 2010): 36-60, 38.
48   LEDs Magazine, "Citiled Installs World's Largest LED Media Façade at King's Road Tower, Jeddah," December 9, 2010, https://www.ledsmagazine.com/company-newsfeed/article/16691210/citiled-installs-worlds-largest-led-media-faade-at-kings-road-tower-jeddah.
49   Wolfgang Leeb, "The Grand Indonesia Tower," Media Architecture Institute blog, October 10, 2006, https://www.mediaarchitecture.org/the-grand-indonesia-tower/.
50   Daniel Oberhaus, "Building the Largest LED Screen in the World," *Vice*, September 19, 2016, https://www.vice.com/en_us/article/qkjw97/building-the-largest-led-screen-in-the-world.
51   Kevin Chung, "Samsung LED Displays at Times Square," *Korea News Plus*, June 19, 2019, https://newsarticleinsiders.com/samsung-led-displays-at-times-square.

displaying advertisements; exude technological "cool"-ness; and continue the exhibition of the building's size and monumentality which otherwise get swallowed into the darkness of night. But these scenes of LED screens also constitute a different urbanscape – not one of building structures, but an urbanscape of visual media in terms of what McLuhan calls "light through," as opposed to "light on." In *Understanding Media*, McLuhan first cites how artist, designer and educator György Kepes, through his experiments with photography, photomontages and photograms, "developed these aerial effects of the city at night as a new art form of 'landscape by light through' rather than 'light on.'" Later in the book, McLuhan again picks up these phrases in his chapter on television, connecting "the ceaselessly forming contour of things" in the TV image as appearing "by light through, not light on," with "the quality of sculpture and icon."[52] McLuhan thus differentiates between two kinds of illumination which constitute media: light *on* – in the sense of a landing or resting on a surface; and light *through* – in the sense of a permeation, such as how Kepes's photographs showed the infusion of light in the city, or, indeed, in the ever-changing, flickering plasticity of television images.

In this "lit-through" urbanscape of giant "screens" appearing out of il-luminated building façades, the contours of a border-less post-screen start to emerge. On one hand, the boundaries of these LED grids-as-screen are clearly defined by way of the building's edges. In that sense, they are really just massively scaled-up common LED computer screens that happen to be attached to huge buildings, displaying a combination of text, still and moving images amplified in width and breadth for maximum attention grab in today's era surfeit with media. As McCullough points out, these screens "belong to an architecture for the age of YouTube," designed for "one-minute video clips going viral on the Internet."[53] They are continuations of the glut of conventional screens all around us – just much bigger.

Yet, even as clearly defined screens of light – the "thing that glows and attracts attention with changing images, sounds, and information"[54] – these "media façades" also evoke two paradoxical senses of the post-screen. The first is in terms of cover and protection as discussed in chapter 2. With LEDs spread over the building's surface constituting its "screen"-ness, its

52 All quotations from McLuhan in this paragraph are from Marshall McLuhan, *Understanding Media: The Extensions of Man* (Berkeley, CA: Gingko Press, 2013), 88 and 213.
53 McCullough, *Ambient Commons*, 149.
54 Charles R. Acland, "The Crack in the Electric Window," *Cinema Journal*, 15:2 (2012): 167-171, 168. See also the discussion on the definitions of screens in chapter 1, text accompanying footnote 16.

construction invariably recalls the idea of a skin or mesh over the wall.[55] The technologies' patented names, such as "MediaMesh®" or "IlluMesh®,"[56] likewise reflect this idea, as do the breathy descriptions of their capabilities: the Iluma building (now known as Bugis+) in Singapore features a digital façade called "The Crystal Mesh" where "LEDs integrate well with tiled skins" to combine visual display with "a breathable mesh of polycarbonate polygons, overlaid outside a conventional structural façade."[57] Already, the boundaries of the media façade appear fragile and on the verge of disintegration: it is a cloak of light which covers and wraps the façade, yet as a skin of mesh covering is also permeable, porous, flimsy, and penetrable, per human skin.

The second paradox of these (post-)screens on buildings is that which more directly heralds the post-screen through light projections and projection mapping in its core argument of interplay between materiality and immateriality; of animation and being inanimate. This interplay ultimately reduces to one idea about these lit-through buildings: *the more the building is lit, the more it disappears*. While the brightness of the LED grids renders the lit parts of the building maximally visible at night, the rest of the building structure, shrouded in darkness, becomes invisible in the night. It both appears and disappears. As with the post-screen and its other paradoxes discussed in earlier chapters, the buildings and their boundaries thus disappear in this interplay of light and night, display and concealment.

In this post-screen-esque fluidity, a sense of animation also arises. Thinking here of Boccioni's city whose light gives rise to its specific energies, the LED lights of the city similarly emit a vitality in the appearance and dis-appearance of its structure and boundaries. As Seitinger et al. write, "[a]t the urban scale, strategically deploying ambient light makes the night-time city landscape editable."[58] While their statement primarily refers to using

---

55   *Cf* Laura Marks's response to discussing the screen as skin in an interview: when asked specifically about how the screen can "be thought of as skin," Marks replies that she "wouldn't over-emphasize the screen, for it is just one part of the material way the image reaches the viewer"; the skin of the film covers a more extensive range of materiality in connecting image and viewer. See Laura U. Marks with Dominique Chateau and José Moure, "The Skin and the Screen – A Dialogue," in *Screens: From Materiality to Spectatorship – A Historical and Theoretical Reassessment*, eds. Dominique Chateau and José Moure (Amsterdam: Amsterdam University Press, 2016): 258-63, 259.

56   Two patented names for LED technologies which can be applied onto steel meshes for day-and-night illumination of large surfaces. See their brochure at GKD UK, https://gkd.uk.com/ services/mediamesh-and-illumesh-media-facade-systems-with-leds/.

57   McCullough, *Ambient Commons*, 149.

58   Susanne Seitinger, Daniel S. Perry, William J. Mitchell, "Urban Pixels: Painting the City with Light," Proceedings of the SIGCHI Conference on Human Factors in Computing Systems, April 4-9, 2009: 839-848, 840.

"programmable points of light"[59] as design opportunities for urban displays of text and images, it also gestures to a quality of fluidity and variability of the city's material buildings animated out of the interplay between the dazzling cover of LED screens and the dissolving cloak of darkness. This is not, or not yet, about how buildings and urban structures "come alive" (which arises in more substantive discussion later in relation to projection mapping). Rather, the quality of the post-screen here refers to a vitality out of which lit buildings become changeable, evanescent and ephemeral between darkness and light, particularly as manifested against their structural solidity and monumentality. Paving the way for the paradoxes of the post-screen through light projections, the material and the immaterial thus become convertible: what appears to be solid becomes deliquesced. In this fluid roil of mass, energy, matter and dynamism, the screen itself in these urban spaces thus also collapses in its boundaries: the post-screen emerges.

In more recent years, as interactive media increasingly takes the fore, this dynamism of light in the energizing of buildings becomes ever more visible, where LED façades take on palpably reactive and interactive properties as "a 'skin' that responds to stimuli from outside."[60] For example, the LED façade of the Hotel WZ Jardins in São Paolo, Brazil, responds in real-time to noise and real-time changes in the local air quality as picked up by microphones and sensors installed around the building, and translates the stimulants into different manifestations (such as warmer tones of red and orange to indicate more polluted air; blues and greens for less). Users may also interact with the building via a smartphone app by voice or finger taps.[61] Self-dubbed by its own architects as "The Light Creature," the tension between the nature of the inanimate and the alive-ness of urban structures becomes literal as its boundaries *qua* a screen fall away into the energy of a living surface, or even urban beast.

59   Seitinger et al., "Urban Pixels," 841.

60   Huhtamo, "Pre-envisioning Mediatecture," 20. The idea of lighting as "skin" recalls Daney's metaphor of the screen as hymen, discussed earlier in this book; it will again receive further discussion later in this chapter, particularly in relation to Bazin's cinema as a "skin" of history. A further note here that the interactive building façade pre-dates the advent of LED technology: in 1992, the architect, engineer, designer and "public art provocateur" Christian Moeller, in collaboration with Rüdiger Kramm and Axel Strigl, created "Kinetic Light Sculpture," which embedded "a Frankfurt office building's perforated aluminium façade with floodlights that shifted colors from yellow to blue in response to temperature and wind conditions." See Hugh Hart, "Big Brother; Armed with a Spotlight," *The New York Times*, March 4, 2007, https://www. nytimes.com/2007/03/04/arts/design/04hart.html.

61   Stu Robarts, "São Paulo LED building façade shines a line on noise and pollution," *New Atlas*, August 6, 2015, https://newatlas.com/hotel-wz-jardins-sao-paulo-led-building-facade/38813/.

The "lit-through" urban-scape of media façades thus portends the post-screen, where its scaled-up illuminations already interrogate boundaries between their paradoxes of disappearance and appearance, immovability and dynamism. In the discussions over the next sections on light projections in terms of "lit-on" media, we will see how the post-screen comes to fruition – a skin and body that, with light, becomes alive, pulsating, energized and ultimately self-devouring.

Light Projections (1): Light that Dissolves and Constructs... and of Latency

> The suddenly stopped rays find their own screen against the darkness.
> ~ E.H.G. Barwell[62]

The section above outlined two perspectives – namely, architectural framing; and being "lit through" – in which urban surfaces may thus re-purpose into screens. The third perspective with which to relate buildings and screens, then, is via the other half of McLuhan's bisectional distinction between "light through" and "light on," whereby building façades turn into screens simply by having light projected or shone *onto* them. Again, this is neither new nor contemporary. Erkki Huhtamo, for example, traces the history of urban large-scale public projections to the nineteenth century use of magic lanterns in the United States, where "slides were projected outdoors on screens, blank walls and even public monuments from the 1860s."[63] Many of these projections took place as advertising: citing E.S. Turner, for instance, Huhtamo describes how "ads for 'pills, blacking, and watches' were projected on the side of Nelson's Column and the pillars of the National Gallery [in London]."[64]

As with LED grids, on one level the boundaries and sense of the screen for such projections are straightforward – the virtual boundaries are the outlines of the projected light; the physical boundaries, if at all relevant, are the edges of the material surface. However, contemporary light projections, particularly in their creative and political messaging, actively *subvert* screen boundaries so that walls and façades become not simply re-purposed

62   E. H. G. Barwell, *The Death Ray Man: The Biography of Grindell Matthews, Inventor and Pioneer* (London: Hutchinson, 1943), 118.

63   Huhtamo, "Pre-envisioning Mediatecture," 24.

64   Huhtamo, "Pre-envisioning Mediatecture," 26.

bordered screens – or brute surfaces for the opportunistic displays of light to hawk and advertise – but a dismantled and unsettled *post-screen*, giving rise to dynamic interrelations between the materiality of urban structures and the immateriality of light which trace and scribe new energies across the city. Here, screen boundaries are subverted not so much by the concealment of their edges, as with the totalization of the VR media environment or by the hidden screens and strategic reflections of holographic projections. Rather, they diminish by being writ large in electrical light across the otherwise blanketing darkness of city walls and building surfaces obscured in the night. In the process, they enable a different city to emerge – one whose materiality becomes fluid, as with Metropolis melting under its dazzling light; whose structures become dismantled into different energies and volatilities which pulse the city, as with Boccioni's city; whose walls fall apart in media's convertibility between light and mass to reveal new media-engineered political and democratized spaces: the border-less city of the post-screen.

The key to the post-screen through light projections is thus the conveyance of this dissolving touch. As usual, cinema – the apparatus par excellence of light *and* projection – had always known this convertibility between light and matter, and demonstrates it in creative self-reflexion. There are many examples; a brief highlight here of a prominent example would suffice. In a celebrated scene from Giuseppe Tornatore's *Cinema Paradiso*,[65] film projectionist Alfredo (played by Philippe Noiret) faced an angry crowd denied entry to their film screening due to the cinema theatre being full. Using his knowledge of film projection and reflection (and to the delight of Salvatore (played by Salvatore Cascio), a child who had struck up a friendship with Alfredo), Alfredo cleverly deflected the image to project the movie onto the wall of the building opposite the theatre. As Alfredo moved the projector, the image drifted and undulated across the walls of the darkened projection room, as if possessed of its own life force. Alfredo's and Salvatore's eyes – and certainly the audience's own as well – could not tear away from the hypnotic vitality of its serpentine movements. As the images of the film finally snaked across to the opposite building, the ecstatic crowd gathered and set up their chairs, their eyes glued to the now-transformed wall-as-screen.

For those moments in the film-within-the-film, the actuality of the building is imperceptible; what completely takes over is the virtuality of the projected light, its energies manifest in its animated movement, to which

65   *Cinema Paradiso*, directed by Giuseppe Tornatore (1988; Shenley: Arrow, 2019), DVD.

the audience is irresistibly drawn. Then, it happened: halfway through the viewing, a resident opened his balcony door which happened to be in the middle of the "screen," only to be befuddled by the light beaming onto him and the audience's ensuing indignation at the disruption. The resident promptly retreated and shut the door behind him, and the viewing resumed. The incident lasts a mere few seconds in a scene which ultimately sets up a disaster to come,[66] but it expresses an important message about screens and projected light: the *screen* under the power of projected light becomes a *post-screen* of capricious boundaries that plays between materialization and de-materialization in its metamorphic magic. The opened door abruptly re-materializes the building by (literally) rupturing the light, exposing and undoing its transformation: on the shredding of the light, the screen reverts to the actuality and real-ness of the wall as an urban structure. As the door shuts, the wall-as-screen manifests once more, its light intact and spellbinding. The post-screen here through light projection thus asserts its knife-edge physics of the material and the immaterial, balanced between latency and assertion in the energies of the moving light, and their equally easy dismantling of one state for the other.

Cinema Paradiso is a toast to cinephilia and the inherent magic of pro-jected immaterial light on material surface via the mediated architecture of cinema. Other light projections similarly manifest this interplay of material-ity and immateriality via the post-screen, if out of different inspirations, motivations and politics. For instance, Shimon Attie's 1991 art project, *The Writing on the Wall*,

> ...[slid] projected portions of pre-world war II photographs of Jewish street life in Berlin onto the same or nearby addresses where the photos were originally taken 60 years earlier... Thus parts of long destroyed Jewish community life were visually simulated, momentarily recreated.[67]

Here, in echoes to *Tribute in Light* for 9/11, the projected light becomes a restorative *im-material* – or, we can say, a material of immateriality – for its post-screen surfaces: the light becomes the bricks, mortar, panes and bodies of nostalgia and pastness, shot through with the horror and tragedy

---

66 Immediately after this scene of the re-directed movie, the film strip in Alfredo's projection room catches fire and the original cinema hall goes up ablaze. As the audience empties the cinema in panic, it is left to the child, Salvatore, to rescue Alfredo, who suffers life-changing injuries.

67 Shimon Attie, "The Writing on the Wall," Berlin, 1991-2, http://shimonattie.net/portfolio/the-writing-on-the-wall/.

to come; after all, this light is *the writing on the wall*. In the meantime, though, the light re-constructs, restores and refurbishes: ruined walls are made whole; abandoned doorways are occupied with people in activity; empty windows are filled with thriving businesses. The light projections here thus differ from *Cinema Paradiso*, where the materiality of the wall gives way to the immateriality of the light, whose hypnotizing energies of animated life dissolve the physical into the virtual. In *The Writing on the Wall*, the projected light *reconstitutes* lives and activities in the virtual, through which material and bodies return with their own energies, borne, of course, by the light.[68] The physical spaces become reanimated, charged with their own life force of historical presence and inexorable strain of their bleak futurity.

In other cases, though, these forces take the form of hope and solidarity. In the many dark and deserted nights of Covid-19 lockdowns across the world for much of the first half of 2020, light projections took on another peculiar life force, their brightness substituting the vitality of still cities emptied of life by enforced lockdown. In late January 2020, across Wuhan, the then-epicentre of the virus and one of the first cities in the world to undergo lockdown, the message "Wuhan Jiayou" (meaning "Wuhan, keep going") and other variations were lit across various buildings in the city.[69] Many similar examples followed across the world in the months to come – again, as a random sampling: in February 2020, "Wuhan Jiayou" was projected on Tehran's Azadi Tower as a message of solidarity;[70] in March 2020, "merci" was lit on the Eiffel Tower in Paris as a thank you to key workers;[71] in April 2020, "thank you" was again flashed on the Burj Khalifa skyscraper in Dubai.[72] Light projections also appeared in the form of artwork, such as "LIVES" by Manoel Enrique which featured an illustrated message of "Clean Hands/Save Lives" projected on the NYU Langone Hospital in Manhattan in April 2020.[73]

---

68   The tragic irony being, of course, that this re-constitution via light also only lights the horrors in store for these people, businesses and lives.

69   CGTN, "Wuhan Light Show Honors Heroes as COVID-19 Lockdown Ends," YouTube Video, 1:24, April 7, 2020, https://www.youtube.com/watch?v=LGu-L5ULByw.

70   Iran Foreign Ministry, Twitter post, February 19, 2020, 8:04 a.m., https://twitter.com/irimfa_en/status/1230040418181664769?lang=nl.

71   Le Parisien, "La Tour Eiffel s'illumine et dit "merci" aux soignants," YouTube Video, 0:51, March 28, 2020, https://www.youtube.com/watch?v=Lo7ZfjZ99uU&feature=emb_title.

72   Burj Khalifa, Twitter post, April 1, 2020, https://twitter.com/BurjKhalifa/status/1245405341636005888?ref_src=twsrc%5Etfw.

73   Amplifierart, "CLEAN HANDS SAVE LIVES by @mane_fissurinha projected outside of the NYU Langone Hospital in Manhattan by @the.illuminator." *Instagram*, April 20, 2020, https://www.instagram.com/p/B_BjW5FJx_R/.

In all these instances, as with *The Writing on the Wall*, immaterial light takes on not representation or re-presentation, but reconstituted *presence* – of the material, of the living, and of filling in the absence of life on the streets. In these projections, the literal boundaries of the "screens" are evident, even if, such as for *The Writing on the Wall*, effort was made to incorporate and merge the images into their surrounding environment. However, the insertions are not seamless, but neither is that the point. The post-screen here is not about melding the virtual with the actual into the same space-time, as is the case, for instance, with holographic projections. Rather, it is a kind of re-animation of the inanimate with projected light, a revitalization of absences with presence, a filling of emptiness with bodies. Here, as part of the convertibility through light between mass and energy, the interplay of the material and the immaterial through the ambiguous screen borders of the post-screen lies in reconstruction, rather than dissolution. This play thus presents another dimension of the transformational convertibility of light *as the matter of light* – not only between materiality and immateriality, but also between presence and absence of the corporeal and the structural.

However, in other kinds of projections, the immateriality of the light itself *does* become the point. In *The Writing on the Wall*, the light serves as an *im-material* of reconstruction which re-writes the definitions, effect and boundaries of the wall-as-screen. In other cases, such as that of digital graffiti, the light serves as the im-material *of immateriality* itself. For instance, in 2007 the design studio Graffiti Research Lab (GRL) set up a large-scale projector to enact what have since become very high-profile digital takes on spray-painted street graffiti. Using a vision tracking camera to pick up laser light as beamed by a user onto an unlit surface (such as a darkened building façade or garage door, or the underside of a bridge), the camera tracks the laser beam's movements as it appears on the physical surface.[74] The camera then feeds the data into a computer, whose software mirrors the laser's movements on its own screen. The projector, connected to the computer, finally casts the computer-generated image as "inked" graffiti writ large on the selected urban surface.[75]

---

74   As with Pepper's Ghost projections which rely heavily on appropriate lighting for the illusion to work, the surface for this likewise has to be dark enough for the set-up to work, as it is critical that the camera picks out the brightness of the laser beam in order to register it accordingly onto the computer.

75   Joshua Yaffa, "The Writing's On the Wall. (The Writing's Off the Wall.)," *The New York Times*, August 12, 2007, https://www.nytimes.com/2007/08/12/nyregion/thecity/12graf.html?ex=1188100800&en=8d3af598019e4c0a&ei=5070. In 2010, the Graffiti Research Lab, along with various other partners, made news again with their development of the Eyewriter, a "low-cost,

As with *Cinema Paradiso* and *The Writing on the Wall*, the projected light of GRL for "digital graffiti" transforms building façades and urban surfaces into screens for this "lit on" play of light. One such transformation is the now-familiar deliquescence of the building façade into its surrounding darkness, rendered as a screen only by the graffiti: as Susik writes in relation to these "projection bombing" graffiti sessions, "the 'screen' *dissolves* into an indexical record of the performative act of reinscribing the city space with the life and consciousness of its dwellers." (emphasis added)[76]

However, the transformation from light in this case goes beyond the dissolution of the material and, for that matter, its restoration. Rather, the light transforms the im-material *as immateriality* – a showcase not of the paradoxes of a surface-screen but of its sheer ephemerality, as the digital graffiti appear and then disappear without a trace, as does the darkened "screen" of the building surface itself correspondingly dis-appear and disappear. The post-screen here is thus apparitional in every way – the light on the building surface demonstrates neither its actuality nor virtuality but its *latency* of creativity and potential; of what *could* be written on it out of the unseen vitality of users' energies and efforts (graffiti itself also a longstanding instantiation of subverted creativity); of where and how the screen *could* emerge. In this space of potentiality between building-façade-screen and light, the post-screen thus not only re-scribes graffiti itself – as tags of ephemeral light from a laser pointer rather than paint from a spray can, applied across much larger distances and scale – but also the spaces on which it appears. It re-forms urban surfaces into the fleeting double-ness of the post-screen between the material and the immaterial, and between the present and the absent, re-drawing the boundaries not only of where and how images appear, but where and how screens *could* appear...and disappear.

Light Projections (2): Walls that Fall Apart... and Re-Form

In more recent years, these re-forming of urban surfaces out of the post-screen through light projections has taken on another level of significance, namely, the signalling of new political spaces whose boundaries navigate

open source eye-tracking system that will allow graffiti writers and artists with paralysis to draw using only their eyes": Tannith Cattermole, "Eyewriter enables paralyzed artists to express themselves with eye-drawn art," *New Atlas*, March 19, 2010, https://newatlas.com/eyewriter-art-paralyzed-artists/14566/.

76   Susik, "Sky Projectors," 88.

different vacillations not between materiality, but the establishment of power and the challenges to it. In what has been called "protest projections," text and images are projected onto significant buildings to register various kinds of dissent. While such projections have gained recent prominence, protest projections can be traced as far back as 1989, where protesters gathered at the then Corcoran Gallery of Art (today the Corcoran School of the Arts and Design) to protest the cancellation of a Robert Mapplethorpe retrospective "over concerns that its content [featuring sexually explicit images with homoerotic and sadomasochistic themes] would affect the museum's government funding."[77] In resonance with the holographic protest marches discussed in chapter 4, the protesters at the Corcoran projected images of Mapplethorpe's photographs on the museum's exterior walls, using the light as a defiant substitution of protested virtual presence for mandated actual absence.[78] Once more, the post-screen through projections emerges in fluid fluctuations of boundaries between wall and screen. Here it is not so much a statement on light in transformative flux between the materiality and immateriality of urban surfaces and spaces, but, rather, a re-drawing of space with light as energized by dissent, objection and challenge. Projected light in this case is thus a charge of politics between the virtual and the actual; the post-screen, in turn, becomes the mutable and adaptable space for it.

As mentioned, such projections for political messaging and protest have only gathered apace in recent years.[79] Just to cite a few examples: in October 2010, the Glass Bead Collective projected a sequence of images and text on the J. Edgar Hoover Building, also known as the headquarters of the US Federal Bureau of Investigation (FBI), in protest of FBI raids and

77   Kriston Capps, "A Museum Cancelled its Robert Mapplethorpe Shows – and Decades Later, It's Finally Trying to Make Amends," *The Washington Post* online, June 12, 2019, https://www.bbc.co.uk/news/blogs-trending-39934436. Notably, in June 2019 or thirty years later, the Corcoran Gallery of Art, now the Corcoran School of the Arts and Design, featured a show, "6.13.89: The Cancelling of the Mapplethorpe Exhibition," to continue that conversation about the navigation of politics by art and museum institutions.

78   Stephanie Williams, "Robin Bell contemplates the power of transparency with 'Open'," *The Washington Post* online, February 14, 2019, https://www.washingtonpost.com/express/2019/02/14/robin-bell-contemplates-power-transparency-with-open/?utm_term=.0d519ae2c999.

79   Although light projections which function as protest may also do the work for the entrenchment of power, as social media shared images of live projections of the current Russian President Vladimir Putin on "several buildings in Moscow and Saint Petersburg" as he made a speech on January 15, 2020: see Soviet Visuals, Twitter post, January 16, 2020, 9:40 a.m., https://twitter.com/sovietvisuals/status/1217743391292784640?lang=en. As social media users noted, the images of Putin's face in such largeness of scale staring out from the buildings inevitably invoke the dystopic surveillance of George Orwell's *1984*.

tactics against activists, including claims of intimidation, imprisonment and murder.[80] In 2011, artist Mark Read, with help from other video projection artists, used a 12,000 lumen projector and dedicated software to project a series of messages supporting the Occupy Wall Street movement onto the side of the Verizon Building in Manhattan, New York.[81] In May 2017, another artist, Robin Bell, projected anti-Trump messages onto the side of the Trump International Hotel in Washington DC;[82] in 2018, he again projected slogans onto a Washington DC courthouse which related to the nomination (and subsequent appointment in November later that year) of Brett Kavanaugh as a US Supreme Court judge in the wake of allegations against him for sexual assault.[83] In October 2019, artist Jenny Holzer, famous for her many light projections of text and poetry on public surfaces from the 1990s, set up *VIGIL*, a two-night display, and also part memorial, of projected light onto landmark buildings of the Rockefeller Centre in New York City which shared "testimonies, responses and poems by people confronting the everyday reality of gun violence."[84] Across the Atlantic to the United Kingdom, in early September 2019, an anti-Brexit activist group called Led by Donkeys projected onto the walls of Edinburgh Castle an image of then-House of Commons leader and Conservative Member of Parliament Jacob Rees-Mogg's "infamous Commons slouch."[85] This "slouch" was a reference to the reclining position in which Rees-Mogg took and was photographed (and rebuked by other MPs) on the front benches of the House of Commons during a significant parliamentary debate on the Brexit process.[86] The projection was accompanied by an appropriately punned slogan: "Lying Tory."

80   See also Susik, "Sky Projectors," 87-88.

81   Dashiell Bennett, "How Did They Project That Occupy Wall Street Message on the Verizon Building?" *The Atlantic*, November 18, 2011, https://www.theatlantic.com/national/archive/2011/11/how-did-they-project-occupy-wall-street-message-verizon-building/335321/.

82   Lamia Estatie, "Protest message projected on Trump hotel in Washington DC," *BBC News* online, May 16, 2017, https://www.bbc.co.uk/news/blogs-trending-39934436.

83   Neuendorf, Henri, "'We're Going to Go Where He Works': An Artist Projected a Message of Protest Against Brett Kavanaugh Onto His DC Courthouse," *ArtNetNews*, September 26, 2018, https://news.artnet.com/art-world/robin-bell-brett-kavanaugh-protest-1357492.

84   Philip Stevens, "Jenny Holzer Confronts the Reality of Gun Violence with Rockefeller Center Light Projections," *designboom.com*, October 11, 2019, https://www.designboom.com/art/jenny-holzer-gun-violence-rockefeller-center-light-projections-vigil-10-11-2019/.

85   Sarah Turnnidge, "Huge Picture of Jacob Rees-Mogg As 'Lying Tory' Projected Onto Edinburgh Castle," *Huffpost*, September 5, 2019, https://www.huffingtonpost.co.uk/entry/jacob-rees-mogg-edinburgh-castle_uk_5d70bc33e4b09bbc9ef9fe47.

86   Kevin Rawlinson, "'Sit Up!' – Jacob Rees-Mogg under fire for slouching in Commons," *The Guardian* online, September 3, 2019, https://www.theguardian.com/politics/2019/sep/03/sit-up-jacob-rees-mogg-under-fire-for-slouching-in-commons.

There are also variations to such projections of slogans and messaging in terms of the nature of the surface used, the content of messages involved and/or the mode of the light's production. One variation projected illuminations not in a city, but on the Alpine mountainside of the Matterhorn in early 2020. Nightly, with weather permitting, for more than three weeks from March 24, 2020 to April 19, 2020, the light artist Gerry Hofstetter, in co-operation with Zermatt Tourism, beamed light projections onto the Zermatt peak as an artistic response to the Covid-19 pandemic which had then been escalating to its gravest heights that year throughout Europe. These projections consisted of succinct text messages, such as "#hope," "solidarität" and "#grazie" (to key workers), and images of various countries' flags to demonstrate solidarity in the face of their respective Covid-19 crises.[87] In a similar vein, "Stay Home" and "Stay Safe" messages in text were also projected onto the Great Pyramid of Giza in Egypt in April 2020. As another variation, one night in the summer of 2019, anti-government protesters in Hong Kong directed laser pointers onto the exterior walls of public buildings, such as the Hong Kong Space Museum, to protest the arrest of a student leader, Keith Fong, for buying laser pointers which the Hong Kong police then labelled "offensive weapons."[88] Unlike the other protest projections, the lights in Hong Kong were not organized projections, but "crowd-sourced" laser lights beamed by protesters in a collective yet uncoordinated effort – a mass of laser light dots dancing across the exterior wall of the Museum that neither spelt out any particular slogan nor put together any coherent image.

Through all these light projections which befuddle site and screen, the post-screen emerges. More importantly, the shuffle and flux of materiality against immateriality via the post-screen here come into being as *screen-sites* of politics, hope, solidarity, media and transformational light: as with the projection of *Il Pompieri di Viggiù* on the building wall in Tornatore's *Cinema Paradiso*, in these moments of light appearing on a physical surface, their virtuality becomes the relevant visual realities – here as political resistance and/or energies of hope and solidarity – while the actuality of the structure

87   Eben Diskin, "Tonight, the American flag is being projected onto the Matterhorn in a display of solidarity," *Matador Network*, April 16, 2020, https://matadornetwork.com/read/messages-hope-solidarity-projected-onto-matterhorn/. Note that the Graffiti Research Lab has also projected their "laser tagging" onto different kinds of surfaces, including "miniature pyramids in Italy" and "snow-covered mountains…in Utah": see S. James Snyder, "Graffiti 2.0: Gone by Morning," *Time*, April 14, 2008, http://content.time.com/time/arts/article/0,8599,1730645,00.html.
88   Christy Choi, "'No Tears, No Blood': Hongkongers Stage Huge Laser Show to Protest Against Arrests," *The Guardian* online, August 8, 2019, https://www.theguardian.com/world/2019/aug/08/no-tears-no-blood-hongkongers-stage-huge-laser-show-to-protest-police-arrests.

dissolves into the night. The unstable boundaries of the post-screen through urban "protest projections" thus highlight not only the flux between the material and the immaterial per the post-screen, but also their slippages between the meanings of space and the spaces of meaning. Moreover, the projections are invariably made on buildings which in themselves contain significant meaning. As buildings meld into screens, screens also converge into the buildings. The projections as text (via the slogans of "Pay Trump bribes here," "Brett Kavanaugh is a sexual predator," "Lying Tory" and so on) remind the viewer that the "screen" on which they receive these messages is also a brick construction of symbolic meaning (such as the rule of law and legality from a courthouse) and functional identity (i.e. upholding law, rules and justice as a courthouse). The post-screen through light projections thus also unsettles the boundaries of the *siting* (as against the *screening*) of meaning that vacillate between material structure and lit screen. In turn, these sites finally emerge as *screen-sites* in their simultaneous convergences and contrasts against each other.

Nor need this reminder take only the form of projected text; the resonances of meaning against screen and surface may also take place through image and sound. For instance, in *Hiroshima Projection* (1999), artist Krzysztof Wodiczko projected the hands of atomic bomb survivors onto a wall of a river bank next to the Atomic Bomb Dome memorial in Hiroshima. This memorial building, originally known as the Hiroshima Prefectural Industrial Promotion Hall building with a distinctive dome at its highest part, is historically significant as one of the few structures left standing in Hiroshima near the bomb's hypocentre. Wodiczko's projections were further accompanied by voice recordings of the bombing's survivors relating their stories.

As with the protest projections, the post-screen via the bank wall converges the meanings of place (i.e. the symbolism of the Dome as memorial, as well as the building's indexical traces of the historic bombing) with the meanings from the audiovisual projections per their filmed images and recorded stories. In this case, though, the post-screen takes on an additional form – not just wall or screen, but a further layer of reality, more in the sense of a thin coating. Architect and architectural historian Eran Neuman gestures to this additional layer in the following way: "the [Hiroshima Projection] did not derive from the memorial's formal or geometrical contexts. Instead it *suggested an addition* to the site's physical presence… and *added another layer* to the memorial's content." (emphasis added)[89] Neuman does

89   Eran Neuman, "Inside Out: Video Mapping and the Architectural Façade," *Leonardo*, 51(3) (2018): 258-264, 261.

not elaborate on what he means by this "added layer," although, in the context of the article as a piece on architectural façades, he presumably refers to the projections as adding architectural (visual and aural) content to the memorial building. At the same time, the artist also refers to the idea of layering in how he places image over surface. To Wodiczko, the projections become a "skin" of separation, specifically to signify the disengagement of empathy:

> The images are not projected on the white screen, but on the facades that are carved. They have their own iconic arrangements or texture, made of bricks or mortar. And this is important. There is an image; there is a building. There is a body of the person, projected; and there is a body of the building or the monument, animated. *But it is also the skin of the building, the surface, which is seen as something in between.* And that's a very important protective layer—that separates the overly confessional aspect of the speech of those who animate the building and our overly empathetic approach towards the speakers [emphasis added].[90]

However, another way to look at this layering of light is in terms of the boundaries of the post-screen within which the image rests. Or, rather, the multiple transmuting iterations of those boundaries – the surface that becomes the screen that becomes a skin. By virtue of its siting/screening, the post-screen here through this projection amalgamates surface, wall *and* its coating of light as an organic membrane that fuses between viewer, place and history; a layer, moreover, that is weighted by the sheer heft of the events' historicity. This characterization of the post-screen as skin also recalls Serge Daney's description of the cinema screen as hymen – another piece of skin charged with significance and meaning, and, as discussed in chapter 3, is paradoxically protective while breakable. The post-screen through the Hiroshima projections beckons to those same paradoxes, if in an entirely different context of vulnerability and exposure. It is not a skin in the sense of a palpable – safeguarding yet fragile, all with tremendous significance – surface between image and viewer; the post-screen, in its fluid transmutations, does not take that literal iteration here. Rather, the skin of the post-screen here is the skin of history itself – taking "skin" not from Daney, but from André Bazin's metaphor of recorded images as "the

---

90 Krzysztof Wodiczko, "Hiroshima Projection," an interview, originally published on PBS.org in September 2005 and republished on Art21.org, November 2011, https://art21.org/read/krzysztof-wodiczko-hiroshima-projection/.

extraordinary shedding" of the world by "tens of thousands of cameras" each day. The result of Bazin's metaphor is a remarkable re-characterization of the recorded image as skin, particularly recordings of events of war and history: "As soon as it forms, the skin of history peels off as a thin film."[91] The light projections in Hiroshima can thus also be read as that thin film par excellence that is the skin of history; in turn, its screen, as the post-screen, is not just a surface for an image, but the very lodging of that skin of history in the sheer historicity of place. The post-screen here becomes something temporally organic – not simply a sterile façade for an image, but a living and ephemeral membrane conjoining image, place, history and time.

In all these light projections, surface, screen and image come together as a tripartite convergence energized by the light. The heat of their amalgamation also reveals the hidden natures of the surface. As also cited in chapter 2,[92] David Theo Goldberg writes of political walls as dual-facing: on one side "is to be found the shaping of conduct, commercial and social, the social regulation of circuits of mobility both of people and their products"; on the other, walls are also "potentially screens for projecting commercial and political messages both propagandistic and critical or resistant."[93] The transformative energies of light in these projections thus also facilitate the continual passage between such dual sides of the wall's surface – a constant interchanging between engineering and functionality, politics and information, materiality and immateriality, establishment and resistance, power and challenge.

Indeed, that flux continues in perpetuity: even as the light transforms – deterritorializes – the meanings and political spaces of the post-screen surface, the surface also re-asserts itself – re-territorializes – in interminable fluctuation. As the projected lights get turned off, the walls are again structures of material bricks, cement and concrete. More insidiously, the power structures inherent in material infrastructure endure and re-assert, such as the increasing enforcement of law by police and governments over light projections, deeming

---

91 "A peine formée, la peau de l'Histoire tombe en pellicule," as quoted from David Forgacs, *Rome Open City (Roma Città Aperta)* (London: BFI, 2000), 23; the English translation in the text above is my own, and emphasizes the implication of the "thin film" (playing on both senses of film strip and membrane) from "pellicule." *Cf* Bert Cardulloso's translation of Bazin's essay, "On *Why We Fight*: History, Documentation, and the Newsreel (1946)," as it appears in *Film & History: An Interdisciplinary Journal of Film and Television Studies*, 31(1) (2001): 60-62, which reads, at 61: "As soon as it forms, history's skin peels off again."

92 See footnote 33 of chapter 2.

93 David Theo Goldberg, "Wallcraft: The Politics of Walling," *Theory, Culture & Society*, February 27, 2015, https://www.theoryculturesociety.org/david-theo-goldberg-on-wallcraft-the-politics-of-walling/.

them to be, for example, "unlawful posting of advertisements."[94] In 2016, police arrested three members of a New York-based art-activist collective on that basis of unlawful advertising after they projected the message, "KOCH = CLIMATE CHAOS," onto the exterior of the Metropolitan Museum of Art. The Museum had recently named its redesigned plaza the David H. Koch Plaza, and the protest was made in reference to Koch's role in Koch Industries, a conglomerate heavily involved in fossil fuels and with alleged records for breaking environmental regulations and illegal pollution. In March 2019, police arrested the projectionist Robby Diesu (and also one of Robin Bell's collaborators) for projecting protest slogans (including "discrimination is wrong") onto the side of the Rayburn House Office Building in Washington[95] on the basis that the building is regulated as part of Capitol Grounds, protected by US Capitol Police, with its own regulations governing protests. Even as the wall becomes the post-screen as a materialized/dematerialized screen-site of diminished boundaries between media, image, text and politics, the post-screen also reverts into a wall as a material structure of power. The only state of plausibility for the post-screen is thus its unrelenting flux between its tenets of meanings, media, materiality and political power.

Light Projections (3): Particles that Gain a Body... and Transform

A segue, then: where light transforms concrete flat surfaces into immaterial screens, it may also be projected onto amorphous particulate materiality, such as masses of ash, clouds and sheets of water droplets, to be turned into surfaces of screens. As ever, this is neither a new nor contemporary phenomenon. As Abigail Susik writes, in the late 1920s, British inventor Harry Grindell Matthews put together what was called the Sky Projector, an "oversized" projector that "had to be transported like a wartime cannon on the back of an industrial truck." Consisting of a powerful arc lamp, a focusing lens and a plane mirror, "the Sky Projector would project an image

---

94 Per the wording of the legal charge for the "KOCH = CLIMATE CHAOS" projection, as quoted from Corinne Segal, "Projection artists bring light to social issues with attention-grabbing protests," *PBS news*, September 17, 2017, https://www.pbs.org/newshour/arts/projection-light-artists-protest.

95 Peter Hermann and Clarence Williams, "Police arrest man projecting 'discrimination is wrong' onto outside of Rayburn House Office Building," *The Washington Post* online, March 15, 2019, https://www.washingtonpost.com/local/public-safety/police-arrest-man-projecting-discrimination-is-wrong-onto-outside-of-rayburn-house-office-building/2019/03/14/1dd44f0a-469c-11e9-8aab-95b8d80a1e4f_story.html.

high into the sky [onto clouds] without the need for a screen to project the image onto." Of course, other projections onto clouds have since followed. For instance, on four evenings in late 2005 as part of the "Interesting Times" exhibition at the Museum of Contemporary Art (MCA), Australian activist-artist Deborah Kelly projected a large white beam of light from the roof of the MCA spelling out "BEWARE OF THE GOD" in the sky for "one to two hours, depending on the weather."[96]

In more recent years, such inchoate "screens" have morphed out of various kinds of particulate matter, generally leveraging their novelty for publicity and marketing purposes. Sheets or "curtains" of water, for instance – consisting of a layer of fine water droplets up to twenty-five metres in height, discharged downward through nozzles or upwards from pumps, onto which light is then projected to display images – are particularly popular.[97] As with LED screens on buildings, a seemingly ceaseless array of images on Pinterest demonstrates the innumerable instances around the world of projected light on water screens. They tend to be large and intricately designed outdoor displays, held for maximum visual excitement and buzz. Two relatively high-profile examples will suffice here as illustration. In 2009, Paramount Pictures UK projected images of Dr. Manhattan, a major character in the *Watchmen* comics series,[98] onto a water screen twenty-two metres in height and thirty metres across in the middle of the River Thames in London to publicize their then-newly released film of the same name.[99] In September 2014, Polo Ralph Lauren launched a light show at the Cherry Hill section of Central Park in New York City as part of the city's fashion week, projecting images of models against an eighteen metre high "wall of water" (really more like a sheet of water droplets),[100] formed from a sprinkler head. A variation of this is the "fog screen," whereby water is pumped into a

---

96   All quotations in this paragraph are from Susik, "Sky Projectors," 80-1.

97   Through precise control of the water droplets, these projections are even becoming 3D displays: see Michelle Bryner, "New Technology Turns Water Drops into 3-D Display," *Live Science*, July 12, 2010, https://www.livescience.com/8396-technology-turns-water-drops-3-display.html#:~:text=A%20new%20display%20%E2%80%9Cscreen%E2%80%9D%20made,be%20viewed%20without%20special%20glasses.&text=But%20instead%20of%20pixels%2C%20this,water%2C%20othe%20higher%20othe%20resolution.

98   Albeit the projected image of Dr. Manhattan is from the film adaptation *Watchmen*, directed by Zack Snyder (2009; Hollywood, CA: Paramount Home Entertainment, 2009), DVD.

99   Bruce Simmons, "Watchmen's Dr. Manhattan on the Thames," *Screenrant*, March 5, 2009, https://screenrant.com/watchmens-dr-manhattan-thames/.

100   Lauren Cochrane, "Ralph Lauren models walk on water at Central Park fashion show," *The Guardian* online, September 9, 2014, https://www.theguardian.com/fashion/2014/sep/09/ralph-lauren-polo-models-central-park-fashion-show.

fog tank which converts it into "a thick fog made of tiny water particles 2-3 microns in diameter."[101] Once incorporated with fans to spread the particles, the screen essentially becomes a wall of mist not dissimilar to the water screens per the other examples. Nor are such inchoate screens limited to water or water particles. In April 2015, a group called The Illuminator Art Collective projected an image of Edward Snowden's face onto a cloud of ash they had scattered above a column in "a Brooklyn park."[102] The projection occupied the empty space of what should have been a four-feet tall bust of Snowden that had been erected on the column by anonymous artists a few days prior, but had since been removed by the city park's employees.

While clouds in the sky and clouds of ash may constitute a more amorphous "screen" than elaborately designed and calibrated jets of water, the key in this segue is that these inchoate screens are formed not from a solid surface, but a collective mass of atomized matter by way of water droplets or ash specks. The post-screen here thus appears in perhaps its most definitive state: a display surface for images with no literal boundaries, nor, in the case of clouds and ash, even a firm or enduring shape. The images do not so much merge or totalize with their environment as they *are* already a constituent part of it in both material and immaterial affiliations. As with screens out of urban surfaces, the post-screen of inchoate matter is caught in similar flux between materiality and immateriality, if on the opposite directional vector: the post-screen in this case does not render the material surface as immaterial per the previous examples, whereby light dissolves the materiality of building structures. Rather, it is the other way around so that, with light, *the particulate gains a body*. With that body, the inchoate acquires coherence; the collective mass contains sense. Where the post-screen had previously turned from being seen to becoming hidden, here the post-screen emerges from being hidden to becoming seen, with the boundaries of the screen thus roiled in that flux. As quoted earlier, Abigail Susik comments that "the Sky Projector would project an image high into the sky [onto clouds] without the need for a screen to project the image onto."[103] Susik refers to the absence of a screen in the sense of a designated screen or a solid surface on which light may appear. But the Sky Projector, with its vision well ahead of its time of visual culture and display, was not for the screen;

101 Cara Reynolds, "Behind the FogScreen: TechRepublic Interviews Company President Jorden Woods," *TechRepublic*, November 7, 2007, https://www.techrepublic.com/blog/geekend/behind-the-fogscreen-techrepublic-interviews-company-president-jorden-woods/.
102 Segal, "Projection artists," np.
103 Susik, "Sky Projectors," 80.

it was for a more contemporary amalgamation of image and surface, and of where and how images appear today against their environments. Namely, it was for the post-screen, and for all the comments by the post-screen, as through light projections, on the increasingly complex appearances and dis-appearances of images in designating realities, in giving voice, and in propping up power.

A final point: the post-screen through the projection of light is a transformation of not only surface and matter, but also transformation itself. The light comes on and goes off, rhythmic against the movements of light and day as in the rising and setting of the sun; or the dispersal and dissipation of the masses of particles; or the shifting of the clouds in the formation of different shapes, but also of different matter, such as the precipitation of water vapour into rain and evaporation back into water vapour. The state of change in the transformation of light is thus constant and relentless, so that change *itself* changes: no longer merely a transition from one state to another, but a non-stop loop of ephemerality in the turns of becoming and un-becoming, appearance and dis-appearance. If McLuhan had envisaged the shifting of mediated meaning from text to medium in terms of how "the medium is the message," here we might think of another deviation, where meaning emerges from this ceaseless flux between materiality and immateriality, between matter and ephemerality, between form and shape. Here, *transformation* is the message. The projected light, and the words and images it forms, is ephemeral, even spontaneous. However, what matters is neither the text nor even the light, but its sheer appearance out of nowhere and then its inevitable disappearance into somewhere. What matters is the flux and contestations in the amorphous boundaries of the post-screen. It is in the friction of that transformed transformation that the energy of the projected light truly manifests itself in the post-screen, or in finding, per Barwell's words in the opening quotation of this section, "their own screen against the darkness."

Projection Mapping (1): The Image that Devours Structure; the Voracity that is a Media History

In the examples discussed above, light is projected onto a relatively flat wall or building façade, whose display of image and text on a two-dimensional surface transforms it into a (post-)screen. However, from the 2000s, a method known as "projection mapping" or "video mapping"[104] enables the projection of images onto *irregularly* shaped objects so that the light superimposes a visual output onto a *three-dimensional* surface. By mapping a spatial replica of the object with specialized software, an image is then produced by computer graphics programs that take into account the elements and attributes of the surface, such as colour, shadow, angles and unevenness.[105] By aligning the projection with the features of the source object or façade, the result is the presentation of an image which conforms to, or "wraps" around, the uneven surfaces of the object. The object thus appears with a virtual "skin" of light which not only covers it, but *is mapped directly onto its surface*,[106] so that the whole object appears transformed with the illumination of this "skin" that irradiates the object, otherwise appearing neutral in ordinary light, with a laser-like brilliance. More spectacularly, the virtual "skin" may also be animated, so that the image's movements induce the perception that the object *itself* is moving even as it is grounded solidly as a stationary mass, such as a building. However, unlike Wodiczko's "Hiroshima," this "skin" of light is not so much a *pellicule* of history as shed by the camera onto the historicity of place. Rather, it is a skin in a more constitutional sense – as "mapped" to the object, it forms or is bound to it as an organic component. The post-screen here in terms of this "skin" thus merges with the object whose amalgamation, in turn, dramatically changes the notion of the object's solidity or object-ness. As such, the monumental turns enlivened; the material becomes zoetic.

As with every new media discussed here, projections of images onto three-dimensional objects likewise have their pre-millennial precedents and prototypes. In 1969, Disneyland premiered its Haunted Mansion

104 The terminology is still in flux, and there are various other names for the technology, such as "urban projection mapping," "3d video mapping projection," "3d architectural projection mapping," or "spatial augmented reality," with the last being apparently the first term of reference: see Oliver Bimber and Ramesh Raskar, *Spatial Augmented Reality: Merging Real and Virtual Worlds* (Natick, MA: A.K. Peters, 2005).

105 Ramesh Raskar, Greg Welch, Kok-Lim Low and Deepak Bandyopadhyay, "Shader Lamps: Animating Real Objects with Image-Based Illumination," Proceedings of the 12th Eurographics Workshop on Rendering Techniques, London, UK, June 25-27, 2001: 89-102, 89, Berlin; Heidelberg: Springer-Verlag.

106 This effect of augmentation of objects can be traced to the technique of Spatial Augmented Reality (SAR) first proposed in the late 1990s, whereby "the user's physical environment is augmented with images *that are integrated directly in the user's environment*, not simply in their visual field [emphasis added]": Henry Fuchs, Ramesh Raskar and Greg Welch, "Spatial Augmented Reality," First International Workshop on Augmented Reality, San Francisco, November 1, 1998, MIT Media Lab, 1, http://citeseerx.ist.psu.edu/viewdoc/download?doi=10.1.1.439.7783&rep=rep1&type=pdf.

attraction with a projection of a looped film featuring the face of Madame Leota onto a static neutral-coloured object inside a crystal ball. Madame Leota, the character of a psychic medium who resides as a disembodied head inside a crystal ball in the Haunted Mansion, thus appears on the attraction as an appropriately dislodged head, glowing and chanting from inside its bauble.[107] In the 1980s, artist Michael Naimark experimented with superimposed projections on objects and furniture for various iterations of an installation titled *Displacements*. Here, Naimark first used a rotating movie camera to film a living room with all its furniture and occupants in it, including the latter's various movements in the room – a mother who bends over her teenage child with a hug; the teenager who picks up a guitar, and so on. The room's walls and furniture were then painted white as a neutral colour, and the recorded imagery was projected back onto its walls "using a rotating projector that was precisely registered with the original camera."[108] That precise coordination is important, for the subsequent projection of light as rotated onto the white walls and furniture of the room effectively "layers" them with a "skin" of projected images. Appearing with physical and positional coherence, this "skin" over the surfaces of the room thus appears to not only correspond but also *belong* to its walls and objects. Yet the virtuality of the images – their second order reality – is also clearly distinct and separate from the physical actuality of the room: as with biological skin, the layer of images becomes a constituent of the now whitewashed room, yet also sheddable. Just as significantly, as with projection mapping, this layer of imagery also triggers "enlivenment" as objects in the room acquire colour and the moving images of the room's occupants, in precise correspondence to the objects and the space in the room, animate the otherwise empty and static space.

The post-screen through such projections onto three-dimensional surfaces or objects thus emerges in its most nuanced iteration. With projections on flat surfaces, the screen of the surface presupposes between appearance and dis-appearance; it materializes and dematerializes in the oppositional terms of darkness and illumination. Here, the screen merges into the "skin" of images over the object; it *is* the object. The iteration of the post-screen in three-dimensional projections is thus the simultaneous enfoldment, through

107 Jeff Baham, *The Unauthorized Story of Walt Disney's Haunted Mansion* (USA: Theme Park Press, 2016).

108 See Michael Naimark, *Displacements*, as exhibited at the San Francisco Museum of Modern Art, San Francisco, CA, 1984; and M. Naimark, "Spatial Correspondence in Motion Picture Display," *Optics in Entertainment II SPIC* 462 (1984): 78-91.

transformative light, of image, object (as subject of the image) *and* object (as the surface on which the image forms). Where boundaries were blurred or hidden in previous instantiations of the post-screen, the case of projection mapping presents a step beyond that level of obscuring: here, the image *is* the frame; the frame *is* the image. Where Baudrillard's hyperreality of images envisaged the dominance of an order of simulacra stemming from the loss of connection to or differential from their origins, these projections of images herald a different order of simulacra, *whose images directly correspond to* and *overlay their origins*. Where in conventional light projections, the screen shares nebulous boundaries with its surroundings through the interplay of light and darkness, here the screen is encased entirely in the object itself; it forms its organic skin: the screen is the object as the object is also the screen.

In some ways, the post-screen here is not only the screen which subsumes the object; it also appropriates the screen itself. For instance, in 2013, researchers at Microsoft developed a proof-of-concept system to project what they call "peripheral projected illusions."[109] Using a wide field-of-view projector and a Microsoft Kinect sensor, the system calibrates the furniture and surfaces of the wall against which the conventional television or computer screen, connected to the Kinect, is placed. The system then projects images onto the three-dimensional surfaces along the wall so that the wall and its furniture re-appear to the user in various effects. One of these, harking to the totalizing effect of the panorama, extends the images of the game beyond the boundaries of the conventional screen onto the wall surfaces around it, so that the images fill the user's visual field. Another effect uses projection mapping to "augment" the surfaces of the wall so that the objects in the room appear "wrapped" in an animated "skin" of imagery coherent with what is shown on the screen. Here, the post-screen broadens into the furniture along the wall. Not only does the image spread from beyond the confines of the screen, it also consumes the conventional screen itself, which disappears visually and functionally into its surroundings thus "skinned": the image now pulses from across the entire wall, rather than simply from within the boundaries of the television screen. In this way, *the post-screen absorbs its own screen*: even the secondary layer of virtual images in the screen gets devoured into the virtualization of its larger, more encompassing post-screen version.

109 Brett R. Jones, Hrvoje Benko et al., "IllumiRoom: Peripheral Projected Illusions for Interactive Experiences," *CHI*, Paris France, April 27-May 2, 2013, https://www.microsoft.com/en-us/research/wp-content/uploads/2013/04/illumiroom-illumiroom_chi2013_bjones.pdf.

Finally, the post-screen through projection mapping reaches monumental heights when applied to city buildings and other large-scale urban façades. The last decade from the early 2010s featured a rash of 3D projection mapping displays across the world on large building surfaces, usually in the context of high-profile events such as marketing launches of new products or light night festivals held to promote cities. In every case, the projection mapping is set closely to the architectural features of the façade and encases the building surface in a "skin" of lit imagery. The skin illuminates the surface with uncanny brilliance so that the building appears virtualized, endowed with a cartoon-like effect. As with conventional light projections, the light transforms the boundaries of the surface into a flux between wall and screen, in the process also muddying between their being artefacts of architecture and of media.

However, as alluded above, there is more. The "skin" of images may also be animated, inducing perceptions of movement so that the building façades themselves appear to change, pulsate and morph, even fold and collapse, creating a dramatic shock of the object's light and lightness against stability and solidity. As usual, there are many examples of projection mapped light displays; a few here will suffice as illustration. One of the earliest displays of large-scale projection mapping is the project *Perspective Lyrique*, a light presentation shown in 2010 at the annual Festival of Lights (*Fêtes des lumières*) in Lyon, France. The display featured projection mapping which applied an animated "skin" of light to the uneven façade of a former lyrical theatre, transforming it into a humanoid face that twisted and pulsated, and moreover morphed in response to spectators' voices via a microphone (and an audio analysis algorithm).[110] In the same year, Nokia and Windows promoted the launch of the Nokia "Lumia 800 with Windows" phone by projecting a light show on the Millbank building in London. Close-mapped to the (not coincidentally) 800 windows of the 120 metre building,[111] the projection created multiple illusions of the windows – and thus the building itself – shifting, collapsing and coalescing. At one point, all the windows of the building were lit only to have the Windows logo "crash" through them, with the projection presenting flying shards of glass and a dark "hole" in the middle of the illuminated building. In 2011, a 6"05' minute projection

110   No author, "Perspective Lyrique by: 1024 Architecture," February 7, 2012, http://www.mappingaround.com/perspective-lyrique-by-1024-architecture/.

111   John Chapman, "Nokia's smartphone opens new Windows in world of mobiles," *Express*, November 16, 2011, https://www.express.co.uk/news/uk/283966/Nokia-s-smartphone-opens-new-Windows-in-world-of-mobiles.

mapping installation was presented at the opening ceremony of the Tel Aviv Museum of Art's new wing, the Herta and Paul Amir Building. As with *Perspective Lyrique*, the projection "addressed, played with and subverted the notion of the façade" that the architect Preston Scott Cohen had created, particularly in the warping and thwarting of geometrical shapes that make up the building's "twisted and tessellated surfaces."[112] As Neuman describes, the animation of the display as mapped to the façade confronted the physical materiality of the building, producing

> ...a discrepancy that challenged the relationship between the façade's physical dimension and the virtual screening. The façade in its physical dimension was present in the background of the screening, but the projection's virtual dimension disavowed their mutual logic.[113]

Prominent projection mapping light displays also featured in Shanghai during the city's Western New Year countdowns to 2012 and 2013, attracting much global attention out of coverage by international media outlets such as CNN, BBC and NHK.[114] The light show was projected on two historical buildings – the Customs House and the SPD Bank Building – along the Bund, a waterfront area by Shanghai's Huangpu River on which are located dozens of historical buildings housing banks and trading houses; the Bund itself used to be a British settlement. As with the other projections, the building façades were closely mapped so that, in being illuminated with its "skin" of imagery and with the illusion of movement induced by the animation of light, the buildings themselves appeared to be in motion – morphing, shifting, folding and collapsing to dizzying effect.

The projection mapped imagery thus fuses with the three-dimensionality of the uneven building façade, and, more significantly, *in*fuses it with the vitality of its animated movements. In turn, the animation co-opts the physical bulk of the structure into the manipulability and lability of light – the buildings appear to warp, flex, contort and stretch in defiance of their rigidity and mass: the walls pulsate, bricks fly off their foundations and

---

112 Robert Levit, "Geometry('s) Rules: Preston Scott Cohen's Tel Aviv Museum," No. 35, no date, http://www.harvarddesignmagazine.org/issues/35/geometry-s-rules-preston-scott-cohen-s-tel-aviv-museum.
113 Neuman, "Inside Out," 262.
114 These celebrations along the Shanghai Bund have since stopped after a tragic stampede due to overcrowding at the New Year's Eve celebrations of 2014: see Reuters, "A Year After Stampede, Shanghai Opts Out of New Year Celebration," December 29, 2015, https://uk.reuters.com/article/us-new-year-shanghai/a-year-after-stampede-shanghai-opts-out-of-new-year-celebration-idUSKBN0UCoDS20151229.

columns fold down like giant origami. The light temporarily transforms the building wall into an animated creation, warping the boundaries not simply between the inanimate and the animate, but also between the concrete and its disintegration. The building does not disappear as with conventional projections, but is amalgamated with the image as the post-screen in a virtualized vitality of disintegrating matter and atomization of structure – a contradictorily edge-less and impossibly weightless mass.

Connecting movement, matter and life as theoretical strands through media is, of course, not a new combination. Sergei Eisenstein, for instance, had proposed that movement in cinema and in architecture share theoretical provenance, where dynamism in the architecture of buildings can also be seen in cinematic terms.[115] The movement of cinema's moving images likewise renders its subject as animated and alive. However, the combination of virtualized animation, induced movement and the "skin" of the building in the post-screen through projection mapping asserts another kind of vitality – one closer to Boccioni's sense of transformative light that splinters and disintegrates matter towards a peculiar sense of life. The post-screen here effectively colours a specific triangulation of light, material solidity and movement into a statement on the nature of matter itself as between affiliation and disconnection, substantiation and atomization. It is a statement which links to the very relational essence of matter as fundamentally particulate, as assemblages of connections and disassociations that relate the foundational matter of matter – particles, subatomic particles, quanta, microbes, microbiomes. In the process, the post-screen through these large-scale projection mappings depicting the virtualized collapsing of building façades also becomes a visual proclamation of the breakdown or annihilation of those connections under the power of the projected image through it. With that destruction thus also emerges the Boccioni-esque sense of vitality with which these projection mapping images are enlivened to the point of voracity.

In other words, the resonance of the post-screen here is neither with architecture (whose dynamism is essentially imagined out of its structures) nor cinema (whose presentation of images is about a *passing through*, such as the movement of film through its gate or a DVD through the laser beam that reads it). Rather, its reverberations are of Boccioni's *The City Rises* and the Futurist manifesto of vitality and light, where, taking on Le Bon, matter transforms and dissociates into energy, in turn stamping out a statement

---

115 Sergei M. Eisenstein, "Montage and Architecture," trans. Michael Glenny, *Assemblage*, 10 (Dec 1989): 110-31.

about their desired society. Here, *media* transforms matter and dissociates into energy. This sense of volatile weightlessness in media has been noted before, such as Vivian Sobchack's comments in relation to electronic media as she writes of the electronic "instant" creating electronic space as "abstract, ungrounded, and flat – a site for play and display," where "its flatness" is for "spectator interest at the surface." To Sobchack, electronic space "disembodies," or rather, it orientates the body with "a purely spectacular, kinetically exciting, often dizzying sense of bodily freedom."[116] The "bodiless exultation of cyberspace" of William Gibson's cyberpunk classic *Neuromancer* also comes to mind as its protagonist recalls his dizzying exploits as a "cyberspace cowboy" – "all the speed he took, all the turns he'd taken and the corners he'd cut in Night City."[117] In this sense, the immateriality of the digital and the flatness of electronic media presage the effortless transformations of projection mapping, whose contours of light transmogrify the solidity and three-dimensionality of objects into the same sense of disembodied and weightless energy.

However, in the contemporary account of the projection mapped image, such transformational atomization of building structures characterizes another kind of energy – not simply one of disembodiment and bodily freedom, but a wholly more perverse voracity of rendering matter into image, and the devouring of all structure into the spectacle and immateriality of digitized media. The vitality of the projection mapped light display thus signals another kind of virtualized viability, namely, the energy of the image which consumes, demolishes and otherwise swallows whole, whereby the (large) object (of the entire building façade) is co-opted with the image to itself move, collapse and transform. On this post-screen, the image is not displayed, but wraps around, merges with and thereby consumed by the screen. Where the indiscernible relations between the image and the screen signal their convertibility is also the point of a devouring where one becomes absorbed or swallowed up by the other.

The post-screen as screen fused to the object – or where matter, with this grotesque vitality, becomes the image – thus stamps out a statement of another kind of transformation. This is not a statement *qua* Boccioni of a society shot in the veins with the energy of Futurist youth and violence. Rather, it is one whose gluttony for media, and in particular for images,

---

116 All quotations in this paragraph are from Vivian Sobchack, "The Scene of the Screen: Envisioning Photographic, Cinematic, and Electronic 'Presence'," in *Technology and Culture, The Film Reader*, ed. Andrew Utterson (New York; Abingdon: Routledge, 2005): 127-142, 137.
117   William Gibson, *Neuromancer* (New York: ACE Publishing, [1984]; 2003), 4-6.

corrupts the object so utterly that its image dissociates completely from the reality of the object while still *being* the object. The virtuality of the image takes over the actuality of physical materiality – and in that sense draws out new grounds of their contestation, a theme reflected throughout this book – to become a vitality which encompasses all. Bazin had similarly identified such a comprehensive totalization of reality via his vision of "no more cinema" (mentioned previously in the introduction), which stemmed from the desire to clarify cinema's ontology as the seeing of reality – or "the pre-existence of the narrative," "the 'integral' of reality" – through the window of the cinema screen. Bazin – via De Sica's *Ladri di Biciclette*, his tutor text – thus sought reality in that unadulterated sense of the image through what he calls "pure cinema," with the "cinematographic dialectic capable of transcending the contradiction between the action of a 'spectacle' and of an event."[118] The post-screen here through projection mapping thus echoes that sense of the all-consuming image over event or spectacle – where the image *is* the spectacle; it *is* the event; and it *is* the object itself. The projection-mapped image through the post-screen encompasses them all.

Yet, this gluttony is also more than merely a summation of greed in the takeover by the virtual of the actual. The dominating consumption of the image is not just about its totality; it is also about the formal implications of this convertibility between media and screen as a crucial chapter of *media history*.[119] As mentioned in the introduction, the fourth argument of this book on constructing media history through this trajectory of screen boundaries is about the modes through which human activity across its spectrum may be thought or construed to be modulated through media. The gluttony of media, signified here through the post-screen's all-consuming domination of the virtual over the actual, is thus one such modulation that specifically signals *the ease of convertibility* that is colouring much else of twenty-first century living and the forming of its histories, namely the slippages of truth against lies; of post-truth and misinformation and

---

118 André Bazin, "An Aesthetic of Reality: Neorealism," in *What Is Cinema?: Vol. II* (Berkeley; Los Angeles, CA: University of California Press, [1971], 2005): 16-92, 60.

119 In my introduction to this book, I had referred to the four arguments I wanted to make in relation to the post-screen: firstly, that the growing imperceptibility of screen boundaries leads to an obscuring of difference between image and surroundings; secondly, that virtual realities of the post-screen re-draw relations between the virtual and the actual, and re-shape imaginations of the real; thirdly, that the changing virtuality out of disappearing screens also changes affect and subjectivity; and fourthly – on media history – that the first three arguments can be threaded into an imagination of the post-screen that is indicative, as a discursive object, of contemporary lives, ways of living and understandings of ourselves.

disinformation; of the accelerating volatility of shifting values and their terms; of the reverberating echoes of social media against all other kinds of discussion; of the dominance of spectacle and the televisual against the event. These are all the signal fires of a moment of media history to which there has yet to be a name or framework, but which we might actually be able to connect to the state of contemporary media and specifically, per this chapter's argument, the convertibility of the post-screen. If we may speak of a post-screen politics, it might just be one shot through with this convertibility between screen and media per the post-screen, as with the translation in physics between mass and energy where the terms of solidity change and the basis of previous values irrevocably shift. In this sense, the post-screen – in relation to its convertibility and the gluttony of the virtual – thus perhaps holds up a reflecting shard of a bigger picture. Media and history – not as an unfolding account of events nor even as the occurrences that we live through, but as that *which lives through us*, as that in which we *are*, perhaps even without our knowing that we are in it – thus also connect to each other in this sense. The former is the canvas for the latter's colour, shape and form; one weathers, erodes and stratifies the ground for the other. The post-screen here might thus be one such moment of history in the politics of the twenty-first century or at least certainly of its second decade; it forms even as I write.

Projection Mapping (2): The Exterior that Reveals; the Permanence that Fades

If the outcome of projection mapping on building walls and façades is the apparent evisceration of solidity and materiality, projection mapping on *human* faces and bodies draws the opposite effect. Deployed thus far primarily in art installations and stage performances, the image as projection mapping onto human skin similarly takes over its object. However, in this case, projection mapping renders the somatic surface of skin into molten yet solid material, so that the skin appears to shed its biological properties to become an unnatural, inorganic yet animated layering over the face or body, appearing as both part of yet distinct from the face or body.

To date, there have only been a few examples of such projection mapping, though they have attracted substantial attraction and publicity. Thus far the most prominent usage of face projection mapping is work by Japanese artist Nobumichi Asai of WOW INC, a visual design studio based in Tokyo, evidenced through his creations of several projection mapping projects over the last five years. In 2014, for instance, Asai's art installation, *Omote* (in collaboration with Hiroto Kuwahara and Paul Lacroix),[120] first tracked and captured a human model's face in detail via numerous sensors on the face to capture and process marker data in operations akin to those by motion capture systems. The software then produced a virtual replica of the model's face rendered with various animated textures by computer graphics (CG). Those rendered images are finally projected back onto the model's physical face, appearing as a "skin" due to the image's close correspondences to and alignments with the face's features, such as its eyes, nose, lips, shadows and contours.

As with other instances of projection mapping, the "skin" of light which virtualizes the face renders it similarly transformative, here mutating the face from biological to technological object. The model's face – and their skin as a *material* of human likeness – disappears into, or gets covered by, its mask of light, which shifts and morphs the face in real-time into different colours, structures and patterns. The meaning of this projection resonates further in terms of how masks or exterior displays ("omote" moreover being the Japanese word for face or mask) are – in the Japanese context or otherwise – culturally specific external vehicles for signifying interiority or inner life.[121] The *exteriority* of this transformation in rendering permeable skin into seemingly impenetrable casings thus takes on a further meaning – an exteriority reflecting not the interior, but an *additional* exteriority doubled down: an encasing of light which removes further from the humanness

---

120 Nobumichi Asai, Hiroto Kuwahara and Paul Lacroix, "Omote / Real-Time Face Tracking and Projection Mapping," http://www.project-omote.com/. Asai has also created several other facial projection mapping projects, albeit all displaying similar effects which use projection mapped images to dramatically change the external appearance of the models' faces and/or bodies. A further example of Asai's work is "Connected Colors," part of an advertising campaign by Intel Corporation: see https://www.wow.co.jp/en/art/connected_colors. Asai also collaborated with Intel and Nile Rogers to create a similar effect for Lady Gaga's tribute performance to David Bowie at the 58th Grammy Awards in February 2016, where the singer opened her performance with projection mapping over her face in the motifs of "lightning, sun and spider make-up" to symbolize Bowie: https://www.nobumichiasai.com/works/136/. Another work, *INORI (Prayer)*, is "a dance performance video that uses real-time facial projection mapping to change the look of the dancer's faces. Over the course of about one minute, the dancers are made to look like skulls with empty eye sockets, big-toothed clowns, and terrifying dolls with their jaws unhinged": see Lizzie Plaugic, "Watch A Dance Performance Change in Real Time with Facial Projection Mapping," *The Verge*, April 1, 2017, https://www.theverge.com/2017/4/1/15135962/inori-prayer-music-video-facial-projection-mapping-how.

121 See Martha Johnson, "Reflections of Inner Life: Masks and Masked Acting in Ancient Greek Tragedy and Japanese Noh Drama," *Modern Drama* 35(1) (Spring 1992): 20-34.

of skin; or which reflects an external environment, such as the ethereal landing of a virtual butterfly on the model's cheek.

In this sense, the post-screen here transacts between the interior and the exterior of humanness, specifically refuting the display of humanness via skin, membrane and face for the technologically rendered human physical form, closely encased almost to perfection. On the last, we may also recall Barthes's ascribing (albeit, in his view, seemingly only to the face of Greta Garbo) no less than a Platonic essence to the face – specifically, "the essence of [a] corporeal person descended from a heaven where all things are formed and perfected in the clearest light."[122] In this respect, facial projection mapping through the post-screen thus conducts its ultimate dissonance, where the face and its features remain – if still with some kind of essence – in residual form even as transformed into virtualized object, encased with its inorganic animated skin. Where projection mapping in *Perspective Lyrique* morphs the infrastructural façade of the lyrical theatre in Lyon into a humanoid face of the building, here the human face transforms into *a humanoid face of the human*. As with the other displays of projection mapping, the face here is not simply a surface for the landing of light. It is a statement of the post-screen and the transformation it heralds, where the image is now capable of absorbing biological epidermis, or even the essence of forms, and where it speaks now of a world of even biological matter, substantive in their own terms of life and of the living, as consumed into image.

As a variation to Asai's facial projection mapping, the work of Oskar & Gaspar, "a collective of visual art specialized in video mapping and motion graphics,"[123] similarly focuses on projection mapping the uneven surfaces of the face and, in particular, the whole human body. To date, they have produced several projects along these lines, including a turn on the competition television show, "America's Got Talent," where they demonstrated projection mapping techniques on the body of one of the show's judges. One of their more remarkable pieces of work, a 2015 project in Lisbon, *Ink Mapping: Video Mapping Projection on Tattoos*,[124] was labelled "the world's first live tattoo

---

122 Roland Barthes, "Face of Garbo," in *Mythologies*, trans. Annette Lavers (New York: Farrar, Straus & Giroux, [1972], 1991): 56-7, 57.

123 Oskar & Gaspar, "Providing the World with New Digital Creative Experiences," https://oskargaspar.com/.

124 Nathaniel Ainley, "Tattoos Transform into Moving Images with Ink Mapping," *Vice.com*, October 23, 2015, https://www.vice.com/en_uk/article/aen45j/tattoos-transform-into-moving-images-with-ink-mapping.

video mapping event."[125] During the event, the artists projected coordinated animated images over models' tattooed bodies so that the projections appeared as animations, or in some cases live continuations, of the models' body markings. Flowers or patterns burst into life with animation or uncanny brilliance; snakes or manta rays appear to slither up torsos or float across thighs. Oskar & Gaspar's work demonstrates the usage of the whole human body as a screen for images, closely mapped and in correspondence not only with its bodily features but also to its markings, enlivening them under the transformational properties of the light.

Projection mapping across the body thus continues the post-screen of the face – here the body becomes a screen in the sense of a surface for the landing of light, but whose boundaries between media and biological organism become particularly muddied. As with facial projections, the post-screen here makes a statement of the copious virtualization of matter by the projection of light. Yet it is also a different kind of virtualization. Where facial projection virtualizes the face as a membrane or revelation of interiority into an opaque surface, the projections on the tattooed body virtualize permanence into impermanence, and in the process wipe out their significance. Like the hymen, tattoos appear with cultural and social significance: in non-western tribal cultures generally, they may be indications of social roles; in western societies, they may also function more as "a mechanism for demonstrating one's disaffection from the mainstream."[126] Yet the virtualizations of these tattoos in the image swallow not only the body and the markings, but also their meanings, embedded as they are in the permanence that body modification tattooing entails but now transformed into the ephemerality of the projected light. Again, invoking Baudrillard's notion of the all-consuming image, this is not a virtualization of an order of simulacra far removed from the original; this is the virtualization directly of the original itself to the point where even forms of enduring signification themselves get caught up in its all-encompassing sweep. What we have here is not an order of simulacra or of the copy, but of a transformation of ontology itself, where faces and bodies lose their biological groundedness and matter – *qua* carbon, water, blood, skin and epidermis – as absorbed into image. The image no longer appears and dis-appears in the post-screen; it consumes all.

125 Oskar & Gaspar, "Ink Mapping: Video Mapping Projection on Tattoos, by Oskar & Gaspar," uploaded October 22, 2015, Vimeo video, 03:12, https://vimeo.com/143296099.
126 Clinton R. Sanders and D. Angus Vail, *Customizing the Body: The Art and Culture of Tattooing* (Philadelphia: Temple University Press, 2008), 2.

The Ground Beneath Our Feet

The ground, the ground beneath our feet. My father the mole could have told Lady
Spenta a thing or two about the unsolidity of solid ground. The tunnels of pipe
and cable, the sunken graveyards, the layered uncertainty of the past. The gaps
in the earth through which our history seeps and is at once lost, and retained in
metamorphosed form. The underworlds at which we dare not guess.
~ Salman Rushdie[127]

Through my long sojourn all these years as a film enthusiast, it was always
the light to which I returned. Barthes wrote of the darkness of the cinema
theatre as "truly a cinematographic cocoon," the blanket of blackness in
which he, the film spectator, is sustained precisely by being shut in it,
swathed in this stuff of alternate reverie and eroticism. The light of the
cinema screen hits the spectator with assertion – an "imperious thrust"
out of the luxury of darkness – which, even in its forcefulness, fascinates
and hypnotizes: "we are entranced by this brilliant, immobile and dancing
surface, without ever confronting it straight on."[128] The assemblage of light
and darkness – light out of darkness; light against darkness – harks to
ancient binaries, those which, like bimetallic strips, necessarily make sense
only with and against each other: good versus evil; fire against the night;
ignorance and knowledge. The light at the end of the proverbial tunnel is
precisely what makes the tunnel a tunnel: its light enables both coherence
and affirmation, whose message, short and profound, is simply "this is what
things are and how they make sense." The light from the cinema screen – as
ratification and revelation – is also a central part of the intense love cinema
solicits: Susan Sontag's accounts of 1960s and 1970s cinephilia, for instance,
describe "the full-time cinephile always hoping to find a seat as close as
possible to the big screen, ideally the third row center."[129] The character of
Matthew (played by Michael Pitt) in Bernardo Bertolucci's romantic tribute
to cinephilia, *The Dreamers*, specifically sought the screen: "Why do we
sit so close [to the screen]? Maybe it was because we wanted to receive
the images first. When they were still new, still fresh."[130] The image for

127 Salman Rushdie, *The Ground Beneath Her Feet* (Canada: Vintage Canada, 1999), 67-68.
128 All quotations in this paragraph from Roland Barthes, "Upon Leaving the Movie Theatre,"
in *Apparatus*, ed. Theresa Hak Kyung Cha (New York: Tanam Press, 1981): 1-4, 2.
129 Susan Sontag, "The Decay of Cinema," *The New York Times*, February 25, 1996, https://www.
nytimes.com/1996/02/25/magazine/the-decay-of-cinema.html.
130 Dialogue line from *The Dreamers*, directed by Bernardo Bertolucci (2004; Burbank, CA:
Walt Disney Studios, 2004), DVD.

Matthew was a balm of complete absorption, the surrender to which was not a mindless capitulation, but the acceptance of a life-giving and world-renewing energy. The cinema – beyond all of its ever-evolving technologies, standards and markers of industry, and apparatus of ideology – ultimately is, if in unabashedly romanticized terms, a grand source of light against the enveloping darkness. Its light is transformational in its invoking of the elemental and its rejuvenating of the primal, with the screen as the magical fount of those unspoken expressions. I sit in the cinema in the midst of that primitive contest between light and darkness, my modern and personal equivalent of a campfire against the blanket of the terrorizing night and its hidden beasts. So many years later, even when the practices and habits of my film consumption expanded to the television screen (and then the computer, and then the tablet), I am still thus captured by the light.

As a cinephile, I gazed on Boccioni's *The City Rises* with deep shivers of recognition of *that light which transforms*, whose rays carried all those hopes of a new city and society borne out of new energies and new futures. Those chills of recognition swelled again when I watched the projection mapping display of the buildings in Shanghai for the first time in 2012, clearly born of cinema in its apparatuses of projector, image and surface, and with unmistakable echoes of its institution in the form of sizable audiences gathered before an enlarged, encompassing image lit against the darkness. But the display was also radically different: with the light mapped to the building, the screen moulded into the object, and the light on it became angry, transforming the building with aggression and a ferocious plasticity. The long history of light projection in art, architecture and media demonstrates the role of light in constant assertion and subversion of solidity and matter, in play between presence and absence, and in the constitution of space between material and immaterial structure. Light itself has always been transformative. Yet, the light of projection mapping emanates from its contemporary screen with a visuality which erupts out of the very object of the screen itself and makes its very point in deriding the object's solidity. As the post-screen begins to take on increasing amounts of contemporary society's images, the regime of the image likewise stirs – first in minuscule degrees, almost imperceptibly but also affectively, but which now threatens a tectonic heave. Where it once bathed my face like an open window of air, the light now asserts itself from the darkness by collapsing the ground beneath my feet.

Conclusion/Coda

Abstract

This Conclusion/Coda summarizes the book's key argument: the post-screen as a visual phenomenon which, through contemporary instantiations via Virtual Reality, holographic and light projections, blurs boundaries between virtuality and actuality, and re-formulates conditions of media, reality, death, life, matter and history. The Conclusion also points the post-screen towards two further ideas which drive its concept: difference, in terms of screen boundaries demarcating image against surroundings, and on difference without positive terms; and gluttony of visual media, specifically in relation to play between the real and the unreal. Both ideas not only serve the diminished boundaries of the post-screen in terms of the book's analyses, but also render the post-screen a framework for today's politics of post-truth, misinformation and deepfakes as a moment of media history. Finally, the Conclusion extends the post-screen to the (as of writing) current Covid-19 pandemic as a mirror of the internalization that is of both post-screen media and virus – both are in us, and inescapable.

Keywords: screen; media history; difference; gluttony; Covid-19; intern

Postscripts to the Post-Screen: The Holiday and the Global Pandemic

Vignette 1

In the midst of finishing this book, I took a week-long family holiday to Dubai in late January 2020. Taking our place in the crush of tourists, we visited the Burj Khalifa, the tallest building in the world at 829.8 metres from ground to tip, by taking the elevator – whose trip alone took three minutes – to its 124th floor observatory deck at 450 meters towards the sky. The deck was an enclosed circular floor covered in wall-to-ceiling glass windows, so that one could walk an orbit to take in a 360-degree view of

Ng, J., *The Post-Screen Through Virtual Reality, Holograms and Light Projections. Where Screen Boundaries Lie*. Amsterdam: Amsterdam University Press 2021
DOI: 10.5117/9789463723541_CONC

the city. The ghosts of the nineteenth-century panorama whispered in my ear as I strolled slowly along the circular deck, gazing out at my own twenty-first century panorama, tracing steps in the same kind of round tour as nineteenth-century Londoners did with my panorama's visual forebear, wide-eyed with the same kind of wonder.

At that height, the famous skyscraper-replete vista of the city of Dubai, boasting twelve of the world's tallest seventy buildings, looked like geometrical saplings. In the distance, the Arabian Desert stretched away into dusty and dusky shimmers. With a constitution happily unaffected by vertigo, I stepped right up to each window pane, gazing out at the view as if at a panel of a continuous panorama painting. At that elevation and across the totalizing glass pane, my visual reference to the scene before me was not only godlike,[1] but, to all intents and purposes, *virtualized*. The view at that height had no spatial or dimensional relation to its reality on the ground – a reality I had only half an hour ago observed and photographed from the tourist bus at ground level before boarding the elevator – other than its vast vertical distance from it. In this respect, too, the Victorian viewer could only relate to the locales and depictions of the nineteenth-century panorama by their foreignness and distance from the actual sites of catastrophe and heat of battles. It was at this point that I realized I was viewing not out of an observation deck, but *at a screen*. Born in Asia, living in Europe, I had come to this point mid-way between the two continents and gotten myself onto the observation deck of the tallest building in the world to look at a screen.

Yet, as I continued to move from pane to pane, border to border, I realized there was something else in the alien-ness of that view, namely, that even while I was gazing out onto something which contained its own order of reality, at the same time *it constituted my actuality*. And this was the difference: where the notion of the screen had always been linked to ideas of spectacle and spectatorship – and therefore to questions of representation – here was a different experience. This was no painting with concealed boundaries; there was neither illusion nor trick of light. Rather, this was an image commandeered through a bodily experience of the visual and the affective which could only make sense of this visual data as an unreal reality. This was not an image in the sense of that which tried to capture what is out there; *this was an image that was in me*. This was not a screen; this was the post-screen.

---

1   See Paula Amad, "From God's eye to Camera-eye: Aerial Photography's Post-humanist and Neo-humanist Visions of the World," *History of Photography*, 36:1 (2012): 66-86.

Vignette 2

I returned to the UK in early February 2020. A month later, the SARS-CoV-2 virus swept through Europe as Covid-19 cases started soaring in Italy. Within weeks of my return, the UK entered a national lockdown on March 26, 2020 that was to end up lasting, in one variation or another, 432 days (and still continuing as of writing). For at least the next seven weeks, we lived almost entirely through onscreen realities as people worked from home and all visits to others outside one's household were banned. Virtually all manner of human interaction, from work meetings to blind dates to Parliamentary sessions,[2] were conducted on Zoom (or Microsoft Teams or Google Meet), and friends and family Skyped to stay in touch, or waved to each other across window panes. The screen, always having been ubiquitous and ambient, even wearable, now asserted its presence in our lives with a new, almost violent forcefulness. Its boundaries became a suddenly cruel clarion call of the vast and imposing distances between the virtual realities of the screen, with its semblances of human warmth and interaction against the actual realities of isolation and confinement in daily lives under lockdown.

The status of the screen in life under Covid-19 conditions, then, heralded an unexpected postscript to the post-screen. The original post-screen, as this book has argued, was to be the state of our visual world as compelled via various contemporary media technologies, and through which images and reality present themselves, blurring critically important boundaries between virtuality and actuality, art and life. In wider terms, the post-screen re-formulates statements of contemporary conditions of media, reality, death, life, matter and history. The heady view from the heights of the Burj Khalifa observatory deck, just before Covid-19 became its own reality, sealed to me the pinnacle of that post-screen: a view of a reality unfettered by boundaries not merely in terms of the floor-to-wall glass windows but, more importantly, by its state of actuality apprehended as so unreal it was, in its essence, an internal virtualization. The post-screen thus signalled a visual regime whose boundaries of reference were so truly deliquesced it compelled the conceptualization of a successor to the screen as our most dominant and ubiquitous visual framework today: hence, the post-screen.

---

2   See Andrew Woodcock, "'Digital Parliament' Details Unveiled: MPs to Ask Questions by Zoom," *The Independent* online, April 16, 2020, https://www.independent.co.uk/news/uk/politics/coronavirus-uk-parliament-zoom-house-commons-video-call-link-a9468476.html; and Sophie Haslett, "You Can Now Go On a Virtual Blind Date Organised by Matchmakers For Singles in Lockdown," *The Daily Mail* online, April 15, 2020, https://www.dailymail.co.uk/femail/article-8219621/You-VIRTUAL-blind-date-organised-matchmakers-singles-lockdown.html.

However, mere weeks after, the world in the grips of the pandemic was swept into Covid-19 lockdown realities of virtualized meetings and ubiquitous communication through screens of computers, laptops, tablets and mobile phones. Their screen boundaries were not only of realities between the virtual and the actual, but also between isolation and companionship, silence and speech. In that sense, the pandemic demonstrated, and continues to do so, a gulf across boundaries of immense distance, cruel and stark.

Yet, in another sense – and of the same kind of paradoxes which have coloured this book – the Covid-19 pandemic is *also* of the post-screen: like the image of the post-screen that is in us, so is the virus. As does the post-screen eliminate boundaries between the actual and the virtual, so does the virus carry across borders and bodies. Screens may have been everywhere during lockdowns across the world, but, per their boundaries against the actual and the virtual, were neither protections nor defences against the viral threat. The status of the screen under Covid-19 conditions, then, turns out to be the more radical post-script to the post-script, whereby the virus itself is not of the screens to which life under Covid-19 retreated; it is actually *of* the post-screen. As with the image of the post-screen, the virus spills across boundaries and renders the same paradoxes of danger and protection in its wake, such as those of our bodies' vulnerability and immunity. What started as an idea of a contemporary visual world thus also became a conceptual anchor of the current pandemic world. Both worlds are inescapable.

***

Twin Obsessions (1): Difference

Two obsessions drive this book. The first is with difference – that quality of limbo which relies on relation, existing not as a thing but as against *some*thing. In that respect, difference is inherent in representation, whose being, by definition, is necessarily of one as against its object. The well-known Saussurean idea of "differences without positive terms" also comes to mind here,[3] whereby difference itself is not about the sign but about those that surround it. The task of the theorist, then – and certainly of this theorist via this particular obsession – is to be clear about difference without positive terms.

---

3   See Paolo Virno and Timothy Campbell, "The Money of Language: Hypotheses on the Role of Negation in Saussure" *Diacritics*, 39(4) (winter 2009): 149-161.

Yet, even as difference exists to mark one against the other, the counter-drive emerges to eliminate it. In this, the language for the visual already contains its own twist, where "representation" ruptures itself both with punctuation and in pronunciation as "re-presentation." André Bazin foresaw this outcome in relation to cinema (and its brand of realism as its aesthetic) as a regime of representation, where his reading of the medium as a "perfect aesthetic illusion of reality" engendered his prophetic vision of "no more cinema."[4] In that sense, to Bazin and the theorists of cinematic realism, cinema – as technological innovation, but also as the ushering of a new level of representation by the moving image and its capture of reality – heralded the effective extinguishing of their difference. Moreover, the seam between image and reality becomes increasingly invisible with technological advance-ment, as Serge Daney, paraphrasing Bazin, puts it: "I know (that the image is not real) but all the same... *With each technical change, the transparency grows, the difference seems to get smaller,* the celluloid becomes the skin of History and the screen a window open to the world." (emphasis added)[5]

As with Baudrillard (if in the more complex context of simulacra), Daney cautions against the perfect seam:

> We should not split up the screen but show the split occurring on it... not break continuity but make a rupture stand out on the conveyor belt of presence... *To intern difference* [taking from Derrida's phrase of sublimation, or idealization, by "interning difference in self-presence"] *means saving representation* [emphasis and quotes in square brackets both in original].[6]

But what does saving representation mean? And, in turn, what is at stake in difference? Or, perhaps, the question might be: what is representation for? As with storytelling, representation belies deep and ancient drives – the cave drawings, tens of thousands years old, of Lascaux, Chauvet, Altamira, Bhimbetka, Laas Gaal and, most recently, in South Sulawesi[7] stand alongside

4   André Bazin, "An Aesthetic of Reality: Neorealism," in *What Is Cinema?: Vol. II* (Berkeley; Los Angeles, CA: University of California Press, [1971], 2005): 16-92, 60.

5   Serge Daney, "The Screen of Fantasy (Bazin and Animals)," in *Rites of Realism: Essays on Corporeal Cinema*, ed. Ivone Margulies, trans. Mark A. Cohen (Durham, NC: Duke University Press, 2002): 32-41, 34.

6   Daney, "The Screen of Fantasy," 33.

7   Hannah Devlin, "Earliest Known Cave Art by Modern Humans Found in Indonesia," *The Guardian* online, December 11, 2019, https://www.theguardian.com/science/2019/dec/11/earliest-known-cave-art-by-modern-humans-found-in-indonesia.

oral myths, performance rituals, lore and legends. To represent in image or story is to make the most primitive sense of world and life, such as allegorizing dangers and threats into monsters to be slain. Representation is also an existential documentation of our presence – a pushback against death, decay and the grinding of long and deep time, where even bones turn to dust. It is an expression of the most primordial kind – to see and hear oneself as an externality in order to affirm the shortest of all possible grammatical linguistic utterances: we are; or, I am. To save representation, then, is also to conserve these expressions in and of the image so intimately bound up with identity, being and selfhood. Difference is not about disagreement but confirmation; it is an affirmation of our realities against what they are not.

But media technologies constantly push against these boundaries of image against reality, so that representation itself is thus also driven by twin obsessions, one extended from the other – to create; and then to create *entirely*. These two obsessions stretch as broad vectors across the history of visual representation from painting to magic lanterns to panoramas to photography to cinema to holograms to virtual reality and so on – not only to construct a *real* external reality in image and sound, but also to eliminate their boundaries of *difference* with the actual real. The desire at stake is the compulsion that drives every creator – to devise and commandeer an order of reality that is theirs, to engender not only the conditions of the present but also its history and memories. It is, perhaps, the ultimate gesture to the renewal myth writ large: the begetting of new beginnings, the rebirth of the prelapsarian, the return of innocence and wholeness.

As it turned out, cinematic representation, with its cumbersome apparatus and eventual baggage of editing, narrative, composition, sound and so on, was not quite the perfect illusion as envisioned by Bazin. As cinema institutionalized, its screens lapsed into industry standards and models which fossilized expectations of the site of the image – the multiplex, the art house theatre, the drive-in with its big outdoor projector screen, and so on. In becoming a set audiovisual system, cinema established its language, institutional norms and social practices; its regime of representation correspondingly became tamed, familiar, bloodless. The image that used to be so big became small, not in its literal dimensions (though much of that has happened, too, with the advent of multiplex screens, and then with sites of consumption of films shifting to domestic screens and then mobile screens), but diminished in what it could say, in itself, about images as against object, about the virtual in relation to the actual, and about what representation as a semiotic system can mean as *an understanding of understanding*.

Instead, as I have shown in this book, other kinds of representational media systems have emerged as the more telling formulations of our states of realities and histories today, formulations which I have thus encapsulated as *the post-screen*. Three of these media technologies in this regard form the corpus for this book: virtual reality (VR); holographic projections; and light projections. They are by no means new phenomena, as painstaking research into their media histories demonstrates critical trajectories from various precedents and predecessors, and highlights the important nuances of understanding the "newness" of new media.

However, that line of scholarship is concerned with, so to speak, media's assemblage – in the Deleuzian and Guattarian sense – in terms of media's forms of content and expression.[8] What I have argued through this book, rather, per this first obsession with difference, is the shifting of the theoretical lens from the newness of media in relation to form to *its increasing lack of discernment of difference* – per its diminishing boundaries – and, in turn, what that might mean as a critical statement of today's conditions of truth, reality and, in turn, today's politics of mediated truth and reality. On its ontology of difference, representation has always balanced, in long lines of theoretical and philosophical thought, affirmation and challenge against reality, be that iconic likeness as in portraiture, or the existential imprint of light in photography and film for its evidentiary properties of documentation and truth, or the challenges thereof. At the turn of the millennia, computer graphics imagery (CGI) upended these relatively established grounds of truth and reality as photorealistic images with no connection to their referents flooded screen and print media. The re-questioning of truth and reality from representation reverberated across the photoshopping of celebrities' bodies to the computer-generated fireworks shown in the live broadcast

8   Deleuze and Guattari first refer to the term "assemblage" in their text, *Kafka: Toward a Minor Literature* (Minneapolis: University of Minnesota Press, 1975), where, in their analyses of assemblages in Kafka's literature, Deleuze and Guattari state that "an assemblage...has two sides: it is a collective assemblage of enunciation; it is a machinic assemblage of desire," 81. Subsequently, they re-affirm their ideas of the formalization of assemblage as a constitution of form of content (via bodies) and form of expression (via acts and statements) in their 1980 work originally in French, *A Thousand Plateaus: Capitalism and Schizophrenia* (Minneapolis: University of Minnesota Press, 1987): "[A]n assemblage comprises two segments, one of content, the other of expression. On the one hand, it is a *machinic assemblage* of bodies, of actions and passions, an intermingling of bodies reacting to one another; on the other hand it is a *collective assemblage of enunciation*, of acts and statements, of incorporeal transformations attributed to bodies [emphasis in original]," 88.

of the 2008 Beijing Olympic opening ceremony, passing off as a filmed record of the actual event.[9]

On one level, then, today's manifestations of VR, holograms and light projections thus continue as the latest evolving statements of those contestations between image, representation, truth and reality. Yet it is more than even that: in light of today's indisputable ubiquity of screen media, the combination of their pervasiveness with their ensuing subversions of difference might also present an amplified statement of where and how we and our times stand today in terms of *media history* (*cf* the history of media, the distinction of which was also discussed in the introduction and chapter 5) – as in, the modes of media and mediation through which we live the times in which we exist and that which lives through us *in* the times in which we exist. History today is thus not simply that which we live through or the tracing of accounts of events; it is also a *media history*, whereby media – images, utterances, sound, screens, the Internet, social forums etc – becomes the sense-maker of the politics, policies, history, power, law and events which turn the billions of axles through which human and non-human lives grind and pass, and collectives formed and shaped.

In this respect, the manifestation of the post-screen today, then, also constitutes one such moment of media history, a moment I have chosen to identify as the twenty-first century ruling political climate, certainly in the Anglo-US context, through which runs a current – or more like an enveloping toxic cloud – of constantly shifting, essentially unreal, almost apparitional basis of truth and values. The Information Age, once heralded as a new dawn for human society, crowned by the Third Wave from the 1980s that swept in the post-industrial society,[10] and crested with the heady liberations of mainstream Internet in the 1990s, is now indisputably corrupted into The Misinformation Age,[11] with fake news and post-truths deftly and ably developed into a whole alternative system of non-values. The issue is not so much about the existence of lies and falsehood (though there is certainly no shortage of those in our politics today), but the non-existent grounds for their existence.

Hence, media, as systems of representation – and as harbingers of *difference* between real and unreal, actual and virtual, art and life – connects

9   Jonathan Watts, "China Faked Footprints of Fire Coverage in Olympics Opening Ceremony," *The Guardian* online, August 11, 2008, https://www.theguardian.com/sport/2008/aug/11/olympics2008.china.

10   Alvin Toffler, *The Third Wave* (New York: William Morrow, 1980).

11   Caitlin O'Connor and James Owen Weatherall, *The Misinformation Age: How False Beliefs Spread* (New Haven, CT: Yale University Press, 2019).

also to history; hence, post-screen media, as eliminators of such difference, itself connects to current histories. Where media is, as it has always been, the systems which push and interrogate how we sense and apprehend, and, in turn, establish the terms on which we receive information, truths and values, they also become the core of the corruption of those information, truths and values – the misinformation, disinformation, mistruths and untruths which colour contemporary times. In other words, our terms of information changed as media itself changed and, *qua* the post-screen, changed *in terms of how it sets, establishes, guards and polices the boundaries of difference.* It is a subtle, multi-faceted and not entirely perceptible process, though I have tried to give an account here of its gradual shades and incremental degrees of change as anchored to the concept of the post-screen. In this account, first across chapters 1 and 2 via the institutional representational system of cinema as a primary exemplar, the winds for the emergence of the post-screen first shifted in how screen boundaries were already moving and porous from the early days of cinema in terms of revelation and concealment, protection and defence; the virtual already leaking into the actual. In turn, the shift gathers apace in the bordering of our representational systems via VR, holographic projections, true holograms and light projections which enable us to re-think key ideas of meaning across ontology, death and matter: the placements of actual and virtual reality in the totalization of virtual environments (chapter 3); the conceptualizations of afterlife, after-life, ghosts and apparitions (chapter 4); the transformations of matter under light as against the material and the immaterial, mass and dematerialization (chapter 5).

The elimination of difference in the post-screen thus shifts not only our thinking of key ideas, but also the ground beneath their feet. At this point, we may also re-visit Daney's provocation, as quoted earlier: "to intern difference means saving representation." To *intern*, then, takes on different kinds of meanings: in the context of his essay, Daney most likely refers to the sense of confinement, as in to intern a prisoner; hence, to capture or delimit difference to show the split on the screen.

However, *to intern* may also refer to internalization, or to render internal. In turn, this tension in the ambiguity of definitions points precisely to the problematics of the change from the screen to the post-screen, which itself signals a shift *from confinement to internalization*. Where the image onscreen had always been on the outside (as in, outside of the spectator) to be captured (*to save representation*), the post-screen erases the difference between spectator and actor, so that action, spectating and image all become internalized unto the spectator themselves. The extinguishment of difference thus becomes the assertion of a new kind of virtuality, one that is neither a second order of the actual nor even simulacra in terms of its distance from the originating referent, but a virtuality that is *internalized* – that meshes with, dissolves or folds into the actual to create different kinds of ghosts. As with the view from the observatory level of the Burj Khalifa, the unreal actuality is also the real virtuality – such is the paradox of the post-screen, and illustrated here by way of a personalized instantiation. But, as with the viral pathogen of SARS-CoV-2, the post-screen is also an internal entity – it is in us. Covid-19, as the most critical analogue to the post-screen, and the real-ness of its system of contagion, borderless across bodies, thus progresses difference in this sense to become something that is more than a lack of the discernment of difference – it is also about the internalization of difference, and the implications thereof. As much as the post-screen is of media technology, it is also of another notion of history, epitomized here in a timely convergence via Covid-19 as pathogen, disease, pain and violence, for which there is no immunity, just as screens are not in themselves any kind of protection. In turn, that points to the issue of media today as one not of representation, but internalization. There is no full circle of, for instance, VR from the panorama in terms of the extinguishment of difference; what we now face (and are hit with) is a mediascape against which, like the viral contagion of our present, there is no escape – because it is internalized. On that same logic, what we thus also have today is not just the contestation of the real against illusion or the unreal, but the disappearance of difference without positive terms *as a moment of media history* – the history of uncertain values and a ground pulled from our feet, in which we are still currently interned and internalizing.

The post-screen thus engenders for reality not only no more cinema, but also no more media in terms of history: we may now think of everything as cinema, or as image, or as virtuality. But the thinking is not in the sense of "no more cinema" as Bazin had meant it, whereby he flipped the illusion of reality to reality itself in cinema's borderless zone of the unreal against the real. Rather, our conclusion today is that the terms of reality and illusion no longer have their old semantic values as they did when shot across the screen's boundaries; *they are no longer related in the ways they used to be related.* In the post-screen, reality and illusion are not counters or opposites to each other. Where the difference without positive terms has disappeared in the internalization of the image is also the emergence not of another reality but *another regime of truth values that has returned on the far side of media history.* But this is no return of, say, the prelapsarian, which only speaks of a nostalgic return; this is a return of another notion of history out of a profound internalization of the shit storms, the dis/misinformation and the post-truth of current politics – a return of the positive in the most scurrilous and outrageous of styles.

This conclusion thus also answers the "why now" question currently so faddish in academia in justifying its enquiries: why now in writing this book, why now for this argument? The answer, if one is at all needed: because we are in a regime today where, to put it succinctly, truth is lost. This is not to say that truth is not around or not in its place and we are in search of it, as in "I have lost my keys," or even as in Proust's seeking of memory *à la temps perdu*. Rather, truth is lost in the sense that it is disoriented, adrift and constantly unsettled in being fudged and muddied against half-truths, speculative facts, exaggerations, omissions, out of context information, misstatements and outright falsehoods, all merging into the kind of virtuality – part simulacra, part virtual, part actual – which drives our realities today. How, then, to make sense of this virtualized reality *sur les stéroïdes anabolisants* – not in terms of discerning the truth, but in simply grasping the shape and state of this discombobulation?

Media, as the representation system par excellence of our realities, thus becomes an obvious candidate of tool. In turn, the post-screen, as a devised concept in thinking through these critical boundaries, becomes a mental map to trace these precarious and volatile grounds today of information, knowledge and knowing. The re-thinking of representation and re-presentation, replacements and re-placements, ghosts and spectrality, virtual and actual, materiality and immateriality is thus enmeshed not only with the currencies of media technologies, but the wider question of how we understand and engage with our values, truths and information. Where media is ubiquitous and screens are omnipresent, something is also happening to their boundaries. Where screens meld into air, buildings, bodies, and where images fold down into their surroundings – *as difference vanishes into us* – something is also happening to our apprehensions of truth and falsehood, and the difference between them. The post-screen is thus also a mental model for signalling this disappearance of difference, if one not quite as with simulacra. It is a different kind of vanishing – a dis-appearance that is constructing, even as I write, this moment of media history.

Twin Obsessions (2): The Gluttony

The second obsession, then, is with gluttony, or, perhaps more precisely, with the mindless devouring and consumption that comes out of not only an unsated but unnamed hunger. Such craving in the culture is not new to the millennia, albeit perhaps different to each era. We might recall here Slavoj Žižek's description, taking from Alain Badiou's "passion for the Real [*la passion du reel*]," of a drive in the twentieth century for "the thing itself," or a (Lacanian) Real that comes from the bare-knuckled face of violent and direct transgression:

> In contrast to the nineteenth century of utopian or 'scientific' projects and ideals, plans for the future, the twentieth century aimed at *delivering the thing itself...* The ultimate and defining moment of the twentieth century was the direct experience of the Real as opposed to everyday social reality – the Real in its extreme violence as the price to be paid for peeling off the deceptive layers of reality [emphasis added.][12]

Now well into the millennia, we might be able to think of this desire for the real – in its *Fight-Club*-esque exposure of harshness, intensity and brutality against suburban indulgence, social proprieties and mindless consumerism – as shape-shifted into a desire for another kind of real: *the real of the unreal.* There was no clear turning point, though perhaps 9/11 – as the millennia's first globally visible violent realization of the real, so much "like a movie"[13] – augured a shift in the wind, as several critics have noted, in terms of the almost uncanny folding of event into media (read as Hollywood films), and vice versa. Certainly, the appearance and appetite for the unreal drive the millennia's culture of image production with particular energy: at some point in the last twenty years, the manipulations of photorealistic CGI imagery, which at the turn of the millennium so radically shattered the evidentiary bases of celluloid film images, crystallized into a cliché, rolled into "Photoshop fails" memes and counter-exposed by assertions of "natural" photographs, such as #NoMakeup and #wokeuplikethis selfies, paraded for no reason other than their *lack* of enhancement and touch-ups. The ubiquitous deployment of the "green screen" (if previously blue for film) used for compositing computer-generated images as backgrounds and foregrounds

---

12 Slavoj Žižek, *Welcome to the Desert of the Real!: Five Essays on September 11 and Related Dates* (London; New York: Verso, 2002), 5-6. The defining cultural fiction of this argument would be The Wachowski Brothers' (as they were then) 1999 film, *The Matrix*, whose plot and theme are precisely about the two realities against each other: the indulgence of a mindlessly accepted state of reality as conjured by puppeteering entities (the computers) as opposed to the cold and harshness of the real, or the thing itself.

13 See, for eg, Bülent Diken and Carsten Bagge Laustsen, "9/11 as a Hollywood Fantasy," *P.O.V.*, No. 22 (December 2005), https://pov.imv.au.dk/Issue_20/section_1/artc4A.html.

to live actors also pumps out vast amounts of computer-manipulated photorealistic imagery, its ceaselessness not unlike the inhumanly generative Moloch machine so memorable in Fritz Lang's 1922 film, *Metropolis*. Actors in various metamorphoses of digital skins appear onscreen as completely unrecognizable. The armies of computer graphic artists and visual effects technicians, evidenced in the increasing lengths of credits on visual effects work (VFX) scrolling at the end of every major blockbuster film, add to the gargantuan assemblages of CGI fakery production.

Like reciprocal presupposition, yottabytes of unreal images which engulf today's visual culture, wearing their fakery and manipulation like a proud brand, drive a gluttony that is, in turn, driven by an unsated hunger. The previous centuries had been marked by a similar appetite for images, hungry for their realism – the central appeal of the Victorian panoramas, for example, was its totalization of a believable virtual environment around the viewer with photorealistic paintings, hidden boundaries, faux terrain, clever lighting and so on. Cinema at the turn of the twentieth century was attractive and (eventually) celebrated because of its capture not only of reality – a rushing train; a line of workers out of factory gates – but also the realism in its smallest and most intimate details: as Kracauer declares, cinema is the "instrument which could capture the slightest incidents of the world about us."[14]

We are still devouring images in monstrous appetites today, if anything more than ever, fanned by the currents of circulation gusting through our connectivity networks: the photographs and videos, the Instagram streams and the TikToks, the more than 500 hours of video around the world being uploaded onto YouTube every minute.[15] But images in the twenty-first century feed a different hunger. These images are not so much unreal in the sense of CGI's photorealism which so uncannily resemble their referents, to which we may point as the last answer to the question of flipping the real for the unreal. Rather, the images in the twenty-first century explicitly play *between* the real and the unreal, so that photorealism is no longer a representation, but an off-balance vacillation between extreme naturalism and unabashed fakery; between recognisability and open alteration; between realism and manifest manipulations. Dozens of photo filter apps available

---

14   Siegfried Kracauer, *Theory of Film: The Redemption of Physical Reality* (Princeton, NJ: Princeton University Press, 1997), 27-8.

15   Statista, "Hours of video uploaded to YouTube every minute as of May 2019," https://www.statista.com/statistics/259477/hours-of-video-uploaded-to-youtube-every-minute/#:~:text=As%20of%20May%202019%2C%20more,for%20online%20video%20has%20grown.

for our smartphones today are specific tools and advertisements for adjusting and enhancing Instagram photographs. CGI itself in the twenty-first century has become extreme fakery in the form of deepfakes, now not just about the computational manipulation of images, but the leveraging of machine learning and Artificial Intelligence in training generative neural network architectures to manipulate or produce photorealistic audiovisual content. This manner of image production is thus no longer about the exact duplication of the real, for there is none – the referent has, in short, moved into the realm of computational parsing and the reduction into patterns. Representation becomes the computational generation of a real that, as a result, is paradoxically so real and recognisable that it is no longer real and recognisable. These images thus ask questions beyond those of the real and the unreal which still plagued the era of CGI and digital manipulation. They point to a new discomfort, one borne out of vacillation rather than a specific positioning in one or the other. It is the discomfort with a new kind of uncanny photorealism – not a return to the real (as still figured by CGI), but an elevation to another level of limbo between the real and the unreal, one that is neither here nor there, and completely unhistorical.

Indeed, as life goes live online in the Covid-19 era restrictions on travel, socializing and meeting, the computational unreal of the real becomes a constant output of video imagery. Video conferencing platforms, now the lifeblood of work, play and socializing under Covid-19 lockdowns, offer real-time enhancements, or fakery in actual time. Features such as Zoom's "touch up my appearance" filter work to "smooth out the skin tone on your face, to present a more polished looking appearance" during one's live conference calls.[16] Chinese video sharing websites, such as Bilibili, are flooded with videos that show young women whose features look dramatically changed *in real-time on live calls*. Their faces thus appear live onscreen in different presentations, whereby the "desired" face is one changed in real-time to feature enlarged eyes, smaller mouths and narrower chins – not without coincidence in line with the appearances of female anime faces: the actualized cartoonization of the real in actual time; or the unreal occupying the real in real time. Other videos show users' faces in real-time calls with subtle changes made to their eyes so as to shift their eyelines, so that the user's line of sight matches their audience's onscreen, an effect that could only be achieved if the user looked straight into their camera rather than their screen, as is normally the case.

16   Francis Kot, "Video Enhancements," on Zoom's Help Center, https://support.zoom.us/hc/en-us/articles/115002595343-Touch-Up-My-Appearance?mobile_site=true#collapsePC.

In turn, the Work From Home (WFH) phenomenon, now embedded across the world as ushered in by governmental, travel and health restrictions in relation to Covid-19,[17] has singlehandedly, if unexpectedly, migrated much of office and social life into the home. Via the webcam, engagements previously actual and embodied also digitize into constant streams of images. In a single tectonic heave as a result of the Covid-19 pandemic, the virtuality of webcam images has substituted much of daily life, driving a visual culture ever more obsessed with consuming the real via the webcam in terms of the unreal by way of video enhancement in actual time, simulated backgrounds which change coherently with the user's movements, deepfake software and, surely in time to come, realistic real-time avatars which completely replace the actual. The vacillation between the real and the unreal is thus extant today not only in the consumption of images of the real while yet in open and blatant image manipulation, but also in real-time and with relentlessly deeper sophistication and firmer implantation. Our times of the real unreal, previously manifest only in print, digital and moving media, is now also live and literally alive, powered by the relentless production and consumption of its images.

As an exercise of deploying theoretical creativity to account for our contemporary culture and phenomena, the post-screen is thus also an experiment in naming this devouring. In transferring a conceptual spotlight onto the increasingly hidden and subverted boundaries between virtuality and actuality, the post-screen also draws attention to the appetites which drive that muddying, and the obsession of the virtual that does not so much replace the actual but merges with it, as with the formation of some kind of mutant hybrid. Boundaries matter here because they are the gateways to the gluttony for this unstable hybrid, constantly vacillating between the real and the unreal – the changed and "enhanced" looks and appearances which yet remain recognisable; the photographs of ourselves in our fattened Instagram accounts both realistic yet unrealistically perfect; the videos of ourselves that have become genuinely viable substitutes for living bodies at work and play complete with artificially corrected lines of sight in real-time. The issue is not of the real against the unreal, but the sense of this new pushing of their borderline, guzzled with consummate ease and efficiency because all natures of their boundaries are obscured and melded.

---

17   Google, Microsoft, Morgan Stanley, JPMorgan, Capital One, Zillow, Slack, Amazon, Pay-Pal, Salesforce and other major companies have allowed work-from-home to continue for the foreseeable future: Jack Kelly, "Here are the Companies Leading the Work-From-Home Revolution," *Forbes*, May 24, 2020, https://www.forbes.com/sites/jackkelly/2020/05/24/the-work-from-home-revolution-is-quickly-gaining-momentum/.

The post-screen thus signals this different order of things – not a replacement of the original as with simulacra, but a compounded and amalgamated hybrid where virtuality melts and folds into the actual, and whose reality is simultaneously real and unreal. A hybrid whose real is as yet ungraspable and incomprehensible. The unadulterated form of the unreal might have crystallized at the turn of the millennia by way of avatars in virtual worlds such as Linden Lab's Second Life, itself a veritable phenomenon from the turn of the millennia and at one point in 2013 boasting thirty-six million accounts and more than one million regular users.[18] The leakage at the time between the onscreen realities of avatarial existence and the actual realities of users was much documented and noted;[19] it was really only a matter of time – a decade later, to be more precise – before that leakage became a full-on rush across screen boundaries into the current hybridization as a more scandalous and madder amalgamation between the actual and the virtual.

In this sense, the consumption of the mutant hybridity is not just of the constant streams, scrolls and walls of real-yet-unreal images, but, in some way, a devouring also of each other – and of ourselves – to the disappearance of ourselves, or at least some level of authenticity of ourselves. The removal of human agency from all manner of human activity, not least creative activity, is itself another and much wider anxiety: where and what is the role of the human if the computers are doing all the film creating, music composing, painting, writing, chess playing, stock brokering, even flying?[20] The post-screen thus also becomes the conceptual space to accommodate this sense of collapsed boundaries between the human and the digital falling apart under the force of the gluttony unleashing the real-of-the-unreal in our ceaseless streams of images which overwhelm even our sense of ourselves as spectators. As Simon Lefebvre puts it, writing in precisely the context of movies being made today in this sense of the unreal: "all screens finally give in and spectators watch themselves disappear, swallowed by some green screen that they cannot see."[21] Notably, he wraps up his point

18   "Infographic: 10 Years of Second Life," June 20, 2013, https://www.lindenlab.com/releases/infographic-10-years-of-second-life.
19   See, for example, Sherry Turkle, *Life on the Screen: Identity in the Age of the Internet* (New York: Simon & Schuster, 1997). Also Steven Morris, "Second Life Affair Leads to Real Life Divorce," *The Guardian* online, November 13, 2008, https://www.theguardian.com/technology/2008/nov/13/second-life-divorce.
20   The only thing AI cannot seem to do is drive: see The Economist, "Driverless Cars Show the Limits of Today's AI," *The Economist: Technology Quarterly*, June 11, 2020, https://www.economist.com/technology-quarterly/2020/06/11/driverless-cars-show-the-limits-of-todays-ai.
21   Simon Lefebvre, "The Disappearance of the Surface," in *Screens: From Materiality to Spectatorship – A Historical and Theoretical Reassessment*, eds. Dominique Chateau and José Moure

in a concluding mention of Steven Spielberg's *Jurassic World*,[22] the fourth film in a famous string of films which are all variations on the theme of humans trying to avoid being eaten by the dinosaurs they have scientifically created. As Lefebvre observes,

> ...the parallel between digital creation and scientific creation is clear and it is interesting to note that the whole movie is based on a system of explosion of separation, through the system of enclosures. Furthermore, all enclosures and protection devices are vitrified bubbles behind which humans are protected, at least for a while.[23]

Lefebvre's point is clear: along the way in the accelerating chain of digital imagery production, some sort of separation exploded and, as allegorized by the dinosaurs of Jurassic Park which eat its human creators, our uncannily photorealistic digital creations are also somehow consuming us. In the same sense of collapse, the beautiful avatars used in Second Life and other virtual worlds as our handles in that virtual reality have spilled into our Instagram feeds, social media accounts and online video calls as beautiful users in real-time, and this spillover is invasive, encroaching and threatening. In the post-screen breakdown of boundaries, we have somehow swallowed all the fakery in our images and spat them out as virtually masticated versions of ourselves, a process accelerated by the enforced migration to online exist-ences due to Covid-19 lockdowns and restrictions. This force of consumption is not quite coercive but mindless, not quite intimidating but a helpless thrall. This gluttony not only drives the space of the post-screen, but also colours it with a distinct aggression. The conduct of this all-encompassing absorption is not the antagonism of the twentieth century's real in its shocks of brutality, disorder and destructiveness for a kind of sought-after clarity. It is an aggression of the ceaseless ingestion of signs which have no material existence out of the wilful oblivion or singular impossibility of knowing or realizing saturation, or a surfeit that is entirely without satisfaction. It is an entirely different kind of violence.

However, the flipside of gluttony is starvation – two extreme ends of a deadly spectrum which also meet up, where binging becomes a kind of response to deep deprivation. To the gluttony of reality television, to the bottomless satisfaction of watching ourselves as Baudrillardian ready-mades

---

(Amsterdam: Amsterdam University Press, 2016): 97-106, 106.

22   *Jurassic World*, directed by Colin Trevorrow (2015; Universal City, CA: Universal, 2015), DVD.

23   Lefebvre, "The Disappearance of the Surface," 106.

("cloned to our own image by high definition, and dedicated by involution into our own image to mediatic stupefaction"),[24] Baudrillard aligned and identified this mass consumption to "the mediatic class... starving on the other side of the screen, in front of an indifferent consuming mass, in front of the teleabsence of the masses."[25] What feeds the voracity is hunger. Yet, the starvation in relation to the gluttony of the mutant unreal is more difficult to pinpoint. Perhaps it is the counter-response to the absolute of the twentieth century real in its brutality, as Žižek identified – a freaked withdrawal from its violence and terror, and from its bare-knuckled reflections of injustices and wars and misery and suffering, once held up as a mirror to the world's truths and now simply a route to overwhelming grief. Perhaps it is a deviated disengagement from the exponentially increased anxieties and stresses of millennial living, from eco-anxiety to the relentless exploitations of the precariate to the inexorably spiralling living costs of Generation Rent to the ever escalating culture wars and shit storms, at least in the Anglo-American sphere, whose extreme nastiness, personal attack and weaponization of social media only intensifies every year, or, seemingly, with every US election. All this with no apparent solution or resolution in sight. The starvation to the gluttony might also very well be a cry for help.

The Post-Screen in the Time of Covid-19

> But the screen is not a mirror, and, while it was some kind of magic to pass beyond the mirror, there is no magic at all in passing beyond the screen. It's impossible anyway – there is no other side of the screen. No depth – just a surface. No hidden face – just an interface.
>
> ~ Jean Baudrillard[26]

> Whoever passes through the screen and meets reality on the other side has gone beyond jouissance.
>
> ~ Serge Daney[27]

Per the opening quotations of this section, both Baudrillard and Daney allude to a certain process of passing through the screen, both with some

---

24 Baudrillard, "The Virtual Illusion," *Theory, Culture & Society*, 12 (1995): 97-107, 100.
25 Baudrillard, "The Virtual Illusion," 100.
26 Baudrillard, "The Virtual Illusion," 101.
27 Daney, "The Screen of Fantasy," 34.

accompanying sense of an access to a beyond. However, they arrive at dia-
metrically opposite conclusions on this viability of passing. Baudrillard takes
the starker position of flatly denying its possibility – to him, the spectating
masses are on both sides of the screen, and so what joins across the divide are
similar voids of panic-inducing simulacra, so that interactivity itself becomes
a farce. In short, there is no reality beyond the reality of the simulacra.

For Daney, though, not only is there the possibility of passing through
the screen into other realities, but its passage, moreover, becomes acutely
profound as one which moves the viewer into territories of the deeply forbidden and prohibited. In the context of the quotation, Daney takes on André
Bazin's declaration of "no more cinema" – whereby reality and image are
finally and completely merged – so that the screen becomes precisely this
viable membrane of transparency, whose viewer may pass from one reality
on one side to another on the other. This "cinema of transparency" thus also
crowns the elimination of difference and distance between the image and
its surroundings, whose resulting reality is now both image and object.

But the most interesting point of Daney's provocation in passing through
the screen is the prospect of the viewer having then entered a kind of utterly
elusive zone – that of "beyond *jouissance*." The translator notes *jouissance*
as meaning "both pleasure and orgasm" (and an allusion to Freud's *Beyond
the Pleasure Principle*); the passing through the screen – the experience of
the reality once beyond the transparency between image and object – thus
takes on not only enjoyment or satisfaction, but also the real of sexual
gratification. In turn, though, sexual pleasure itself is double-edged, for
the orgasm in French also contains connotations of death – *à la le petite
mort* – which extends, in paradox, both the real and its impossibility beyond
the screen. On one hand, *le petite mort*, as with the moment of death – "the
unique moment par excellence," as Bazin puts it – cannot be part of reality
onscreen. As Bazin argues in his essay, "Death Every Afternoon," both death
and the sexual act

> ...[e]ach is in its own way the absolute negation of objective time, the
> qualitative instant in its purest form. Like death, love must be experienced
> and cannot be represented... without violating its nature. This violation
> is called obscenity.[28]

---

28   Both quotations from Bazin in this paragraph are from André Bazin, "Death Every Afternoon,"
in *Rites of Realism: Essays on Corporeal Cinema*, ed. Ivone Margulies, trans. Mark A. Cohen
(Durham, NC: Duke University Press, 2002): 5-9, 9.

On the other hand, there is also no reality more real than death and love; and here we might likewise recall the conclusion to chapter 3, whereby only love is generative and transformative in the re-placed space of John Hull's blindness made symptomatic through the near-darkness of the totalizing screen of its Virtual Reality (VR) project, *Notes on Blindness*. Daney, read with Bazin, thus gestures towards both the real and the unreal in that space beyond the screen, bound up in all the extremities of love, death, sex and pleasure. The extinguishing of screen boundaries itself – in passing through the screen – is thus also about these irreconcilable paradoxes of "beyond *jouissance*": of accessing another level of reality real and unreal, visible and unrepresentable, forbidden and transgressed.

A certain madness in the crosshairs of these paradoxes thus lies in that space beyond the screen, a step we take at our peril. In that sense, the post-screen – in its extinguishing of boundaries and the ultimate disappearance of the fixed screen, as both a literal and figurative statement of the discombobulation of our mediated realities and truths today – also rests in those irrational indeterminacies as an edging onto an ultimate nihilism, or some kind of final apocalypse. The loss of the determination of the screen is precisely about this zone or sense of the forbidden – or to read Baudrillard another way, so prohibited it is impossible – because it heralds a zone of such profoundly irretrievable loss. The ground that shifted beneath the feet of truth with the advent of CGI and digital manipulation is in danger of disappearing altogether into a sinkhole of unrepresentability by virtue of the real descending entirely into the unreal. This, then, is the real peril of the post-screen: it is not just the real against the unreal, or just a question of that which is against representation; *it is the entire loss of representation in the place of real-unreal poles that cannot be grasped in its scandalous and impossible madness.* In this respect, Paul Virilio writes, too, of the disappearance of "the whole universe" in the literal loss of the image – per the quotation below – but he might as well also be speaking about the loss of the boundaries of the image, whose edges, or the "ontological" cut, are the very sense-makers of representation itself:

> In the West, the death of God and the death of art are indissociable [sic]; the zero degree of representation merely fulfilled the prophecy voiced a thousand years earlier by Nicephorus, Patriarch of Constantinople, during the quarrel with the iconoclasts: 'if we remove the image, not only Christ but the whole universe disappears.'[29]

29   Paul Virilio, *The Vision Machine*, trans. Julie Rose (Bloomington/Indianapolis, IN: Indiana University Press, 1994): 16-17.

At the same time – as already alluded above – the post-screen spreads beyond screens and image to other conditions of our world, namely, in current times, the Covid-19 global pandemic. On one hand, life under the pandemic has re-inserted the screen and its boundaries into our lives with almost ferocious violence as travel is restricted; work and study from home is instituted for foreseeable futures; and various degrees of governmentally decreed lockdown, particularly in Europe, relegate socialization back to screens and across screen boundaries. In that sense, the boundaries are never realer, where screen edges are also the symptoms of our "bubble" isolations, across which we reach to others, whose rigidities are cold reminders of our enforced confinements and severely restricted inabilities to travel.

On the other hand, the virus, as with the image of the post-screen, are also internal to us, and here is where the post-screen might augur the new condition of internalization as triangulated between media, environment and bodies. Perhaps, then, the ultimate post-script to the post-screen, for now at least, is how the Covid-evolved boundaries of the post-screen *re-place* to the boundaries of ourselves as human bodies, people and identities, and the questions of where, how and in what ways do we count as *being* in an existence now no longer just replete with screens, but made essential and indeed possible only through them. Via the screens of Covid-19, we have truly become streaming digits, our lives rendered into rushing packets of information which, in turn, form the socially viable and economically useful units operable in twenty-first century capitalism. This is the vision of Gibson's *Neuromancer* writ large – each of us digital bits racing through the network – but without its exhilaration and romance of the cowboy adventure. Or perhaps, we can say, a technologically dystopic version of Jordan Peele's 2019 film, *Us*, also a film of various boundary transgressions, whose binary worlds of rich/poor; have/have-nots; light/shadow similarly cleave across our current versions of ourselves as unstable vacillations between the real and the unreal. The post-screen of the Covid-19 screen landscape is thus the erased boundaries of manipulators and manipulated, a distinction that was relatively clear even in the plasticity of the CGI era. But in reducing ourselves to these digital bits, we have become manipulators of ourselves – our own men behind the curtain, our own Wizards of Oz. This time, though, there is no Toto to pull aside the curtain, and no "back room" of a little old man pulling the levers and pushing the buttons to literalize the scene of manipulation. That scene of manipulation has become us, each of us in front of our computer, laptop or tablet screen – or the post-screen out of the Covid-19 pandemic. Perhaps the true discontinuity of the post-screen is its paradoxical continuity, where the image persists but on this alternative plane of also, as does Covid-19, changing the visibility *and* viability of ourselves as bodies in our media environments.

A final point: the post-screen, both today and/or accelerated or evolved by the Covid-19 pandemic, is thus most marked by this quality of dissolving boundaries: boundaries of screens or otherwise which mark distances and differences; contain and close off; demarcate and differentiate. But boundaries are also and always hubs of ceaseless flux and movement, or temporary spaces of deterritorialization and reterritorialization. In that sense, even conventional screens can no longer stand by themselves, as Casetti also notes: "[Screens] have become transit hubs for the images that circulate in our social space. They serve to capture these images, to make them momentarily available for somebody somewhere – perhaps even in order to rework them – before they embark again on their journey."[30] In the hub of such ceaseless flux, the post-screen thus also heralds a different order of things – not the order of antagonism between simulacra and truth, whereby, as with Spider-Man (discussed in chapter 1), we try to discern with a tingle, to get that prickling of the neck as we seek to differentiate one from the other. In the post-screen, the order is one of no order: it is only about acceleration and volatility through our screens, or of ceaseless movement and activity across its boundaries. And, in a way, this, too, might herald new kinds of existences where, trapped even as we are in our screens in Covid-19 lockdowns, we find new liberations in new movements.

However, in another way, the endless fluidity and change might also only ring in a hopeless and helpless chase, like Orpheus running after Eurydice whom he was never going to get back – a tragic pursuit doomed from beginning to end. Perhaps all our activity and flux across the screens is simply a cover for a race towards something we have lost, to which we cannot give a name but can only vaguely sense amidst the noise and deceptions that constitute our contemporary underworld. It is also in this sense of fumbling through a dim memory that I valiantly and vainly try to recall the movie whose title I have forgotten: the one where, having spent much of it reeling from spectacular accident to accident caused by the interpenetration of two opposite realities – one fictional, one actual – the movie ends with a satisfying reclamation of order: the return of everything into their right places, as it were, on either side of the screen.

---

30   Franco Casetti, "What is a Screen Nowadays," in *Public Space, Media Space*, eds. Chris Berry, Janet Harbord and Rachel O. Moore (Basingstoke: Palgrave Macmillan, 2013): 16-41, 17.

Index

[*murmur*]   28
3D
   hologram fans   25
   cinema   96-97
6x9   147
9/11   213, 225, 264

Acland, Charles   32-33, 54, 56, 122, 220n54
actual: *see* actuality
actuality   41-44, 88-92, 134-136, 139-142,
   172-184, 224-247, 253-254, 262
affect   22, 114, 128, 132, 179, 246n118
afterdeath   77
afterlife   37, 77, 155, 158-168, 185, 204
Al-Joulan Nayef   159
Alberti, Leon Battista   26, 63n43, 68
algorithm   102-105, 242
Alice Cooper   192, 198
Andrew, Dudley   19, 30, 47n105, 61, 62n38,
   70n81, 138n103
Antonioni, Michaelangelo   73-74
Apple 1984 commercial   49
arche-screens   54
Arnheim, Rudolf   61-62, 202, 203n50
Arrival of a Train at La Ciotat   89
art, land   27
Artaud, Antonin   196n21, 198
Assange, Julian   155, 177-178
Atkinson, Sarah   97
Attie, Shimon   225
Augmented Reality   19n11, 52, 104
   Spatial   239n104
Aumont, Jacques   71, 138n103

Badiou, Alain   264
Barker, Jennifer   68n66, 178
Barthes, Roland   74n100, 158, 195n17, 249n122,
   251
Baudrillard, Jean   39, 43-47, 53, 61n34, 129-131,
   136, 191n4, 241, 250, 257, 269-272
Baudry, Jean-Louis   91n39, 109, 110n4
Bauhaus   112-113, 119
Bazin, André   45-46, 61-62, 64, 65-66, 68-69,
   73-74, 85n22, 85, 158n9, 160n24, 170n67, 185,
   197, 246, 257-258, 262, 271-272
Belisle, Brooke   121
Bell, Robin   229n78, 230
Belting, Hans   196
Benjamin, Walter   138, 163n34, 196n21
Bergson, Henri   60
Bertolucci, Bernardo   251n130
Black Mirror
   "Be Right Back" 168
   "Black Museum" 180
Blade Runner 2049   193

Blair Witch Project, The   93, 95, 97n68
Blow-Up   68n68, 74-75, 77n108, 78n109
Boccioni, Umberto   207-208, 211, 214,
   244-245
Bollmer, Grant   149-150
Bolter, Jay   31, 123n55, 126n67, 126n69, 128. *See
   also* Grusin, Richard.
boundaries
   as thresholds   15, 31, 36, 89, 191
   as transformative   24, 51, 67-68, 214, 240,
     248
   between art and life   17, 19, 34, 98, 225,
     260
   breach of   58, 80, 83, 85-88, 97, 134, 153
   diminishing   21, 33-34, 56, 75, 259
   imperceptibility   20, 23, 182, 246
   of literature   28
   of paintings   16, 20, 26-27, 29-30, 34-36, 54,
     61-64, 117-120, 144-145, 254
   of screens: *see* screen
   perceptibility   16, 21, 31, 108, 208. *See also*
     boundaries: imperceptibility.
   physical   56-61
   virtual   56-61
Braidotti, Rosi   39
Brown, Tom   32n62, 63n44, 95
Brown, William   71n86, 140n118
Bruno, Giuliana   19, 138n106, 214n29
buildings, as screens   217-235
Burch, Noël   31n58
Burj Khalifa   219, 226, 253-255, 262
Burke, Chris   74-75

Camera Lucida   158
camera
   "found"   93-95
   diegetic   87
   direct address   63n44, 99-100, 162n30, 266
   dual-facing   104
   eye   62, 70, 139, 254n1
   handheld digital   179
   mobile   71, 72, 99
   obscura   23, 59
   unleashed   71-72, 138n103
   virtual   43, 75-77
   vision tracking   227
Carbone, Mauro   54-55, 60n29, 70, 72, 81-82,
   86n27
Cardinal, Roger   68-70, 77
Carne y Arena   146
Casetti, Franco   82, 167n56, 274
cave, Plato's   72, 81-82
CAVES   122-123
cell phone: 83. *See also* mobile: phone;
   smartphone.

CGI   71, 161, 163-166, 168-169, 259, 264-266,
    272-273
  pornography   179-180
Chamayou, Grégoire   18, 42
Chou, Jay   169, 172
Cinema Paradiso   224-226, 228, 231
cinema
  no more   45-46, 246, 257, 262-263, 271
  marketing of   97-98
cinéphilia: *see* cinephilia
cinephilia   67-68, 161, 225, 251
Cinerama   122, 118
Circarama   122
Citizen Kane   70-71, 78
City Rises, The   207-209, 211, 215, 244, 252
city   98, 120, 124, 207-230, 252, 254
clone   132
Clouds over Sidra   148-149
Cloverfield   95, 179n88
co-location   99-102, 105, 157
Cobb, William Jelani   145, 151
Comment, Bernard   26, 120n44, 137, 138n100-101
compassion   23, 146n132, 150
computer screen films: *see* desktop films
Condition Humaine, La   34-36, 46, 109, 131-132,
    140, 152-153
confinement   116-119, 123
  VR   124-125
Consumer Electronics Show (CES)   25, 26n31
copy   36, 57, 128, 132, 136, 151, 153, 158, 250
Covid-19
  lockdown   18, 101, 226, 255-256, 266-267,
    269, 273-274
  pandemic   18, 38, 101, 143, 231, 255-256, 267,
    273-274
  virus   253, 255-256, 262, 273
Crary, Jonathan   117
CRASH   118
Cubitt, Sean   19n13, 218n45
Cumiskey, Kathleen   168n57
Cushing, Peter   165
cybersickness   135-136

Daney, Serge   84-86, 88, 222, 233, 257, 261,
    270-272
danger paradox, the   133-136
Danto, Arthur   26
De Nachtwatcht   16, 20, 109
de-materialized   213 *See also*
  de-materialization.
de-materialization   225. *See also* de-
  materialized; re-materialize.
DeadSocial   167
dead
  actors   161-165
  singers   165-166
deepfake   266-267. *See also* fakery.
Deleuze, Gilles   60n31, 66n61, 70n80, 96,
    129n76, 181, 197-199, 259n8

Derrida, Jacques   29n48, 163
desktop films   103-104
detail, peripheral   68-69
difference
  disappearance of   130, 262
  extinguishment of   261-262
  intern   257, 261
  internalization   253, 261-263, 273
  without positive terms   256, 262
differentiation   15, 23, 30, 46
digital apparitions   199-204
dis-appear   55-56, 78, 127, 132, 151, 228, 250
dis-appearance: 51, 61, 76, 134-135, 145, 153,
    221, 238, 240, 263; *see also* dis-appear.
disinformation: 21, 247, 261. *See also*
  misinformation.
Displacements   240
District 9   98
Dome of Light   213n25, 214
Dreamers, The   251
drone   18, 42, 52
Duchamp, Marcel   27
Dyer, Geoff   111

$e=mc^2$ 215
Einsteinian physics   213, 215
Eisenstein, Sergei   61-62, 66, 162n31, 196n21,
    212, 244
electronic bulletin board systems (BBS)   100
Eliot, T.S.   91
Elkins, James   60n32
Elsaesser, Thomas   32n60, 41n83, 63n44, 66,
    178n84
empathy   133, 146, 149-151, 233
  machine   116n22, 146, 148
energy   37, 132, 207-209, 211-215, 222, 227, 238,
    244, 247, 252, 264
engulfment   116, 119-123
  VR125
environmental theatre   27
ergon   29, 36. *See also* parergon.
escape
  across screens   49, 262
  VR   133, 137-140, 142, 145
Excursus: Homage to the Square   214
existential   47-49, 111, 131-133, 139-140, 152, 200,
    204, 258-259
Expeditions   140
externalism   196-197. *See also* internalism.

façades
  building   217-218, 220-223, 227-228,
    233-234, 238, 242-245
  media   216, 219-223
Facts in the Case of Mister Hollow, The   76-78
Facts in the Case of M. Valdemar, The   76-77,
    173-174, 185
fake news   21, 260
fakery   78, 265-269. *See also* deepfake.

Farocki, Harun 42
fireside chat 100n75
Flaxman, Gregory 196n21, 198n31, 198n34
Flusser, Vilém 64-65, 68, 200-204
 "crisis of linearity" 189, 201
Foer, Jonathan Safran 15n1, 186-187
FOMO 47-48
Foucault, Michel 181, 183n108
found-footage 92-99, 102-103
frame 109, 111-112, 115-116, 120-122, 175, 216-218,
 241. See also boundaries; frameless.
frameless 120, 122
Friedberg, Anne 20, 31, 56, 63, 70, 89-90,
 141n122, 213n25, 217n42
Futurist 207n2, 208, 244-245

georama 121. See also panorama; panoramic.
ghosts
 amongst the living 170-174
 as memory, daydream, secret 199-205
 in the city 211n19
 in the media 158-168. See also
  Phantasmagoria.
 of the living 174-185
 vitality 163n34, 164, 171
 vivid 160, 164n38, 170n63
 will not starve 152
Gibson, William 245n117, 139n110
glass ceiling 88
gluttony 38, 245-247, 263, 265, 267-270. See
 also hunger; voracity.
Goldberg, David Theo 39n80, 89n33, 234
Google Tilt 144
Gorky, Maxim 158
Gouhier 171-172
Graffiti Research Lab 227, 231n87
Grau, Oliver 19n13, 110, 121n49, 125
Griffiths, Alison 71n84, 122n52, 135n90, 138n102
Grønstad, Asbjørn 74
Grosz, Elizabeth 21, 24, 89
Ground Zero 213
Grusin, Richard 31, 123n55, 126n67, 126n69,
 128. See also Bolter, Jay.
Gunning, Tom 32, 178n84

Hagener, Malte 57, 63n44, 66
hand-screen 83
Hansen, Mark 125, 128
Happn 104
Harbou, Thea von 212, 214
Hatsune Miku 172n69
Haunted Mansion 239-240
Haunting of Hill House, The 137n94, 199, 204
Heilig, Morton 124n59
hierophany 49
Hight, Jeremy 28
Hiroshima Projection 232-234, 239
Hitchcock, Alfred 70n74, 95n56, 96n59, 155n1,
 156

Hjorth, Larissa 168n57
Hlavajova, Maria 39
holograms
 technology of 191-195
 vitality 175, 178. See also ghosts: vitality.
 Princess Leia 53n4, 193, 194n15
holographic projections
 Holocaust survivors 183-184
 protest 181-183. See also projections:
  protest.
 tetravalence 37, 155, 175, 178, 181, 183-184
Hofsetter, Gerry 231
Holodeck 108, 112, 153
Hookway, Branden 19-20, 67
Hubert, Renée Riese 35
Hugo 57
Huhtamo, Erkki 33n66, 51n1, 53-54, 81n6,
 83n18, 117n31, 118n34, 121, 123n56, 137n98,
 216, 222n60, 223
Hull, John 150, 153
human
 being 34, 36, 50, 153, 173
 condition 36, 48, 69, 152, 187
hunger 43, 47-49, 152, 264-265, 270. See also
 gluttony; voracity.
hybrid of actual and virtual 197, 267-268
hymen 85, 88, 222, 233, 250
hyper reality: 142. See also Virtual Reality
 (VR): location-based.
hypermediacy 31. See also immediacy.
hyperreal 43-44, 53, 129-131, 136, 241

Illouz, Eva 106
IMAX 122, 126, 135n90
im-material 225, 227-228
immateriality 22, 38, 175, 180, 208, 211n19,
 212-214, 221, 224-229, 231, 234, 237-238, 245,
 263. See also materiality.
immediacy 31, 126, 128
immersion: see VR: immersion.
InCell VR 140
Incredibles 2, The 17
indifference 21
Ink Mapping 249
interface 19-20, 29, 47, 49, 67, 91, 105, 126, 186,
 215-216, 270
internalism 196-197. See also externalism.
Irom, Bimbisar 149
Irwin, Robert 214

Jackson, Shirley 137, 204
Joi 193
jouissance 270-272

Karen 102-104
Kafka, Franz 152, 153n155, 259n8
Kate Moss 174-175
kingdom of shadows
kino-eye 139

Kircher, Athanasius 118, 159n17
Kiss, The 192, 195
Kittler, Friedrich 40, 202, 203n50-52
Koeck, Richard 98
Krauss, Rosalind 30
Kress, Gunther 81-83
Kroker, Arthur 22, 53n6
Kroker, Marilouise 53n6
Kronhagel, Christoph 216, 218n46, 219n47

land art 27
Lara Croft 72
LED 217-223, 236
Lefebvre, Simon 96-97, 268-269
Liberty Leading the People 141, 144
Liew, Kai Khiun 170, 171n64
Light Creature, The 222
light
    as lighting 212
    as the matter of light 209, 212-214, 227
    electrical 210-211, 213, 216, 219, 224
    on 220, 223
    through 220, 223
    transformation of 208-210, 214, 227-228,
      231, 238
linear perspective 77
Locative Literature 28. *See also* Locative
    Narratives.
Locative Narratives 28
love 108, 110, 115, 123, 153, 156-157, 271-2. *See*
    also cinephilia.
Lunenfeld, Peter 31n56, 49n112, 108n3, 109

McCullough, Malcolm 30, 216-217, 218n46,
    220, 221n57
McLuhan, Marshall 44, 54n9, 139n107, 220, 238
McQuire, Scott 211-213, 216n34, 218n43
Machine To Be Another, The 146n136, 147
magic 59, 86-88, 163n37, 177n80, 194, 203, 225,
    252, 270
    lantern 54, 82, 159, 159n17, 160, 223, 258
Magritte, Rene 34-35, 46, 109, 132, 140, 152. *See*
    also Condition Humaine, La.
Manovich, Lev 41, 42n85, 56n17, 58, 64, 111,
    126-127
Mapplethorpe, Robert 229
Mariah Carey 68n66, 175-176, 183
Marks, Laura U. 178, 221n55
mask 58, 62, 73, 83, 132n87, 159, 248
mass 37, 207, 211, 214, 222, 224, 227, 231, 237,
    239, 243-244, 247, 261
    communication 138
materiality 56-57, 74, 136, 175, 180, 208-209,
    211-214, 219, 221, 224-229, 231, 234-235,
    237-238, 243, 246-247, 263
    formal 51, 57, 59, 62
    not formal 57
Matrix, The 22n23, 264n12

media
    archaeology 26n32, 53
    façades: *see* façades: media
    history 23, 37-38, 99, 207, 238, 246-247,
      253, 260, 262-263
    interactive 36, 72, 74-75, 99, 102, 166-168,
      204, 222
    mobile 18, 19n11, 23, 28, 83, 104-105, 127,
      140n117, 161, 167-168, 242n111, 256
    old and new 31
    post-democratic 78
    social 18, 21, 48, 166-168, 171, 204, 229n79,
      247, 269-270
mediatecture 218-219
Méliès, Georges 85-86
memento mori 159
Merveilleux éventail vivant, Le: see *Wonderful*
    Living Fan, The
Mesdag panorama 139
Metropolis 49, 212, 224, 265
Microsoft
    Kinect 241
    Windows 63-64, 217, 242
Milk, Chris 116, 146, 148,
Minority Report 156, 193
mirror 48, 59, 60n29, 61, 66, 192n8, 194, 235,
    270
Misinformation Age, The 260
misinformation 38, 78, 246, 261, 263
Mitchell, William J. 209n6, 221n58
Mitry, Jean 61, 65, 73, 179n89
Möbius strip 78
mobile
    app 102-104, 123, 140, 144n129, 150, 222
    narratives 28. *See also* Locative
      Narratives.
    phone 18, 23, 83, 123n58, 127, 167, 256. *See*
      also smartphone.
Modi, Narendra 177
Monkey's Paw, The 157n6
Monkhouse, Bob 161-164, 176
movie ride 141
Multitudes of Amys 107-108, 110, 153
Mulvey, Laura 69n73, 159-161
Murray, Janet 20, 99, 109-111, 135n90, 141n121,
    148-149, 153n154
Musser, Charles 53-54

Nachtwatcht, De 16, 20, 109
Naimark, Michael 240,
necromancy 37, 162-164, 166, 168, 176, 203
necrophilia 37, 155-156, 161-162, 166, 168
Neuromancer 139, 245, 273
Nobumichi Asai 247, 248n120
Notes on Blindness 150, 153n156, 272

Oculus Rift 123, 133-134
Oscar & Gaspar 249-250

panorama 71n84, 117, 118-122, 125-126, 137-139, 241, 254, 262. *See also* georama;
panoramic wallpaper.
panoramic wallpaper 121-122
Paranormal Activity 94
Parasite 189-191
parergon 29, 36. *See also* ergon.
Parikka, Jussi 26n32, 40, 53n6
Party, The 146-147
pastness 158-159, 225
peep
   media 118-119, 126
   practice 117
   show 54, 117-118, 137
Pena, Jean 59
Pepper's Ghost 169-170, 174, 227n74
Pepperell, Robert 196-197
Peretz, Eyal 32n59, 66
Perez, Gilberto 158
Peters, Benjamin 99n74
Peters, John Durham 23-24, 24n26, 46-47, 160n22
Phantasmagoria 159, 169
   New, the 169
Phelan, Peggy 171n66, 172n68
Philidor, Paul 159, 169
photograph 76-78, 113, 158-159, 164n38, 194-195, 202n47, 204n55
Pisters, Patricia 197
pixel bleed 109, 128
Polanyi, Michael 39
Polo Ralph Lauren 236
portal, 67, 105
post-, meaning of 39-50
post-screen
   definition 22-23, 39-50
   imagining 22, 24
   media 20-22, 116, 185-186, 253, 261
   politics 247
posthumous 162n30, 163-166, 171-172
   duet 165, 166n49, 169, 172
Pound, Ezra 211, 212n20, 214
presentness 84, 158-159, 170, 172
Project Syria 148
projection mapping
   of buildings 238-246
   of faces 247-249
   of tattoos 249-250
projections
   graffiti 227-228
   of light 223-237
   protest 229-232, 235
   vitality 209, 221-222, 224, 226, 228, 243-246
   vivid 176, 178, 180n96, 181, 184
proto-screens 54, 123

Raft of the Medusa, The 117-118
rape 85-86

re-materialize 225
re-place: *see* re-placement
re-placement 37, 98, 107, 109, 127, 131-133, 135-136, 141, 145-146, 149-150, 152-153, 263
readymades 27
real
   actual 89, 115, 134-136, 258
   definitions of 22
   the 21-22, 35, 37, 39, 42-46, 62, 84-85, 88-89, 91, 109, 114, 126-127, 130, 132-133, 136, 151-152, 166n49, 172, 203, 246n118, 262-268, 271-273
   virtual 89, 108-110, 115, 133-134, 175
Rear Window 95-96
Rembrandt 16n3
remediation 31, 77
replace: *see* replacement
replacement 31, 37, 43, 45, 98, 109, 127-132, 138-141, 144-150, 153, 263, 268
resurrected, the 159
resurrection 37, 159-160, 162, 165n47, 166, 168, 174-175, 180, 186
Ringu 86-89, 92, 157
Rogue One: A Star Wars Story 165
Rombout, Ton 120
rupture, of screens 32n63, 80, 84-85, 88, 92, 95, 102
Rushdie, Salman 251
Ryan, Marie-Laure 49, 110, 118

Salen, Katie 111-112. *See also* Zimmerman, Eric.
sameness 131. *See also* difference.
Samsung 25, 26n31, 123, 140, 219
Saunders, Rebecca 179
Sayad, Cecilia 93-94, 97-98
Schechner, Richard 27
Schivelbusch, Wolfgang 209-212
Sconce, Jeffrey 160, 162n33
scopophilia 69
scotoma 6
screen-sites 231-232
screen
   "forgetting" 37, 107, 109, 114-117, 119, 123-127, 134
   and information 15, 21-22, 33, 51, 53, 56, 63, 70-78, 83, 105, 115, 186-187, 199, 203, 216-218, 220, 234, 261, 263
   as barrier 21, 81-86, 90-91, 144, 149
   as brain 38, 189, 196-199, 203
   as defence 81-85, 88-89, 91, 101, 256, 261
   as keyhole 73
   as mirror. *See* mirror
   as partition 36, 79, 81, 83, 87-88, 90-93, 99, 101, 103-105, 108, 115
   as Plato's Cave. *See* cave, Plato's.
   as portal 48, 67, 104-105
   as protection 79, 81-83, 85, 88, 91-92, 98-99, 101-102, 105, 108, 115, 220, 256, 261-262, 269

as skin   84-85, 208, 214, 221-223, 233-234, 239-242, 244, 247-250, 257
as transgression   69, 86-88, 97
as window   26, 34-35, 42, 61-65, 67-68, 73, 75, 77, 96, 116, 140, 148, 199, 217-218, 246, 257
definition of   33, 81, 220n54
formalization   127
gendered   79, 85-86
looking at   99, 116, 215
looking into   65, 99
looking through   99
of ash   38, 207, 235, 237
of water   38, 207, 235-238
passing beyond   270
spillover   79, 84, 93, 97-98, 102
   television   63, 87-88, 92, 100, 148, 157, 252. *See also* television.
   totalization   37, 107-108, 109, 115, 123, 126-127, 131, 142, 145, 151, 199, 224, 246, 261, 265
   ubiquity of   17-18, 22-23, 260
screenness   55
screenology   33n66, 53
Second Life   268-269
Sensorama   118n36, 124n59
shadow theatre   54
Shanghai, Western New Year countdown   243
Sherlock, Alexandra   158n7, 161-162
Simulacra and Simulation   43n92, 129, 130n78-80. *See also* Baudrillard, Jean.
Sky Projector, The   235, 237
Sky-Pi   213
Slender Man   94
smartphone   42, 57n22, 102, 104, 135n92, 222, 266
Solaris   66n60, 111
Something Awful   94n46
Sondheim, Stephen   107n1, 108
Sontag, Susan   67, 251n129
Snoop Dogg   169, 172
Snowden, as cloud of ash   237
Sobchack, Vivian   33, 34n71, 59n27, 179n87, 245
Spector, Warren   115
Spielberg, Steven   116, 139n113, 156n4, 193n14
Spiderman: Far From Home   51-52, 55, 76, 91
spirituality   50
Star Wars: Secrets of the Empire   142n124, 143
starvation   131, 269-270
Strauven, Wanda   57, 81-82, 85-86, 90
subject
   and object   15, 23, 39, 41
   viewed   170, 175, 184, 241, 244
   viewing   19, 117
subjectivity   22, 44n100, 60n31, 73n97, 74, 145-146, 148-151, 202, 246n119
substitution   109, 127, 141, 186-187, 229
Super Mario 64   72

surround
   affective   37, 114, 119
   democratic   113-114
Susik, Abigail   235, 236n96, 237, 215n32, 228, 230n80
Sutherland, Ivan   123n57, 125

Tao Te Ching   29n49
Tarkovsky, Andrei   111
tattoo   86n27, 249-250
television   18, 24, 30, 36, 43, 48, 59, 64, 79-80, 100, 108, 116, 129, 139, 160-161, 165, 176, 220, 241, 249. *See also* screens: television.
   bezel-less   25, 26n31
   broadcast   49, 79, 100, 166, 259
   frame-less   26
   reality   43, 79, 269
teikhíon   72, 81-82. *See also* cave, Plato's.
Teng, Teresa   169
threat   52, 87-89, 91, 256
transcendence   49, 138, 14
Tribute in Light   213-214, 225
Triple Conjurer and the Living Head, The
trompe-l'oeil   77
Truman Show, The   79-81
Tupac   155, 168-169, 171-172, 175, 184
Turkle, Sherry   42, 47-48, 268n19
Turner, Fred   33, 100n76, 112-113

Uncle Josh at the Moving Picture Show   87, 90-91, 134
unreal   253-255, 260, 262, 264-268, 270, 272-273
Us   273

Väliaho, Pasi   197n26
Valenciennes, Pierre-Henri   26, 138n101
vampires   173
Vertigo   155n1, 156-157, 161
vertigo   254
video conferencing   266, 100-101
video mapping: *see* projection mapping
Virilio, Paul   272
Virtual Reality (VR)
   controller sticks   124, 144
   Google Cardboard   123n58, 127, 135, 140n117
   immersion   31n56, 37, 108-112, 114, 116, 121-123, 125, 128, 133, 137-151
   inversion   37, 133, 146, 148, 151
   Liberty Leading the People   141, 144-145
   location-based   141-143, 146
   roller-coaster ride   141-144
   untethered   125, 143
virtual
   camera. *See* camera: virtual
   messaging   100
virtuality
virtualization   22, 26, 104-105, 241, 250, 255
   of the virtual   22

vividness   112, 175-176, 180n96, 183-184. *See also*
    ghosts: vivid; projections: vivid.
vivification   155, 175-176, 178, 180. *See also*
    vividness.
vivified   37, 155, 175-176, 185, 200. *See also*
    vividness.
voracity   244-245, 270. *See also* gluttony;
    hunger.
voyeurism   69-70

wall-window   83
War of the Worlds radio broadcast   89n34
Watchmen   236
weightless   244-245
Wellner, Galit   83
Whole Earth 'Lectronic Link (WELL)   100
widescreen   58n23, 122

Widodo, Joko   176-177
Willemen, Paul   160, 161n25
Windows: *see* Microsoft Windows
Wizard of Oz   80, 273
Wodiczko, Krzysztof   232-233, 239
Wonderful Living Fan, The   85-86
Work From Home (WFH)   267
World Wide Web   63, 100
Writing on the Wall, The   225-228

zero dimensionality   201-205
Zhilyaeva, Anna   144-145
Zimmerman, Eric   111-112. *See also* Salen,
    Katie.
Žižek, Slavoj   167, 264, 270
zombies   173